T0298551

Control Applications of Vehicle Dynamics

GROUND VEHICLE ENGINEERING BOOK SERIES

Dr. Vladimir V. Vantsevich, Editor
Professor and AVMI Director
The University of Alabama at Birmingham

Control Applications of Vehicle Dynamics
Jingsheng Yu, Vladimir Vantsevich

Driveline Systems of Ground Vehicles
Theory and Design
Alexandr F. Andreev, Viachaslau I. Kabanau, Vladimir V. Vantsevich

Dynamics of Wheel-Soil Systems
A Soil Stress and Deformaiton-Based Approach
Jaroslaw A. Pytka

Design and Simulation of Rail Vehicles
*Maksym Spiryagin, Colin Cole, Yan Quan Sun,
Mitchell McClanachan, Valentyn Spiryagin, Tim McSweeney*

Design and Simulation of Heavy Haul Locomotives and Trains
*Maksym Spiryagin, Peter Wolfs, Colin Cole, Valentyn Spiryagin,
Yan Quan Sun, Tim McSweeney*

Road Vehicle Dynamics
Fundamentals and Modeling with MATLAB®, Second Edition
Georg Rill, Abel Arrieta Castro

Automotive Accident Reconstruction: Practices and Principles, Second Edition
Donald Struble

Control Applications of Vehicle Dynamics

Jingsheng Yu, Vladimir Vantsevich

CRC Press
Taylor & Francis Group
Boca Raton London New York

CRC Press is an imprint of the
Taylor & Francis Group, an **informa** business

First edition published 2022
by CRC Press
6000 Broken Sound Parkway NW, Suite 300, Boca Raton, FL 33487-2742

and by CRC Press
2 Park Square, Milton Park, Abingdon, Oxon, OX14 4RN

© 2022 Jingsheng Yu and Vladimir Vantsevich

CRC Press is an imprint of Taylor & Francis Group, LLC

Reasonable efforts have been made to publish reliable data and information, but the author and publisher cannot assume responsibility for the validity of all materials or the consequences of their use. The authors and publishers have attempted to trace the copyright holders of all material reproduced in this publication and apologize to copyright holders if permission to publish in this form has not been obtained. If any copyright material has not been acknowledged please write and let us know so we may rectify in any future reprint.

Except as permitted under U.S. Copyright Law, no part of this book may be reprinted, reproduced, transmitted, or utilized in any form by any electronic, mechanical, or other means, now known or hereafter invented, including photocopying, microfilming, and recording, or in any information storage or retrieval system, without written permission from the publishers.

For permission to photocopy or use material electronically from this work, access www.copyright.com or contact the Copyright Clearance Center, Inc. (CCC), 222 Rosewood Drive, Danvers, MA 01923, 978-750-8400. For works that are not available on CCC please contact mpkbookspermissions@tandf.co.uk

Trademark notice: Product or corporate names may be trademarks or registered trademarks and are used only for identification and explanation without intent to infringe.

ISBN: [9780367681050] (hbk)
ISBN: [9780367681180] (pbk)
ISBN: [9781003134305] (ebk)

DOI: 10.1201/9781003134305

Typeset in Times LT Std
by KnowledgeWorks Global Ltd.

Contents

PART I   *Modeling of Vehicle Dynamics*

PART II   Control Design

Series Preface

Ground vehicle engineering formed and shaped as an engineering discipline in the 20th century and became the foundation for significant advancements and achievements, from personal transportation and agriculture machinery to planet exploration. As we have stepped into the 21st century and facing new economic, climate and other challenges, there is a need for developing fundamentally novel vehicle engineering technologies and proficiently training future generation of engineers in the areas of both conventional (with a driver) vehicles and autonomous vehicles. This timely launched new Ground Vehicle Engineering Series unites high caliber professionals from the industry and academia for producing top-quality technical books and undergraduate- and graduate-level texts on the engineering of various types of vehicles including conventional and autonomous mobile machines, terrain and highway vehicles, road/off-road and rail vehicles, and ground vehicles with novel concepts of motion.

The Series concentrates on conceptually new methodologies of vehicle dynamics and operation performance analysis and control, advanced vehicle and system design, experimental research and tests, and manufacture technologies. Applications include, but not limited to, heavy-duty and pickup trucks, farm tractors and agriculture machinery, earth-moving machines, passenger cars, human assist robotic vehicles, planet rovers, military conventional and unmanned wheeled and track vehicles, and reconnaissance vehicles.

Preface

The success of model-based control design of ground vehicles depends fundamentally on the integration of two key subject-matter areas – the modeling of the vehicle dynamics and the design of the control algorithms. Over the last few decades, a number of intermediate and advanced vehicle dynamics books focusing on the physics and principles of vehicles in motion have been published. These works offer in-depth theoretical and experimental understanding of dynamics of various types of vehicles; however, when it comes to selecting appropriate models of vehicles for control design, these works do not provide clear guidelines. Without such guidelines, highly nonlinear and complex 3D-space vehicle dynamics models may become problematic for real-time applications in controls of both conventional (with a driver) vehicles and autonomous vehicles. Furthermore, current and near-future artificial intelligence-based, real-time decision-making needs and will need appropriate model-based controls. While there are excellent books on advanced methods for control design, they do not offer practical skills in selecting a method most suitable for particular vehicle model.

The primary purpose of this book is to provide advanced students and young engineers with analytical fundamentals and practical skills in selecting the proper vehicle models for designing model-based controls of various vehicle dynamics properties and characteristics. This book also offers different methods of controls that are specifically designed with the use of the presented vehicle models. Thus, the reader will learn how to meet the practicability requirements of controls with sufficient analytical fundamentals of vehicle dynamics models. The reader will also find concrete recommendations on application of the control algorithms to autonomous vehicles.

The authors used their vast experience in industry and academia to lay out the mechanical basics of vehicle dynamics that are needed for designing sophisticated control algorithms and then designed and analyzed the controls to help alleviate core challenges that students and engineers face when designing model-based controls. Such controls, as recent studies of the authors outside of the scope of this book have proven, work efficiently in their integration with machine-learning algorithms that allows for advancing the control applications, especially for severe driving conditions (off-road driving, path following, etc.).

The content of the book is divided into two parts. In Part I, Modeling of Vehicle Dynamics, the reader is introduced to a wide range of dynamic models of vehicles suitable for different control applications. All models are derived and analyzed for purposes of advanced control design. These models are then utilized in Part II, Control Design, which presents and illustrates the model-based controls.

More specifically, Part I (Chapters 1–10) offers an overview and analysis of history of pivotal studies in the vehicle dynamics field that establish vehicle dynamics fundamentals for control application as an essential element of successful vehicle design. Key kinematics and dynamics topics on vector description and transformation, as well as Newton's and Euler's equations with application to vehicle motion, provide the reader with technical information that is necessary for understanding the dynamic models in the book, such as tire and vehicle models of longitudinal and lateral dynamics, normal and roll dynamics, and acceleration and braking mechanics, including regenerating braking and its impact on the braking performance of electric and hybrid-electric vehicles. Various types of steering configurations in combination with wheel driving and braking torque vectoring are explained for the purposes of control.

Part II (Chapters 11–14) is dedicated to control design and control applications in different driving scenarios. When some sensors are not available, engineers have to estimate signals. This book considers different methods for signal estimation and their use in controls. They include state observers, Kalman filter, linear quadratic optimal control and nonlinear control, adaptive and robust control. These methods are illustrated in various applications, including tire slippage control, traction control, vehicle speed control and path following control, yaw stability control, rollover control, and vehicle-trailer stabilization. The designed controls are applicable not only to conventional vehicles, but also to autonomous vehicles. Some control applications in the book were designed specifically for autonomous vehicles.

In addition, LabVIEW examples are provided for simulating the system models and designed control applications. The LabVIEW software platform can be combined with different hardware systems to implement the designed controls. LabVIEW can also be used to run standalone programs; examples are given in the book by applying LabVIEW to simulate some vehicle dynamics models and then to analyze the system characteristics of the models. In the presented controls of vehicle dynamics, LabVIEW was used to set up the simulation environment, in which both the vehicle model and the controller were programmed as VI modules.

Jingsheng Yu   Vladimir Vantsevich
Novi, Michigan   Hoover, Alabama

MATLAB® is a trademark of The MathWorks, Inc. and is used with permission. The MathWorks does not warrant the accuracy of the text or exercises in this book. This book's use or discussion of MATLAB® software or related products does not constitute endorsement or sponsorship by The MathWorks of a particular pedagogical approach or a particular use of the MATLAB® software.

Acknowledgments

Dr. Jingsheng Yu would like to thank Professor Peter C. Müller for introducing him to the field of robotics and mentoring him during research and doctoral studies in the Department of Safety Control Engineering at the University of Wuppertal. Dr. Yu is also grateful to all his colleagues from the Department who helped him gain knowledge of robotics and control theory. Dr. Yu's research work from Wuppertal continued to influence his later studies in control of vehicle dynamics throughout the years.

Dr. Yu is very grateful to his wife, Jia, and their two children, Tiffanie and Kevin for supporting him during the writing of this book, which required extensive research studies outside of his daily job. This book wouldn't have been possible without their encouragement and patience, allowing Dr. Yu to spend significant time during evenings, weekends and holidays.

Additionally, Dr. Yu sends his special thanks to a few individuals who have contributed their spare time to help him with this book. Mohammad Alqawasmeh has done excellent work creating LabVIEW simulation programs. Dr. Mostafa A. Salama has helped with providing LabVIEW simulations for Chapter 4 and Section 14.3 and assisted with writing these chapters. Aaron Showers and Rick Robinson read some of the chapters in the early drafts and proposed many improvements.

Dr. Vladimir Vantsevich sends all of his love to his wife, Svetlana, and their children, Michael, Anna, and Ivan – with deep gratitude for their love, care, and continuous support of his research endeavors that took time away from the family.

Special thanks to his lifelong friend and colleague Dr. Alexandre Opeiko for his professional review, comments and recommended edits. Dr. Vantsevich will be always thankful to his current and former students – Dr. Jeremy Gray, Dr. Mostafa Salama, Dr. Siyuan Zhang, Mr. Jesse Paldan, and many others – for being wonderful people and engineers who supported him through many years of work.

The authors would like to thank Acquiring Editor Nicola Sharpe at CRC Press, Taylor & Francis Group for her patience and professional help during the manuscript preparation and publishing, as well as several anonymous reviewers.

Author Biographies

Jingsheng Yu earned his Dipl.-Ing. degree in Electrical Engineering from Ruhr University Bochum, Germany. He obtained his Ph.D. in robotics from the Department of Safety Control Engineering at the University of Wuppertal, where he was engaged in research and teaching for five years. Dr. Yu has worked as senior engineer, manager and director of engineering in the automotive industry at firms including TEMIC GmbH, Robert Bosch Corporation and JSJ Corporation. He has more than 20 years of experience in development of automotive control systems, such as steering and braking control systems, electronic stability controls, advanced driver assistance systems, and other mechatronic systems including software and hardware electronics. While working in the industry, he has taught university graduate classes and has been active with research in the areas of robotics, modeling of vehicle dynamics, vehicle control and autonomous driving. Currently, Dr. Yu is senior director of design and process engineering at Flex Ltd.

Vladimir Vantsevich is a professor at the University of Alabama at Birmingham. He also serves as Director and Principal Investigator of the Autonomous Vehicle Mobility Institute Program. His research and engineering work with industry and research agencies on conventional and autonomous vehicles is on coupled and interactive dynamics of vehicle systems, agile tire dynamics, vehicle dynamics and control of multi-wheel vehicles for road and off-road applications, vehicle energy efficiency, terrain mobility and maneuver, driveline system modeling and engineering design, modeling of virtual driveline systems for fully electric and hybrid vehicles, and AI-based morphing of autonomous vehicles. Dr. Vantsevich is author of 6 technical books and 170 research articles and has delivered more than 160 seminars and invited lectures to industry, academic institutions and professional societies across 18 countries. He is a registered inventor of the U.S.S.R. with 30 certified inventions. Dr. Vantsevich is the Founder and Editor of two book series with CRC Press and ASME.

Part I

Modeling of Vehicle Dynamics

1 Introduction

1.1 VEHICLE SYSTEM DYNAMICS: BRIEF HISTORY AND FUTURE RESEARCH DIRECTIONS

BEGINNING

Methods and equations of kinematics and dynamics of rigid and flexible bodies are general and can be applied to model any mechanical system, including ground vehicles. The very beginning of theoretical investigations started in the first part of the 19th century. Ackermann steering is a great example of a corner stone of today's vehicle dynamics. Similar results were achieved in the last quarter on the 19th century by Jeantaud (known as "Jeantaud diagram" or "Jeantaud four-bar-linkage").

First formal and systematic publications on vehicle dynamics began in the early 20th century when the Society of Automotive Engineers (United States) and the Institution of Automobile Engineers (United Kingdom) started their regular journal proceedings: *SAE Transactions* and *Proceedings of the Institution of Automobile Engineers* [1, 2]. The passion for research work increased drastically after the lateral tire skid effect was discovered by G. Broulhiet in 1925 [3]. For the first time dependences between the lateral tire force and its slip angle were established in experiments, and research studies on vehicle handling were initiated [4–6].

The next pivotal moment in vehicle dynamics development is understood to be Maurice Olley's works at General Motors in 1930s; *oversteer, understeer and critical velocity* were first formally introduced in engineering practice at that time [6, 7] (even though "oversteer" was first used by F. W. Lanchester in 1907, [8]). In the early 1900s, rigorous research studies of vehicle handling were done by F. W. Lanchester in the United Kingdom [9], I. Gratzmuller and de Seze in France [10, 11] and L. Huber and O. Dietz in Germany [12, 13]. Intensive analytical research work of vehicle aerodynamics and experimental research and tests in air tunnels also began in the 1930s [14, 15]. Properties of tires received much attention in conjunction with their impact on ride and handling [16]. The establishment of the *traction-lateral tire force circle* became a key factor in understanding the mechanic foundation of vehicle motion [17]. The described beginning of vehicle dynamics is illustrated in Figure 1.1.

Concurrently in Russia, a research paper by N. E. Zhukovsky, which was published in 1905, analyzed a particular wheel locomotion technical problem – the effect of a driveline system on vehicle motion [18]. Later in 1923, Prof. Zhukovsky published a study of curvilinear motion of a vehicle with rigid wheels [19]. The decade of the 1930s witnessed one of energetic research in Russia. During those years, E. A. Chudakov published his first vehicle engineering books, including his "Theory of Automobile" [20] (note: the Russian term "theory of automobile" is an equivalent of "vehicle dynamics"). Studies of on-road vehicle motion, in particular vehicle *handling* and *stability of motion*, effect of "Shimmy" and *steering system geometry* were conducted by Ya. M. Pevzner, V. A. Gluh and V. Yu. Gittis [21–24].

This initial period of research work was filled with new discoveries and studies of peculiarities of vehicle movement, and it was becoming more and more clear that vehicle dynamics was being formed and established as a new branch in applied science.

DOI: 10.1201/9781003134305-1

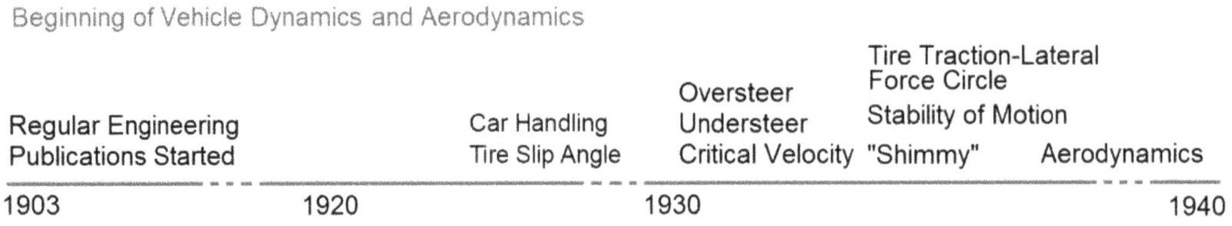

Beginning of Vehicle Dynamics and Aerodynamics

Regular Engineering Publications Started	Car Handling Tire Slip Angle	Oversteer Understeer Critical Velocity	Tire Traction-Lateral Force Circle Stability of Motion "Shimmy"	Aerodynamics
1903	1920	1930		1940

FIGURE 1.1   Vehicle dynamics and aerodynamics beginning.

Going from Steady Movements to Transient Modes, to Off-Road and More

In mid-1940s, active research investigations of *vehicle transient motion* were undertaken (acceleration and deceleration, various maneuvers under variable steering angle); vehicles began to be considered as *multi-body nonlinear systems*; vehicle *dynamic behavior*, including *ride*, was analyzed by taking parameters and characteristics of vehicle systems into consideration. In 1950s, *vehicle handing* and *stability of motion* associated with *transient velocity* and *steering* received a lot of attention and technical consideration in literature; William F. Milliken was a pioneer of this new research and engineering direction in the United States [25–31]. All this is illustrated in Figure 1.2.

It should also be emphasized that from 1945 to 1950s (see Figure 1.2), developments in vehicle dynamics and design were significantly facilitated by experiences gathered during World War II: the shortage of arterial and high-type roads and poor terrain mobility for vehicles continued during postwar years. It was at this time that meanings and definitions of *vehicle operational properties* began to take shape [32, 33]. This was a pivotal moment in the vehicle dynamics history. It allowed dynamics of vehicles to be characterized by a set of *vehicle operational properties*, such as *mobility, handling, stability of motion*, etc., to estimate various features of vehicle movement and effectiveness of the movement in different road and off-road environments while a vehicle is fulfilling its task/mission (e.g., passenger and cargo transportation, aggregation with farm machinery and soil tillage, earthmoving work, interaction with weaponry in military affairs, etc.). Research Institutes established by E. A. Chudakov in 1930s continued research and design work on *parasitic power circulation in all-wheel drive terrain vehicles* [34–36], *stability against lateral skid* [37], etc. E. A. Chudakov published new books on theory of automobile [38, 39], in which *operational properties (terrain mobility, traction and velocity properties, ride, stability against lateral skid and handling and fuel economy)* were discussed in depth. In the same years, technical books on *theory of military vehicles* by G. V. Zimelev and N. I. Korotonoshko were also published [40, 41].

Starting in the second half of 1950s (see Figure 1.2), *elastic/stiffness and damping properties* of vehicle systems and vehicle components that connect the systems such as sprung and unsprung masses, rotating masses of transmission and driveline, steering system and brakes were introduced and actively researched. *Dynamics of deformable bodies* began getting incorporated in vehicle dynamics analysis and system design process [42].

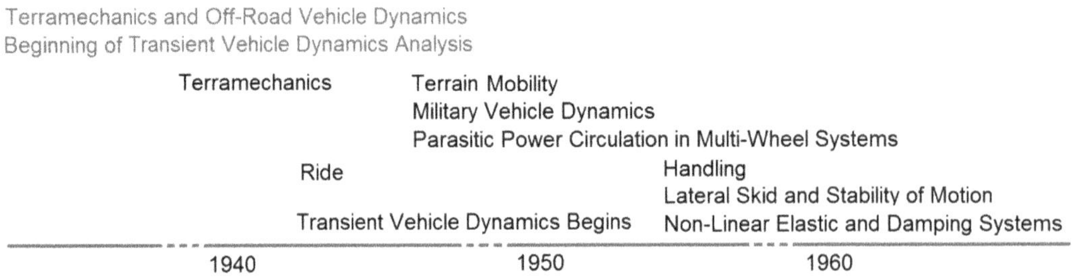

Terramechanics and Off-Road Vehicle Dynamics
Beginning of Transient Vehicle Dynamics Analysis

Terramechanics	Terrain Mobility Military Vehicle Dynamics Parasitic Power Circulation in Multi-Wheel Systems	
Ride	Handling Lateral Skid and Stability of Motion	
Transient Vehicle Dynamics Begins	Non-Linear Elastic and Damping Systems	
1940	1950	1960

FIGURE 1.2   Terramechanics and off-road vehicle dynamics; beginning of transient vehicle dynamics.

It should be mentioned that *terramechanics*, which had been intensely researched since 1930s (Figure 1.2), was notably shaped and profiled in M. G. Bekker's classical work of "Theory of Locomotion" in 1956 [43]. An enormous advancement in research and growth of publications on *terramechanics* and *theory of terrain/military vehicles* happened in 1960–80s. In 1960s, *terramechanics* received its considerable expansion through the work of M. G. Bekker [44], A. R. Reece [45], W. Gill and G. E. Vanden Berg [46]; in these years, first publications by J. Y. Wong [47–49] were issued, and his prolific research studies then resulted in his works [50, 51].

The *soil-farm machine interaction* and its impact on vehicle traction were studied and developed through decades of intensive research of tillage in agriculture, earthmoving and construction operations [46, 52].

The contribution of Russian researchers to *terramechanics* before 1970 was thoroughly analyzed and summarized by M. G. Bekker in his report [53]. (English speaking readers can download the report using the URL link given at reference [53]). The following review can add more specifics on publications that connect *terramechanics* and *vehicle dynamics* studies.

The peak of research on *terramechanics* and *theory of terrain and road vehicles* (*vehicle dynamics*) in Russia was also in 1960–1980s (see Figure 1.3). Among the main works on terramechanics, publications of Y. S. Agiekin should be emphasized [54, 55]. Notable research in various fields of vehicle dynamics were contributed during that time: D. A. Antonov developed the nonlinear theory of stability of motion of multi-axle terrain and road vehicles [56], A. S. Litvinov cultivated the linear and nonlinear handing and stability of road vehicles [57], and Y. E. Farobin presented a theory of turnability of multi-axle vehicles (linear and nonlinear approaches) [58]. Also, V. A. Petrushov developed a linear theory to estimate power losses in locomotion systems of multi-axle vehicles [59] and Y. V. Pirkovsky offered a theory of power losses in terrain multi-wheel locomotion systems [60]. Furthermore, significant inputs to terrain mobility studies were contributed by N. I. Korotonoshko [61], N. F. Bocharov and G. A. Smirnov [62–64] investigated traction dynamics of multi-wheel vehicles, taking into consideration torque distributions between the drive wheels, elasticity of tires and suspension systems and properties of soil. A. S. Antonov developed a theory of power flow management in vehicle systems [65]. P. V. Aksenov presented a system approach to the selection of main design parameters of multi-wheel vehicle systems including vehicle platform general package, steering system, driveline and suspension system [66]. The research works and engineering designs by A. Kh. Lefarov regarding terrain vehicles based on specifications of vehicle operational properties deserves special attention [67, 68]. Simultaneously, in these years, D. A. Chudakov, V. V. Guskov and N. A. Ul'yanov researched dynamics of farm tractors and construction equipment [69–71] following up research studies by E. D. L'vov that were conducted back in the 1940–60s [72].

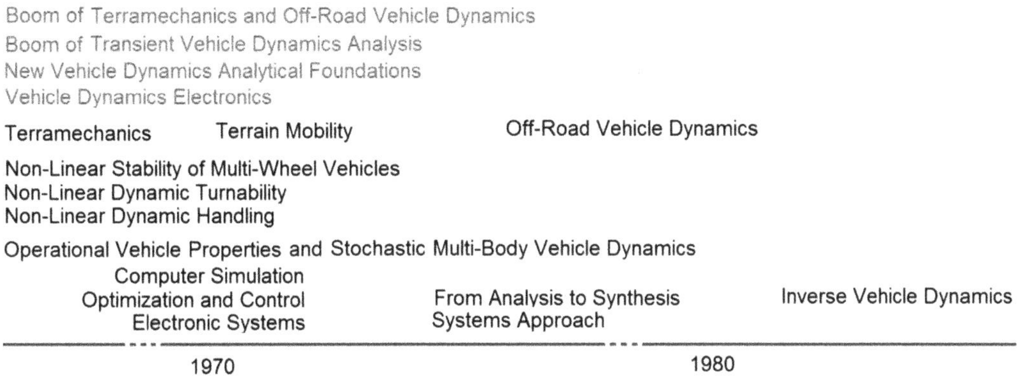

Boom of Terramechanics and Off-Road Vehicle Dynamics
Boom of Transient Vehicle Dynamics Analysis
New Vehicle Dynamics Analytical Foundations
Vehicle Dynamics Electronics

| Terramechanics | Terrain Mobility | Off-Road Vehicle Dynamics |

Non-Linear Stability of Multi-Wheel Vehicles
Non-Linear Dynamic Turnability
Non-Linear Dynamic Handling

Operational Vehicle Properties and Stochastic Multi-Body Vehicle Dynamics
Computer Simulation
Optimization and Control
Electronic Systems
From Analysis to Synthesis
Systems Approach
Inverse Vehicle Dynamics

1970 1980

FIGURE 1.3 Boom of vehicle dynamics modeling and computational simulation; transitioning to new analytical foundations.

V. V. Katsigin and his research group summarized their significant contribution to terramechanics and terrain vehicle dynamics in reference [73].

It is to be emphasized that the listed nonlinear theories of vehicle stability, turnability and handling in Figure 1.3 were very much built upon extensive research work of tire-road interaction and modeling. The works of H. Pacejka, L. Segel, P. Lugner and their successors, and many other researchers have founded a solid basis of tire-road models for many research studies in today's vehicle dynamics [28, 74–79].

Aerodynamics of Vehicles

After the physical phenomenon of aerodynamic drag was discovered in the 1930s [80–82] (Figure 1.1), vehicle aerodynamics has formed as a strong constituent of vehicle dynamics/design [83–85]. It was established that the air flow impacts (i) *vehicle productivity and energy efficiency* and (ii) *safety of vehicles.*

Productivity and energy efficiency of vehicles are strongly inter-connected [86]. An increase of the velocity to surge productivity produces a steep rise of the aero drag, which is proportional to the square of the velocity. This leads to a significant drop of energy efficiency and increases fuel consumption. Over the past 40 years, the influence of vehicle systems and bluff body geometry on the aero drag was studied [87–90], including air resistance produced by wheels of trucks [91], the cabin design [92, 93], the shape of the truck bottom and rear surface of the trailer [88], the height of a tractor/semitrailer and the gap between them [88, 91, 94]. Approximately 25% of the air resistance comes from the air-cabin interaction, 30% is added by the aero drag beneath the trailer, the gap between the cabin and trailer makes 20% and 25% comes from the air vortex behind the trailer [95]. Circling/leading radii of cabin edges and trailer edges were investigated [91]. The side view mirrors generate up to 2% of the total aero drag of a class 8 truck [96]. External air filters and lights also increase air resistance [88, 91]. Based on this research, add-on passive aerodynamic devices were introduced to decrease wake zones, reduce the aero drag and save more fuel. These devices include passive head/side deflectors, frontal air dam [93, 97–99] and spoilers [100–106], half-balloon-shaped airfoils, air shields and trailer skirts to prevent the air flow from entering the space under the trailer (saves of 3–7% of fuel), and rear-end taper [88, 97, 99, 107–110]. Truck platooning is studied to reduce the aero drag [111]. Hydrophobic coating is researched to decrease the aero drag/fuel consumption [112]. It was found that the tire rolling resistance power loss goes up to 30–40% and the air resistance power loss makes 35–55% of the power needed to move a class 8 truck [113].

Starting in 1938 [114], a substantial research work was dedicated to *vehicle directional stability* in straight-ahead driving, which is affected by the lift and pitching moment [88, 94, 115–117]. The air resistance forces were found to be coupled with tire-road forces. *Aerodynamic stability* was then introduced by studying vehicles in the presence of a crosswind [88]. It is important that aerodynamic stability was related to vehicle safety issues in research studies [100, 118, 119]. Passive spoilers were proven to improve stability in turns [88, 98, 120]. Passive spoilers were also studied for fin-type vortex stabilizers, front-end shape and even some bio-inspired surfaces were designed to improve vehicle stability in steady/unsteady crosswinds [121–125]. The influence of aerodynamic forces on braking performance was researched [88, 120, 125–128].

Era of Computer Simulations, Nonlinear, Multi-Body and Stochastic Dynamics

The development of the linear theory of automobile was complete to a large degree in 1960–1970s (Figure 1.3); practically all studies in the following years considered motion of vehicles as *nonlinear multi-body systems* [129]; conditions of motion were more often introduced as *stochastic* [130]. By the mid-1980s, a sustainable set of *vehicle operational properties* had been established and quantitative estimation of dynamics of different vehicle applications could now be done; applications include *terrain mobility, traction and acceleration performance* (this is known as *tractive and velocity properties* in Russian technical literature), *energy efficiency, turnability, stability of motion* and *handling, ride* and *braking properties* [131–133]. The extensive research on terrain vehicle dynamics in 1960–80s assisted developments in battle/military vehicle dynamics [134–139]. At this time, new techniques were developed and put into practice for studying complex vehicle-terrain systems including *parameter* and *system identification* and *model reduction* [139].

Advances in nonlinear theories of vehicle motion were largely enabled by the advances in *computer modeling* and *simulation*. Analog and some digital mainframe computer simulations of vehicle dynamics were first applied by research organizations and vehicle manufacturing companies during 1950s and early 1960s [140]. Those simulations were based on linear and mildly nonlinear equations. Computer simulations soon became a natural attribute in practically all research centers on ground vehicle engineering during 1970s (Figure 1.3).

Numerical methods for computational simulations were specifically advanced for vehicle dynamics needs [141, 142]. This process continues today and modern computational tools are applied to model dynamics of vehicles as well [143].

Computational advantages also inspired a new methodological basis, *theory of optimization* and *system control,* to further research vehicle dynamics and vehicle system design. Since 1970s, the word combination of "vehicle dynamics" and "control" integrated into "*vehicle dynamics control*", which became the "it" factor in vehicle dynamics developments for the following decades [144–147]. As a first product of dynamics-control integration, the Anti-Block System can be mentioned here. An advanced control of vehicle stability and new development methods in vehicle dynamics control followed next [148–150]. New methods led to designs of semi-active systems; for example, a simple and practical "on-off" semi-active damping [151].

The above-mentioned computational advantages and introduction of control and optimization facilitated another move in advancing the analytical foundation of vehicle dynamics and vehicle design: *optimal synthesis*. Previously *analysis* was used as a methodology to study mechanical fundamentals of vehicle motion; in 1970–80s; however, technical problems in vehicle dynamics were formulated using the *optimal motion synthesis* approach. The transition to the new analytical foundation took about 15–20 years (see Figure 1.3: from analysis to synthesis). A characteristic example of this transition period in vehicle dynamics research is A. Kh. Lefarov's work on terrain multi-wheel vehicle design. Unlike the majority of other works, in which one or several *vehicle operational properties* were studied as functions of characteristics of many vehicle systems, his work was concerned with only *one vehicle system*, the driveline system. He designed this system based on its impact on several operational properties, including vehicle mobility, traction, energy efficiency and turnability [68, 131]. Such approach fully complies with a vehicle design engineer's intent to achieve a better design of a system. Another example is passive suspension design that was considered to better satisfy vehicle ride, stability, turnability and handling [57, 152, 153].

In 1980–1990s, the *systems approach* [154–156] was fully integrated in the vehicle dynamics research and design areas; vehicle dynamics and control became more connected with vehicle system design [157, 158]. The vehicle began to be considered as a component of a more complex system, e.g., a vehicle-environment-driver system, a traffic flow, etc. [159, 160]. Each component of the complex systems received rigorous studies. Modeling the driver was a completely new research problem. Most likely, the first steering control model of a driver was pioneered by M. Kondo. The model was presented at the 1956 JSAE Annual Congress and published in Transactions of JSAE in 1958 [161]. This was a preview-predictive model in which the driver fed back lateral position and angular error from the required course. Yamakawa analyzed driver-vehicle system stability employing the root locus method to the preview-first order predictive model [162]. Ohno proposed a program control model of the driver in 1966 [163]. Yoshimoto then proposed the second order predictive model of the driver in 1971 [164].

As a result, the *vehicle system design, nonlinear optimization and different-level controls of vehicle motion and systems* evolved into the *entire vehicle-environment-human interaction* study which used *synthesis* as the new research methodology.

FOUNDING OF MODERN STAGE AND SEEKING OF FUTURE DIRECTIONS IN VEHICLE DYNAMICS

Meanwhile, during the 1980s–2000s, within the context of the *synthesis methodology* and *systems approach*, a novel method of the formulation and solving of problems originated: the *inverse vehicle dynamics approach* [165–168] (see Figure 1.3). There are three major types of inverse problems that are usually studied today [169–171]: (*i*) Recover the force field for a given time history of a system with known parameters; (*ii*) Determine parameters of a system if its time history and force field are given (system identification or parameter estimation problems), and (*iii*) Create the optimal time history of a system and recover the optimal force field for the given or unknown parameters of the system and initial conditions. Several of the listed problems were solved in the ground vehicle dynamics and performance optimization area and described in references [172, 173]. The methods for recovering optimal force fields to provide the given time history of motion and optimal objective functions (performance functionals) of a mechanical holonomic system were developed. Specifically, a closed-form method was developed for determining the force field vector of a mechanical system in unsteady motion, which corresponds to the power distribution between the components of the system at minimal power losses and also leads to appropriate parameters of the system.

New methodological foundations in vehicle dynamics (*systems approach*, etc.) resulted in conceptually new system designs. In the suspension area, a semi-active damping control concept was presented in reference [174] that later found industrial applications (MR-dampers, ER-damper, etc.). An active suspension structure with cross-connected hydro-pneumatic layouts was presented in reference [175]. A systematic survey of active and semi-active suspensions until mid-1990s can be found in reference [176].

Therefore, one can say that, by 2000, modern vehicle dynamics had been shaped in terms of its major technical directions and research topics; the following three books well represent the state-of-the-art in road vehicle dynamics [83], race car vehicle dynamics [85] and terrain vehicle dynamics [50]. Additionally, the following references can be listed to reflect specific areas of vehicle dynamics and design: multi-body vehicle dynamics [177], stability of vehicles [178] and vehicle handling [179–181], off-road vehicle dynamics and regulation of wheel torques [182], dynamics-based design of vehicle systems [183], optimization of vehicle operational properties and inverse dynamics approach [184]. Modern dynamics fundamentals were combined with vehicle dynamics control [185–187]. It should be emphasized that successful developments in vehicle dynamics and control also resulted in new vehicle technologies – namely intelligent or smart vehicles – that enable transport efficiency and safety [188–190].

Continuing vehicle dynamics research in-depth, researchers have also focused on developing vehicle dynamics-based knowledge and technologies to design new vehicle systems. Thus, the development of *inverse vehicle dynamics* continued in the 2000s and resulted in the establishment of *multi-criterion optimization of multi-drive wheel vehicle operational properties* with conflicting criteria [86, 191–197] (see Figure 1.4). Multi-objective optimization approach based on conflicting performance measures was developed for suspension system design [198–200].

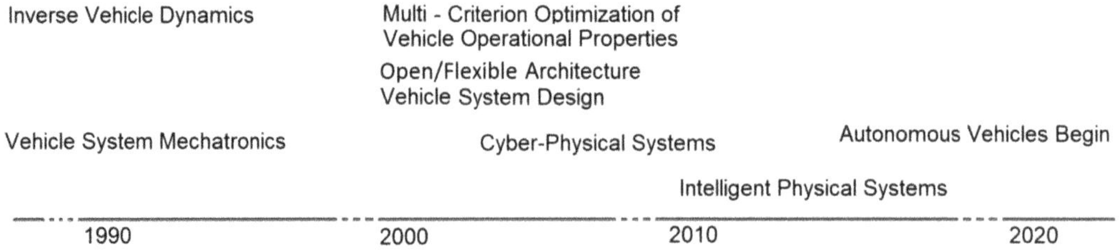

New Vehicle Dynamics Analytical Foundations
New Foundations of Vehicle System Design

Inverse Vehicle Dynamics	Multi - Criterion Optimization of Vehicle Operational Properties	
	Open/Flexible Architecture Vehicle System Design	
Vehicle System Mechatronics	Cyber-Physical Systems	Autonomous Vehicles Begin
	Intelligent Physical Systems	
1990	2000	2010	2020

FIGURE 1.4   Modern state; origins of future directions and trends in vehicle dynamics.

Vehicle *active chassis control* and *integrated/coordinated chassis and powertrain control* are evident examples of multi-criterion approaches to compromise conflicting vehicle systems and their impacts on vehicle normal, longitudinal and lateral dynamics with the purpose to improve different operational properties including ride comfort, handling, braking performance and vehicle stability of motion [186, 187, 201–205]. Fault detection became a new and important feature in control systems; reference [206] presents fault detection with application to roll stability control. A comprehensive summary on contemporary topics in vehicle dynamics and control is given in reference [207].

The mentioned multi-criterion optimization, integrated control and inverse dynamics led to a novel approach to vehicle system design: *open* or *flexible architecture approach* (see Figure 1.4). *Open architecture vehicle electrically driven and suspended corner* is applied to commercial vehicles [208, 209]. The *open architecture approach* is implemented in an *open-link locomotion module*, comprising a driven wheel with an electric motor, a system of electro-hydraulic suspension and an electro-hydraulic power steering system as the basis for the modular design of *manned* and *autonomous ground vehicles* [210, 211]. *Driveline systems* at Volkswagen are becoming *modular component* and *flexible architecture-type systems* [212]. Paper [213] presents a detailed analysis of system architecture in terms of

- relations and interconnections of subsystems and components;
- interfaces to other systems;
- system environment;
- data flow in the system; and
- data and software architecture.

In the 1990s, the above-mentioned practices were keeping pace with the *electrification and computerization* of vehicles and with the growth of new methods for the real-time modeling and controlling of *multiple-physics domain systems*, and designing new types of sensors and actuators. Thus, the mechatronics-based approach to vehicle system modeling, design and control came to practice and is still a critical factor in vehicle design in 2000s [214–218]. Mechatronic systems began gaining more advanced formulation as *cyber-physical systems*, which (by NSF definition) integrate computation and physical components, and *intelligent physical systems* (see Figure 1.4). This steady process particularly enhanced with the introduction of wireless communication between systems of a vehicle, between vehicles, and vehicles and structures/clouds. After launching *cybernetics* in the second part of 1940s [219], recent boom of research in *artificial intelligence* (AI), resulted in the designing of *automated driving systems* and the beginning of the *actual era of autonomous vehicles* (notice: autonomous vehicles and various levels of autonomy are out of the scope).

The change of the vehicle design paradigm to autonomous vehicles requires a fresh look at vehicle dynamics. The advancements in vehicle dynamics and design before 2010 that were discussed and reflected in Figure 1.4 have led to new technologies and novel systems that were not known before. It will be sufficient to mention traction control, electronic stability program, wheel torque vectoring systems, various active suspension, active stabilizers to control roll moment, active steering, active safety systems (e.g., lane departure warning), advanced hybrid-electric and fully electric vehicles, etc. All the above-listed systems are mechatronics-based by their nature, i.e., cyber-physical systems and, some of them, intelligent physical systems. They can respond within the range of 100–120 ms and provide fast interactions with the environment and driving wheel. However, with such time response the actual control of the wheel occurs after the vehicle is reaching to or has already reached a critical situation of losing its traction and mobility [220]. A study presented in reference [221] established a control time response within the boundaries of 40–60 ms that is close to the tire relaxation time constant. By advancing AI-based controls to this time response, an actuation can begin before a wheel starts spinning. Real-time simulations of a Reinforcement Learning – Fuzzy Logic controller demonstrated improved performance of the locomotion module in stochastic terrain conditions within the established boundaries of the time response [222].

The modeling of *fast* (i.e., real-time) and *precise* and, to some degree, *preemptive (i.e., faster-than-real-time) dynamic interactions* of the tire with surface of motion is becoming known today as *agile tire dynamics* and *mobility*, and dynamics of vehicles is evolving to *agile vehicle dynamics* [223–227]. This is another new stage of manned vehicle dynamics and autonomous vehicle dynamics, which requires fast control decisions and implementations [228, 229]. *Model Predictive Control* (MPC) has been becoming one of the most promising controls by incorporating real-time models with feedback and optimization [230]. MPC found applications in active steering, traction control, etc. [231–234].

The above-listed new directions in manned and autonomous vehicle dynamics have been resulting in further innovations in vehicle system dynamics and design. New research topics are raised and approaches developed in practically all research areas over the past 15–20 years. Much effort is paid to developing stochastic tire math models [235]; dynamic modeling of the tire-road interaction and tire-road friction estimation [236–238]; new topics are being established in sensitivity analysis of vehicle dynamics [239] and new suspension design and modeling for vehicle dynamics studies [240, 241], etc. However, in general, autonomous vehicle dynamics is in its infancy, which requires establishing new research areas in modeling autonomous vehicles and their intelligent physical systems, exteroceptive and proprioceptive sensor systems. A special attention should be paid to researching real-time simulations with human/natural intelligence (NI)-based and AI-based decision making and intelligent controls when vehicles operates in unknown environments. Dynamics of both manned and autonomous vehicles, and vehicle system dynamics have been enriched with research and design of observers that reduce the number of sensors and can be well-integrated in AI-based data management and decision making.

The following section generalizes and summarizes the new directions in vehicle system dynamics.

FUTURE RESEARCH DIRECTIONS OF VEHICLE DYNAMICS

A careful analysis of Figure 1.1 through 1.4 shows that more than 100-years of vehicle dynamics research and design work can be illustrated in the following three phases shown in Figure 1.5.

Phase I, named "*Establishing Vehicle Dynamics*", which started with research and explanation of *steady* and *transient vehicle motion* using *traditional methods of Newton's mechanics of a rigid body*, was sustained up to 1980s. This phase resulted in many methods of *vehicle dynamics optimization* and *control*.

Phase II, named "*Advancing Vehicle Dynamics*", *includes multi-criterion optimization* and *mechatronics-based control of vehicle operational properties*; this research and engineering was formed at the end of 20th to beginning of 21st century. Precisely in this period, inverse vehicle dynamics approach was created and open architecture-type systems were introduced. Correspondingly, vehicle system design methods essentially changed toward design of open architecture systems.

The history of vehicle theory of motion, analysis of technical directions in vehicle system design, innovative technical trends and analytical methods employed in recent vehicle dynamics and vehicle system control – *Phases I* and *II* – all show that the performance of vehicles has been significantly enhanced yet vehicle-road/terrain environments become more complex and severe. Today,

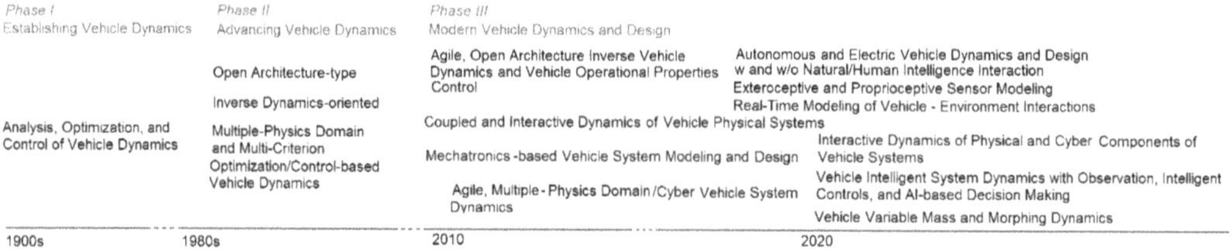

Phase I	Phase II	Phase III	
Establishing Vehicle Dynamics	Advancing Vehicle Dynamics	Modern Vehicle Dynamics and Design	
		Agile, Open Architecture Inverse Vehicle Dynamics and Vehicle Operational Properties Control	Autonomous and Electric Vehicle Dynamics and Design w and w/o Natural/Human Intelligence Interaction Exteroceptive and Proprioceptive Sensor Modeling Real-Time Modeling of Vehicle - Environment Interactions
	Open Architecture-type		
	Inverse Dynamics-oriented		
Analysis, Optimization, and Control of Vehicle Dynamics	Multiple-Physics Domain and Multi-Criterion Optimization/Control-based Vehicle Dynamics	Coupled and Interactive Dynamics of Vehicle Physical Systems	
			Interactive Dynamics of Physical and Cyber Components of Vehicle Systems
		Mechatronics -based Vehicle System Modeling and Design	Vehicle Intelligent System Dynamics with Observation, Intelligent Controls, and AI-based Decision Making
		Agile, Multiple-Physics Domain/Cyber Vehicle System Dynamics	Vehicle Variable Mass and Morphing Dynamics
1900s	1980s	2010	2020

FIGURE 1.5 Main development phases of vehicle dynamics and vehicle system dynamics.

new and growing requirements regarding vehicle energy efficiency, operational properties, safety and even vehicle security relating to road and off-road vehicle applications, manned and autonomous vehicles are being enforced. Therefore, the framework of research work that was established in *Phases I* and *II* cannot provide novel vehicle solutions, which require new *technological paradigms* in vehicle dynamics as the theoretical foundation of vehicle design and engineering. Indeed, establishing novel research domains in vehicle dynamics, i.e., new technological paradigms, can lead to fundamentally new vehicle system designs instead of simply providing more incremental advances in vehicle systems. Nevertheless, *Phases I* and *II* have prepared the ground for *Phase III*, a new period in vehicle dynamics history, which is emerging now ("*Modern Vehicle Dynamics and Design*", Figure 1.5).

A conceptually new research direction in *Phase III* is an extension of the "*agility*" to dynamics of the tire-surface interaction that can be formulated as a new technical paradigm – *agile tire dynamics and mobility*, which targets not only extremely *fast* and *precise,* but also *pre-emptive* 3D-vehicle-tire-surface parameter identification and control within the tire/soil relaxation time constants [221, 242]. This addresses a core problem in wheel dynamics – control of tire slippage to minimize tire-soil power losses, increase wheel mobility and reduce tire wear and soil damage. Indeed, in regards of tire slippage control, *agile tire dynamics* is in demand because it provides enough time for a tire control system to make a control decision and activate a wheel actuator while the tire and soil are deflecting. New *agile tire-soil characteristics* that should be measured and controlled are presented in reference [243].

Another direction for developing pre-emptively *agile* vehicle responses is dynamics of *variable mass and morphing systems* (Figure 1.5). Autonomous ground vehicles (e.g., reconfigurable payload dynamics of rovers and military vehicle applications), passenger cars (e.g., accident reconstruction dynamics), military vehicles (e.g., restoration of required motion in battle and severe terrain conditions, etc.), trucks (e.g., combined aerodynamics and road dynamics) can be named as potential applications. In the project [244], *morphing dynamics* of an autonomous utility truck is studied through as a two-domain interactive problem, comprising of (i) the truck aero-dynamics, and (ii) the truck-road dynamics. This interactive nature will allow for determining the optimal morphological characteristics of the truck, and counteractive flow characteristics of the active add-on flow devices. This will provide the truck-aero-road forces to withstand severe weather and roadway impacts and, thus, to preserve the safety of the truck on roads.

Modern and in-coming mechatronics-based novelties in vehicle technology can facilitate the implementation of *agile vehicle system dynamics* in new *cyber-physical* and *intelligent vehicle system designs* (see Figure 1.5) to be more efficient in occupant safety, energy saving and performance enhancement of the human-vehicle-environment. Thus, *agility* and *agile vehicle system dynamics* is continuing to develop as a key area in the emerging *Phase III* of vehicle dynamics development.

Still another new direction in vehicle dynamics is emerging today that manifests itself through the *interdependence and inter-influence of vehicle systems and their influence on vehicle operational properties* (see coupled and interactive dynamics of vehicle physical systems in Figure 1.5). As shown in the above-analysis in this section of the book, two fundamental but different approaches were established in the past (i.e., during *Phases I* and *II*) to study vehicle system dynamics and design vehicle systems:

i. Each vehicle operational property is studied by taking into consideration separate and independent impacts of various vehicle systems on this operational property.
ii. Dynamics and design of a vehicle system is studied by taking into consideration the impact of the vehicle system on various vehicle operational properties.

Later, along with the development of the *synthesis methodology* and *systems approach*, these two approaches were complemented by the third approach, in which an operational vehicle property is studied and improved by an integration of independent inputs of two systems. In reference [245], the third approach was utilized for a vehicle with two independent drive wheels and steerable non-driven wheels.

Historically, the above-listed approaches (i) and (ii) played significant roles in vehicle dynamics development and vehicle system design. Over the decades of research and design work, however, it has become more and more clear that vehicle systems may have inherited design conflicts (coupled and conflicting actions), and this is why they can interfere with each other when influencing vehicle operational properties. When designing vehicle systems, engineers usually are not concerned about a potential merged action of different systems on the dynamics of a vehicle. Coupled action typically goes undetected in the design process. Thus, coupled dynamics of the driveline and steering systems impacts turnability of the tractor, which is its operational property to make turns. The physics of this coupled dynamics is presented in reference [246], in which interrelationships between *vehicle operational properties* and *vehicle systems* were analyzed; this analysis established an analytical foundation to study coupled dynamics of vehicle systems.

In regards to coupled dynamics, modern integrated control may be considered as a technical solution to reconcile the interference between systems that results in their intrusive and coupled impact on vehicle properties [201]. However, improving of vehicle dynamics and vehicle operational properties, integrated control cannot truly resolve the technical problem of coupled and conflicting actions of systems.

Therefore, new studies of vehicle system dynamics are in demand today in order to thoroughly understand the *interdependence and inter-influence of vehicle system dynamics*. The latter can be achieved by introducing and executing *coupled* and *interactive* vehicle system dynamics when studying dynamics of vehicles [246–251]. Unlike *coupled dynamics*, *interactive dynamics* decouples dynamical effects of systems and then deliberately and purposely incorporates individual actions into an interactive action to enhance vehicle dynamics and performance, improve energy/fuel efficiency and other properties.

With the introduction of *cyber-physical systems* and *intelligent physical systems* (Figure 1.5), couplings extended to the cyber and physical processes in the systems, but they did not receive a proper study yet. Dynamic couplings in physical systems, concurrency and lack of coordination of cyber and physical processes between intelligent physical systems of autonomous ground vehicles require new approaches to de-coupling and establishing interactive dynamics of the systems. An autonomous vehicle should understand the couplings between both its systems and system components for two main reasons:

1. Not to learn and not to make any decisions from its dynamics when it is degraded by coupled systems of the vehicle.
2. Be capable to de-couple the systems and establish their interactive dynamics to facilitate and enhance autonomous movements.

These two challenges require extending autonomy and intelligence from the vehicle level to the level of vehicle physical and cyber-physical system dynamics. These dynamics problems become even more complex when natural/human intelligence is added to in the loop with AI-based controls (Figure 1.5).

There is one more emerging research direction, which relates to *cyber-security* of in-vehicle, vehicle-to-vehicle and vehicle-to-traffic information [252]. These new technical problems are really in today's and tomorrow's research agenda since many vehicle mechatronic systems, including tele-operated systems, belong to the cluster of *cyber-physical* and *intelligent physical systems*. Dynamics of these systems employed at manned and autonomous ground vehicles needs to be secured in *severe electronic/cyber* and *electro-magnetic environment*. In this regards, *real-time reachability*, *verification* and *protection* of the *information exchange* are becoming crucial for the vehicle task/mission fulfillment, vehicle motion safety and security, and for providing the desired level of vehicle operational properties and energy efficiency.

With the vehicle engineering paradigm shift to autonomous vehicles, the modeling of *exteroceptive* and *proprioceptive sensor systems* (i.e., vehicle autonomy related sensors and vehicle system sensors) becomes an essential fundamental of vehicle dynamics to virtually sense in real-time the vehicle-environment interaction and provide information for NI- and AI-based controls. New observation techniques and observers are becoming an important component of the controls. The listed features are illustrated in Figure 1.5.

In summary of the above-discussed new research and engineering directions, Figure 1.5 presents *Phase III* in manned and autonomous vehicle dynamics developments. This phase is the wrapping up of the present dynamics of vehicles and the beginning of future vehicle dynamics, which will continue its development and enhancement as

- inverse dynamics-oriented,
- open architecture system-type, and
- multi-domain and multi-criterion optimization/control-based dynamics.

At the same time, modern vehicle dynamics is acquiring the following principally new features that can result in novel research directions in future vehicle dynamics:

- Agile-to-real-time (or faster-than-real-time) dynamic modeling of vehicles and vehicle systems in interactions with environment.
- Multi-physics mechatronic, cyber-physical and intelligent physical system dynamics.
- Coupled and interactive dynamics of physical and cyber components of vehicle systems.
- Variable mass and morphing aero- and road dynamics of vehicles.
- Modeling of exteroceptive and proprioceptive sensor systems, and developing advanced observation methods integrated with vehicle models.
- AI-based decision making and intelligent controls with or without NI-based interactions integrated with vehicle models.

The above-introduced features of modern vehicle dynamics are emerging today; at the same time, the list of new research directions is not limited to the discussed ones. Indeed, new research areas are coming as vehicle dynamics evolves, especially with autonomous vehicle dynamics receiving further developments. In this regard, the fundamentals of vehicle dynamics for control applications become an essential basis for successful vehicle design. It is these fundamentals that are considered in the book and controls are designed and analyzed.

1.2 MODELING OF VEHICLE DYNAMICS

As seen from Section 1.1, vehicle dynamics as a branch of dynamics of bodies describes fundamentally vehicle motion characteristics, including vehicle and vehicle system positions and displacements, linear and angular velocities and accelerations along with forces and torques acting on vehicles in their interaction with the environment. The description results in a mathematical model obtained from the basic laws of physics governing the vehicle motion. The model can be used to predict the motion caused by given forces, moments and torques, and also recover field of the forces, the moments and the torques that can provide a given history of motion, i.e., a given set of vehicle kinematic characteristics.

A complete dynamic analysis of the vehicle generality to cope with the countless vehicle actual movements is highly complex for vehicle control purposes. Indeed, there are numerous physical and cyber-physical vehicle components connected to each other and dynamically interacting with each other. Including all components and their interactions in a model increases the model complexity and often makes impossible to use it for control design. Therefore, the basic approach to vehicle modeling in this book is to create models for particular vehicle dynamics applications instead of a complete dynamic model. The models are designed for various driving maneuvers, such as acceleration and deceleration, straight-line motion and turning on a curve, path following, and towing a trailer. These models study different aspects of vehicle dynamics, such as wheel dynamics and wheel torque distribution (longitudinal dynamics), braking performance, vehicle lateral and normal dynamics, and trailer dynamics. In order to make the models more effective and complement to general understanding of vehicle dynamics, reasonable assumptions and approximations are made for each model. Thus, the models do not conflict mathematically with each other, but allow for studying various aspects of the dynamics. In general, those individual models can only be accurate to a certain degree comparing to a complete dynamic model including all dynamic effects since the couplings between dynamics of the models are not taken into account. In some situations, however, the unmodeled dynamics can be estimated and then compensated by the control design, for example in control of wheel dynamics. In other situations, the models are sufficient in their accuracy since the actual stability of the vehicle in experimental tests is in the foreground rather than the accuracy of the control target, for example, in control of vehicle lateral and normal dynamics. Also, a robust controller design improves the tracking performance in presence of unmodeled dynamics, such as in longitudinal and lateral dynamics for vehicle motion control.

It is also useful to start the modeling of vehicle dynamics and control design by considering planar vehicle movements within an xy-coordinate system, since the interpretation of the vehicle behavior conventionally begins with analysis of vehicle steering, longitudinal and lateral movements. These models are effective when considering small operating range of the roll angle, the sideslip angle and the hitch angle. The accuracy is sufficient within the small range if the stability of the control is foregrounded.

In Chapter 2, basic kinematic and dynamic equations are considered to form the basis for dynamic equations of vehicle motion in later chapters. Chapter 3 presents primary vehicle motion equations to describe the vehicle longitudinal dynamics as well as pneumatic wheel dynamics. The wheel dynamics is an essential aspect of the general vehicle dynamics. The tire characteristics play an important role in the driving behavior of the vehicle, and they are summarized in Chapter 4. In Chapter 5, longitudinal acceleration performance of a vehicle is viewed where the engine power distribution among the wheels and traction performance are discussed. In Chapter 6, kinetics and kinematics of braking mechanisms as well as the braking force distribution among the wheels is investigated. Chapter 7 provides an overview on regenerative braking related to electric vehicles and hybrid-electric vehicles. Chapter 8 focuses on modeling lateral dynamics of vehicle and then characteristics of the lateral dynamics are analyzed in Chapter 9. In Chapter 10, normal dynamics and roll dynamics are introduced. The normal movements are influenced by suspension behavior as well as tire/wheel excitation while roll dynamics extends beyond normal dynamics and includes also lateral forces and moments.

1.3   CONTROL OF VEHICLE DYNAMICS

This book mainly concerns with control aspects of vehicle dynamics, i.e., the book presents sub-models of vehicle dynamics that were specifically developed for control design purposes and then studies various controls built on those sub-models. The control of vehicle dynamics is yet a developing area for research, development and manufacturing. Due to the highly nonlinear nature and high degrees-of-freedom of vehicle dynamics models, the control design can be quite difficult to implement. A model-based controller requires a vehicle mathematical model as a basis and an algorithm to act on the model. In order to do so, sensory capabilities and actuators for decision making and acting/reacting are required. An integral to the controller design process is issues involving choice and location of sensors as well as actuators. Fortunately, sensor technologies have been significantly advanced in recent years, so that various control methods have become possible for implementation.

Once dynamic models are obtained and presented in Chapter 5 through Chapter 10, automatic control theory can be employed to modify vehicle movements to desired motions. In Chapter 11, a review of concepts of control theory is provided that is needed for controller design for both linear and nonlinear control systems. The chapter begins with a review of second-order linear systems and the state-space description. The key methods such as state observer, Kalman filter, Lyapunov stability theory and linear quadratic optimal control follow through the chapter.

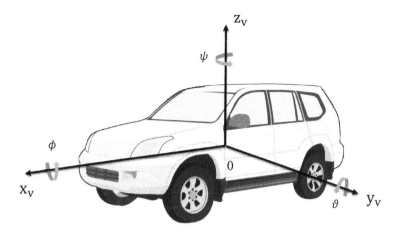

FIGURE 1.6   Vehicle-fixed coordinate system.

In Chapter 12 through 14, each control design application is developed based on the vehicle models presented in the previous chapters. The chapter begins with a tire slip control, which primarily deals with the wheel dynamics. Then a speed control and path-following controls are designed to direct vehicle motions. Different methods of yaw stability control are presented in the subsequent sections, which primarily address lateral dynamics and improve vehicle stability during a turn movement. Based on the computed-torque control that is used in industrial robotic engineering, a state-feedback controller is designed and then extended to the robust controller to deal with model uncertainties or unknown parameters. Rollover is an important topic in the vehicle safety engineering. A three-dimensional dynamic model is developed to detect possible rollover while an open-loop controller is defined to prevent rollover. At the end of Chapter 14, control stability of vehicle-trailer systems is discussed where a bicycle dynamic model is used. Although self-driving technology is not the focus in the book, some controllers are designed for autonomous vehicle applications, for example the path-following control and the yaw stability control of autonomous vehicles.

1.4   COORDINATE SYSTEMS

To represent dynamics of a vehicle as a system of multiple rigid bodies, it is necessary to define a *coordinate system,* or *frame*, in order to measure motion variables of the vehicle. The right-handed axis system with the definition based on the Z-Up orientation is used throughout this book. The definition can be seen in SAE J670 [253]. This newly revised SAE standard recognizes both axis systems with Z-Up and Z-Down orientations, where the Z-Up orientation aligns with the definitions given in ISO 8855 as well as ISO 4130. Based on the axis system with the Z-Up orientation, the following coordinate systems are defined for vehicle, tire, wheel and inertial systems.

Rigid body is typically described with position and orientation in a coordinate system. In mechanics literature, a coordinate system is called a *frame* when it is considered with regard to its position as well as the orientation relative to another coordinate system.

VEHICLE-FIXED COORDINATE SYSTEM

Figure 1.6 shows that the coordinate system, denoted as V_{xyz}[1], is attached to the vehicle. It is called *vehicle-fixed coordinate system*, and normally originates at the CG (vehicle's center of gravity) and

---

[1] *V* presents the vehicle coordinate system while x,y,z denote the coordinate components.

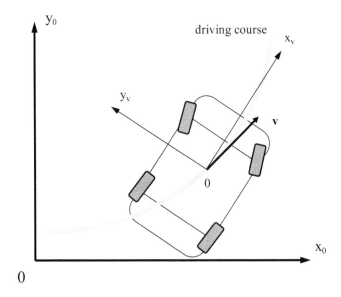

FIGURE 1.7   Inertial coordinate system.

moves with the vehicle. The vehicle motions can be defined with reference to this coordinate system. According to SAE, the x_v − axis is in the center plane pointing forward. The y_v − axis points to the driver's left, and z_v − axis is upward to give a right-hand orthogonal system. Since the vehicle has some acceleration in general, the vehicle-fixed coordinate system is a non-inertial system. Three rotations are defined by SAE as

- yaw rotation angle ψ about the z_v − axis
- pitch rotation angle θ about the y_v − axis
- roll rotation angle ϕ about the x_v − axis

INERTIAL COORDINATE SYSTEM

The inertial coordinate system has axes that do not rotate and they are either fixed or translate with a constant linear velocity; therefore, index "0" is used to identify such system. The inertial coordinate system is often assumed to be fixed on the ground, which the vehicle path and velocity can be referenced to. It has the same axis system, in which the x − and y − axes are in a horizontal plane and the z − axis is directed upward. At the time instant under consideration, the inertial coordinate system is often selected to coincide with the vehicle-fixed coordinate system at the point where a maneuver starts. Figure 1.7 shows the inertial coordinate system, denoted as 0_{xyz}, in the $x − y$ plan projection. To have a better view, a vehicle attached with vehicle-fixed coordinate system V_{xyz} moving with velocity $\mathbf{v}$ is projected in the $x − y$ plane.

TIRE AND WHEEL COORDINATE SYSTEMS

Figure 1.8 shows the *tire coordinate system* used in this book. The tire coordinate system, denoted as T_{xyz}, is based on the right-handed coordinate system with the Z-Up orientation. Its origin is fixed at the tire-road patch center. The x_T − axis is the intersection of the wheel plane and the road

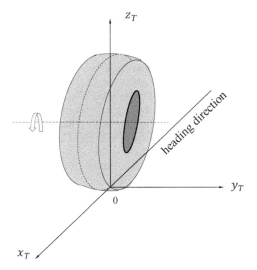

FIGURE 1.8 Tire coordinate system.

surface with a positive direction forward (heading direction). The wheel plane is the plane normal to the wheel-spin axis, which is located halfway between the rim flanges. The y_T − axis is in the road surface. The z_T − axis is perpendicular to the road surface with a positive direction upward.

Figure 1.9 shows the *wheel coordinate system* used throughout the book. The wheel coordinate system denoted as W_{xyz}, is based on the right-handed coordinate system with the Z-Up orientation. Its origin is fixed at the wheel center. The x_w − axis going through the origin is parallel to the intersection of the wheel plane and the road surface with a positive direction forward (heading direction). The wheel plane is the plane normal to the wheel-spin axis, which is located halfway between the rim flanges. The z_w − axis is perpendicular to the road surface with a positive direction upward. The y_w − axis is aligned with the wheel-spin axis so that the coordinate system orthogonal and right-hand.

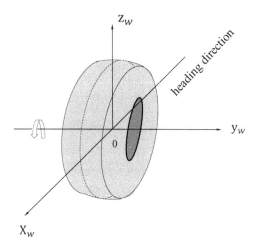

FIGURE 1.9 Wheel coordinate system.

REFERENCES

1. http://www.sae.org/about/general/history/
2. http://www.uk.sagepub.com/journals/Journal202018
3. Broulhiet, G., "La Suspension de la Direction de la voiture Automobile: Shimmy et Dandinement", Société des Ingénieurs Civils de France, Bulletin, Vol. 78, 1925.
4. De, Lavaud, "The Problem of Independent Rear Wheels", Technique Moderne, Vol. 20, N 71, 1928.
5. Becker, G., Fromm, H., Marunu, H., "Schwingungen in Automobillenkungen", Berline, 1931.
6. Olley, M., "Stable and Unstable Steering". General Motors, Report, 1934.
7. Olley, M., "National Influences on American Passenger Car Designs", Institution of Automobile Engineers, Vol. 32, 1937–1938.
8. Lanchester, F. W., "Some Problems Peculiar to the Design of the Automobile", Proceedings of Institution of Automobile Engineers, Vol. 2, 1907–1908.
9. Lanchester, F. W., "Motor Car Suspension and Independent Springing", Proceedings of Institution of Automobile Engineers, N 32, 1937–1938.
10. Gratzmuller, I., "Theorie de la Tenue de Route?", SIA Journal, Vol. 15, N 11, 1942.
11. De Seze, V. G., "Stabilité de Route des Voitures à Traction Avant ou à Propulsion Arrière", Société des Ingénieurs Civils de France, Vol. 10, N 8, 1937.
12. Huber, L., Dietz, O., "Fahrversuche zum Vergleich Zwischen Fahrzeugmodelle und Wirklichem Fahrzeug", Deutsche Kraftfahrforschung, N 2, 1935.
13. Rickert, P., Schunck, T., "Zur Fahrmechanik des Gummibereiften Kraftfahrzeuges", Deutsche Kraftfahrforschung, N 89, 1940.
14. Irving, I. S., "Golden Arrow and the World's Speed Record", Institution of Automobile Engineers, N 24, 1929.
15. Hausen, H., Schlor, K., "Aerodynamische Modellmessungen an Verschidenen Kraftwagenformen und Verhalten des Wirlichen Fahrzeuges bei Seitenwind", Deutsche Kraftfahrforschung, N 2, 1938.
16. Evans, R. D., "Properties of Tires Affecting Riding, Steering and Handling", SAE Technical Paper 350082, 1935.
17. Kamm, W., Das Kraftfahrzeug. Betriebsgrundlagen, Berechnung, Gestaltung und Versuch. Springer; 1936 (in German).
18. Zhukovsky, N. E., Theory of a Device Designed by Dipl.-Eng. Romeiko-Gurko, 1905. Also published in Prof. Zhukovsky's Complete Works, Vol. 8, Moscow, 1937 (Теория прибора инженера Ромейко-Гурко, in Russian).
19. Zhukovsky, N. E., "To Automobile Dynamics", Vol. VII, 1950 (Динамика Автомобиля, in Russian).
20. Chudakov, E. A., "Theory of Automobile", ONTI NKTP Publishing House, 1935 (Теория автомобиля, in Russian).
21. Pevzner, Ya M, "Stabilization of Steered Wheels of an Automobile", NATI News, N 4, 1934 (Стабилизация управляемых колес автомобиля, in Russian).
22. Pevzner, Ya M, "An Investigation of Automobile Stability in Skid", Machine Building. Moscow: Automobile and Tractor Principal Publishing House; 1937 (Исследование устойчивости автомобиля при заносе, in Russian).
23. Gluh, V. A., "Experimental Research of Steering System", Motor, N 11–12, 1936 (Экспериментальное исследование систему управления, in Russian).
24. Gittis, V. Yu, "Automobiles. Theory and Design", Gostransizdat. Moscow; 1931 (Автомобили. Теория и конструирование, in Russian).
25. Shilling, K., "Directional Control of Automobiles", Industrial Mathematics Society, 1953.
26. Julien, M. A., "Convergence des theories francaises et des etudes et réalisations anglosaxonnes concervant la stabilité de route", La Technique Automobile et Aerienne, N 224, 1948.
27. Bastow, D., "Independent Rear Suspension", Institution of Mechanical Engineers, 1951–1952.
28. Segel, L., "Research in Fundamentals of Automobile Control and Stability", Society of Automotive Engineers, National Summer Meeting, 1956.
29. Rocard, I., L'instabilité en mécanique; automobiles, avions, ponts suspendus. Paris: Masson; 1954.
30. Milliken, W. F., Jr. et al., "Research in Automobile Stability and Control and in Tire Performance", Collection of Papers. London: The Institution of Mechanical Engineers; 1956.
31. Milliken, W. F., Jr., "Long-Range Automotive Stability and Control Research Program", Cornell Aeronautic Laboratory, 1952.

32. Chudakov, E. A., "Theory of Automobile", Mashgiz. Moscow; 1940, (Теория автомобиля, in Russian).

33. Taborek, Jaroslav J., Mechanics of Vehicles. Cleveland: Towmotor Corp.; 1957.

34. Chudakov, E. A., "Motion of a Non-Differential Carriage with Rigid Wheels", Academy of Sciences Publishing House, 1946, (Движение бездифференциальной тележки с жесткими колесами, in Russian).

35. Chudakov, E. A., "Motion of a Non-Differential Carriage with Elastic Wheels", Academy of Sciences Publishing House, 1946, (Движение бездифференциальной тележки с эластичными колесами, in Russian).

36. Chudakov, E. A., "Circulation of Parasitic Power in Mechanisms of a Non-Differential Automobile", Mashgiz. Moscow; 1950 (Циркуляция паразитной мощности в механизмах бездифференциального автомобиля, in Russian).

37. Chudakov, E. A., "Stability of an Automobile at Lateral Skid", USSR Academy of Sciences News, N 6, 1944 (Устойчивость автомобиля при боковом заносе, in Russian).

38. Chudakov, E. A., "Theory of Automobile", Academy of Sciences Publishing House, Vol. 1, 1944 (Теория автомобиля, in Russian).

39. Chudakov, E. A., Theory of Automobile, 3rd Edition, Moscow: State Publishing House of Machine-Building Literature.; 1950, (Теория автомобиля, in Russian).

40. Zimelev, G. V., Theory of Automobile, 2nd Edition, Moscow: Military Publishing House; 1957 (Теория автомобиля, in Russian).

41. Korotonoshko, N. I., High-Terrain Mobility Automobiles. Moscow: Mashgiz Publishing House; 1957 (Автомобили высокой проходимости, in Russian).

42. Bohm, F., "Uber die Fahrrichtungsstabilitat und die Seitenwindempfindlichkeit des Kraftwagens bei Geradeaus Fahrt", ATZ, N 5; 1961.

43. Bekker, M. G., Theory of Locomotion. Ann Arbor, MI: The University of Michigan Press; 1956.

44. Bekker, M. G., Introduction to Terrain-Vehicle Systems. Ann Arbor, MI: The University of Michigan Press; 1969.

45. Reece, A. R., "Principles of Soil-Vehicle Mechanics", Proceedings of the Institution of Mechanical Engineers, Vol. 180, N Part 2A2, 1965, pp. 45–67.

46. Gill, W., Vanden Berg, G. E., "Soil Dynamics in Tillage and Traction", Agriculture Handbook N 316. Agricultural Research Service, US Department of Agriculture; 1968.

47. Wong, J. Y., "Behavior of Soil beneath Rigid Wheels", Journal of Agricultural Engineering Research, Vol. 12, N 4, 1967, pp. 257–269.

48. Wong, J. Y., "Optimization of the Tractive Performance of Four-Wheel Drive Off-Road Vehicles", SAE Transactions, Vol. 79, 1970, pp. 2238–2246.

49. Wong, J. Y., "Discussion on "Stress Field under Slipping Rigid Wheels", Journal of Soil Mechanical and Foundation Division, Proceedings of American Society of Civil Engineers, Vol. 98, N SM9, 1972, pp. 977–981.

50. Wong, J. Y., Theory of Ground Vehicles, 4th Edition, New York: John Wiley; 2008 (the latest edition).

51. Wong, J. Y., Terramechanics and Off-Road Vehicle Engineering, 2nd Edition, Elsevier; 2010 (the latest edition).

52. Kushwaha, R. L., Shen, J., Soil-Machine Interaction: A Finite Element Perspective. CRC Press; 1998.

53. Bekker, M. G., "Russian Approach to Terrain-Vehicle Systems (An Exercise in Pragmatism and Continuity)", US Army Research Office, Technical Report No AD730341, 393 pages, 1971, http://www.dtic.mil/dtic/tr/fulltext/u2/730341.pdf

54. Ageikin, Y. S., All-Terrain Wheeled and Combined Locomotions. Moscow: Mashinostroenie Publishing House; 1972 (Вездеходные колесные и комбинорованные движители, in Russian).

55. Ageikin, Y. S., Mobility of Automobiles. Moscow: Mashinostroenie Publishing House; 1981 (Проходимость автомобилей, in Russian).

56. Antonov, D. A., Theory of Stability of Motion of Multi-Axle Automobiles. Moscow: Mashinostroenie Publishing House; 1978 (Теория устойчивости движения многоосных автомобилей, in Russian).

57. Litvinov, A. S., Handling and Stability of Automobile. Moscow: Machine-Building Publishing House; 1970 (Управляемость и устойчивость автомобиля, in Russian).

58. Farobin, Y. E., Theory of Transport Vehicles in Turn. Moscow: Mashinostroenie Publishing House; 1970 (Теория поворота транспортных машин, in Russian).

59. Petrushov, V. A., Shuklin, S. A., Moskovkin, V. V., Rolling Resistance of Automobiles and Trucks. Moscow: Mashinostroenie Publishing House; 1975 (Сопротивление качению автомобилей и автопоездов, in Russian).

60. Pirkovskij, Y. V., "General Formula of Rolling Resistance Power of an All-Wheel Drive Automobile", Avtomobil'naya Promishlennost' (Automobile Industry), N 1, pp. 34–35, 1973 (Общая формула мощности сопротивления качению полноприводного автомобиля, in Russian).

61. Korotonoshko, N. I., Shuklin, S. A., "Influence of Tire Design and Self-Locking Differentials on Terrain Mobility of Ural-375", Avtomobil'naya Promishlennost' (Automobile Industry), N 7, 1968 (Влияние конструкции шин и самоблокирующихся дифференциалов на проходимость Урал-375, in Russian).

62. Bocharov, N. F., Gusev, V. I., "Torque Distribution between the Drive Axles of an Automobile with Positive Engagement and Efficiency of Driveline Mechanisms", USSR University News - Machinebuilding", N 9, 1972 (Циркуляция паразитной мощности в механизмах бездифференциального автомобиля, in Russian).

63. Smirnov, G. A., "Influence of the Number and Location of Axles on Traction Performance of All-Wheel Drive Automobiles", Avtomobil'naya Promishlennost' (Automobile Industry), N 12, 1965 (Влияние числа и расположения мостов на тяговые свойства полноприводных автомобилей, in Russian).

64. Smirnov, G. A., Lelikov, O. P., "Influence of Driveline System on Traction Performance of 8x8 Automobiles", Avtomobil'naya Promishlennost' (Automobile Industry), N 6, 1967 (Влияние схемы силового привода на тягово-сцепные качества автомобилей типа 8x8, in Russian).

65. Antonov, A. S., Complex Powertrains. Theory of Power Flow and Transmitting System Computation. Leningrad: Mashinostroenie Publishing House; 1981 (Комплексные силовые приводы. Теория силового потока и расчет передающих систем, in Russian).

66. Aksenov, P. V., Multi-Axle Automobiles, 2nd Edition, Moscow: Mashinostroenie Publishing House; 1989 (Многоосные автомобили, in Russian).

67. Stepanova, E. A., Lefarov, A. K., Locking Differentials of Trucks. Moscow: State Publishing House of Machine-Building Literature; 1960 (Блокирующиеся дифференциалы грузовых автомобилей, in Russian).

68. Lefarov, A. K., Differentials of Automobiles and Trucks. Moscow: Mashinostroenie Publishing House; 1972 (Дифференциалы автомобилей и тягячей, in Russian).

69. Chudakov, D. A., Fundamentals of Theory and Calculations of Tractor and Automobile. Moscow: Kolos Publishing House; 1972 (Основы теории и расчета трактора и автомобиля, in Russian).

70. Guskov, V. V., Optimal Parameters of Farm Tractors. Moscow: Mashinostroenie Publishing House; 1967 (Оптимальные параметры сельскохозяйственных тракторов, in Russian).

71. Ul'yanov, N. A., Wheel Systems of Earth-Moving and Road Machines. Moscow: Mashinostroenie Publishing House; 1982 (Колесные движители землеройных и дорожных машин, in Russian).

72. L'vov, E. D., "Theory of Tractor", Mashgiz. Moscow; 1946–1960 (Теория трактора, in Russian).

73. Katsigin, V. V., Gorin, G. S., Zenkovich, A. A., Kidalinskaya, G. V., Neverov, A. I., Orda, A. N., Perspective Mobile Energy Vehicles for Agriculture. Minsk: Nauka i Technika Publishing House; 1982 (Перспективные мобильные энергетические средства (МЭС) для сельскохозяйственного производства, in Russian).

74. Segel, L., "The Tire as a Vehicle Component", Edited by B. Paul, K. Ullman, H. Richardson, Mechanics of Transportation Suspension Systems, ASME, AMD, Vol. 15, 1975.

75. Segel, L., "Force and Moment Response of Pneumatic Tires to Lateral Motion Inputs", Transactions ASME, Journal of Engineering for Industry, Vol. 88B, 1966.

76. Pacejka, H. B., "Tire and Vehicle Dynamics", SAE, 2002.

77. Lugner, P., Pacejka, H., Plochl, M., "Recent Advances in Tyre Models and Testing Procedures", Vehicle System Dynamics, Vol. 43, N 6-7, 2005, pp. 413–426.

78. Lugner, P., Plochl, M., "Tyre Model Performance Test: First Experiences and Results", Vehicle System Dynamics, Vol. 43, N Supplement 1, 2005, pp. 48–62.

79. Guo, K., Lu, D., "UniTire: Unified Tire Model for Vehicle Dynamics Simulation", Vehicle System Dynamics, Vol. 45, N Supplement 1, 2007, pp. 79–99.

80. "*Advanced Autonomous Vehicle Design for Severe Environments*", Editors: V. V. Vantsevich and M. V. Blundell, NATO Science for Peace and Security Series, D: Information and Communication Security - Vol. 44, IOS Press, 394 pages, 2015.

81. Irving, I. S., "Golden Arrow and the World's Speed Record", Institution of Automobile Engineers, N 24, 1929.

82. Hausen, H. Schlor, K., Aerodynamische Modellmessungen an Verschidenen Kraftwagenformen und Verhalten des Wirlichen Fahrzeuges bei Seitenwind, Deutsche Kraftfahrtforschung, N 2, 1938.
83. Gillespie, T. D., Fundamentals of Vehicle Dynamics. Warrendale, PA: SAE Press; 1992.
84. Heisler, H., Advanced Vehicle Technology, 2nd Edition, Butterworth-Heinemann, 2002, 653 pages.
85. Milliken, W. F., Milliken, D. L., Race Car Vehicle Dynamics. Warrendale, PA: SAE Press; 1995.
86. Andreev, A. F., Kabanau, V. I., Vantsevich, V. V., "Driveline of Ground Vehicles: Theory and Design", Edited by V.V. Vantsevich, Scientific and Engineering. Taylor & Francis, 2010, 792 pages.
87. Burgin, K., Adey, P. C., Beatham, J. P. "Wind Tunnel Tests on Road Vehicle Models Using a Moving Belt Simulation on Ground Effect", Journal of Wind Engineering and Industrial Aerodynamics, Vol. 22, 1986, pp. 227–236.
88. Aerodynamics of Road Vehicles, Edited by: Wolf-Heinrich Hucho, 4th Edition, SAE, 1998.
89. Kieselbach, R. J. F., Streamline Cars in Germany – Aerodynamics in the Construction of Passenger Vehicles, 1900 – 1945, Stuttgart: Kohlhammer, 1982.
90. Kieselbach, R. J. F., Aerodynamically Designed Commercial Vehicles, 1931 – 1961, Built on Chassis of Daimler-Benz, Krupp, Opel, Ford. Stuttgart: Kohlhammer, 1983.
91. Evgrafov A. N., Vysotsky, M. S., Aerodynamics of Wheeled Vehicles. Minsk: BelAutoTractorMashino-stroenie; 363 pages, 2001 (in Russian).
92. Bouferrouk, A., "On the Applicability of Trapped Vortices to Ground Vehicles", International Vehicle Aerodynamics Conference, 2014, pp. 101–111, http://eprints.uwe.ac.uk/24705/1/Bouferrouk-IMECHE-FINAL.pdf
93. Gilhaus, A., "The Influence of Cab Shape on Air Drag of Trucks", Journal of Wind Engineering and Industrial Aerodynamics, 9, 1981, pp. 77–87.
94. "Road and Off-Road Vehicle System Dynamics", Edited by G. Mastinu and M. Ploechl, Handbook, 2014, CRC Press, 1694 pages.
95. Wood, R. M., Bauer, S. X. S., Simple and Low-Cost Aerodynamic Drag Reduction Devices for Tractor-Trailer Trucks, SAE 2003-01-3377, SAE International, 2003.
96. Leuschen, J., Cooper, K., "Full-Scale Wind Tunnel Tests of Production and Prototype, Second-Generation Aerodynamic Drag-Reduction Devices for Tractor-Trailers", SAE2006-06CV-222, SAE International, 2006.
97. Qi X. -N., Liu Y. -Q., Du G. –S., "Experimental and Numerical Studies of Aerodynamic Performance of Trucks", Journal of Hydrodynamics, Vol. 23, N 6, 2011, pp. 752–758.
98. Chowdhury H., Moria, H., Ali, A., Khan, I., Alam F., Watkins, S., "A Study on Aerodynamic Drag of a Semi-Trailer Truck", 5th BSME International Conference on Thermal Engineering, Vol. 56, 2013, pp. 201–205.
99. Hyams, D. G., Sreenivas, K., Pankajakshan, R., Nichols, D. S., Briley, W. R., Whitfield, D. L., "Computational Simulation of Model and Full Scale Class 8 Trucks with Drag Reduction Devices", Computers & Fluids, Vol. 41, 2011, pp. 27–40.
100. Nakashima, T., Tsubokura, M., Vázquez, M., Owen, H., Doi, Y., "Coupled Analysis of Unsteady Aerodynamics and Vehicle Motion of a Road Vehicle in Windy Conditions", Computers & Fluids, Vol. 80, 2013, pp. 1–9.
101. Han, M.-W., Rodrigue, H., Cho, S., Song, S.-H., Wang, W., Chu, W.-S., Ahn, S.-H., "Woven Type Smart Soft Composite for Soft Morphing Car Spoiler", Composites, Part B, Vol. 86, 2016, pp. 285–298.
102. Wang, W., Liu, P., Tian, Y., Qu Q., "Numerical Study of the Aerodynamic Characteristics of High-Lift Droop Nose with the Deflection of Fowler Flap and Spoiler", Aerospace Science and Technology, Vol. 48, 2016, pp. 75–85.
103. McNabola, A., "Spoiling Air Pollution Dispersion: A Numerical Investigation of Exhaust Plume Dispersion from Cars with Rear Spoilers", Transportation Research Part D, Vol. 16, 2011, pp. 296–301.
104. Stansby, P. K., Pinchbeck, J. N., Henderson, T., "Spoilers for the Suppression of Vortex-Induced Oscillations (Technical Note)". Applied Ocean Research, Vol. 8, N 3, 1986, pp. 169–173.
105. Fukuda, H., Yanagimoto, K., China, H., Nakagawa, K., "Improvement of Vehicle Aerodynamics by Wake Control", JSAE, Review, JSAE9532236, Vol. 16, 1995, pp. 151–155.
106. Kim, M. H., Kuk, J. Y., Chyun, I. B., A Numerical Simulation on the Drag Reduction of Large-Sized Bus Using Rear-Spoiler, SAE2002-01-3070.
107. Patten, P., McAuliffe, B., Mayda, W., Tanguay, B., "Review of Aerodynamic Drag Reduction Devices for Heavy Trucks and Buses, Centre for Surface Transportation Technology, Project 54-A3578", Technical Report, May 11, 2012, 120 pages.
108. Roshko, A., Koeing, K., "Interaction Effects on the Drag of Bluff Bodies in Tandem", Aerodynamics Drag Mechanisms of Bluff Bodies and Road Vehicles, New York: Plenum Press; 1978, pp. 253–286.

109. Im Kampf dem Luftwiderstand, Zeitschrift Nutzfahrzeuge, August 1978, pp. 39-41.
110. Gohring, E., Kramer, W., "Seitliche Fahrgestellverkleidungen fur Nutzfahrzeuge", ATZ, Vol. 89, 1987, pp. 481–488.
111. Bonnet, C., Fritz, H., Fuel Consumption Reduction in a Platoon: Experimental Results with Two electronically Coupled Trucks at Close Spacing; SAE 2000-01-3056, August 2000.
112. Brinker, C. J., Superhydrophobic Coating, Sandia National Laboratories, 2008, 27 pages, http://www.sandia.gov/research/research_development_100_awards/_assets/documents/2008_winners/Superhydrophobic_SAND2008-2215W.pdf
113. National Academy of Sciences. "Technologies and Approaches to Reducing the Fuel Consumption of Medium- and Heavy-Duty Vehicles", Committee to Assess Fuel Economy Technologies for Medium- and Heavy-Duty Vehicles. The National Academic Press, 2010.
114. Faerber, H., Das Autobuch, Offenburg, Burda, 1956.
115. Kramer, C., Gerhard, H., Jaeger, E., Stein, H., "Windkanalstudien zur Aerodynamik der Fahrzeugunterseite", Aahhen, 1974, pp. 71–83.
116. Mitschke, M., Dynamik der Kraftfahrzeuge, 2nd Edition, Berlin: Springer; 1990.
117. Okumura, K., Kuriyama, T., Transient Aerodynamic Simulation in Crosswind and Passing an Automobile, SAE 970404.
118. Baker, C. J., Reynolds, S., "Wind-Induced Accidents of Road Vehicles", Accidents Analysis & Prevention, Vol. 24, N 6, 1992, pp. 559–575.
119. Baker, C. J., "High Sided Articulated Road Vehicles in Strong Cross Winds", Journal of Wind Engineering and Industrial Aerodynamics, Vol. 31, 1988, pp. 67–85.
120. Braess, H., Burst, H., Hamm, L., Hanner, R., "Verbesseung der Fahreigenschafften durch Verringerung des Aerodynamischen Auftriebs", ATZ, Vol. 77, 1975, pp. 119–124.
121. Luo, Y.,Yuan L., Li, J., Wang, J., "Boundary Layer Drag Reduction Research Hypotheses Derived from Bio-Inspired Surface and Recent Advanced Applications", Micron, Vol. 79, 2015, pp. 59–73.
122. Cheng, S., Tsubokura, M., Nakashima, T., Nouzawa T, Okada Y., "A Numerical Analysis of Transient Flow Past Road Vehicles Subjected to Pitching Oscillation", Journal of Wind Engineering and Industrial Aerodynamics, Vol. 99, 2011, pp. 511–22.
123. Beauvais, F. N., Transient Nature of Wind Gust Effect on an Automobile. SAE67068, 1967.
124. Milliken, W. F., Dell'Amico, F., Rice, R. S., The Static Directional Stability and Control of an Automobile, SAE 760712, 1976.
125. Otto, H., Lastwechselreaktionen von PKW bei Kurvenfahrt, Dissertation TU Braunschweig, 1987.
126. Zomotor, A., Fahrwerketechnik, Fahrverhalten. Wurzburg: Vogel Verlag; 1987.
127. Burkhard, M., Burg, H., Berechnung und Rekpnstrktion des Bremsverhaltens von PKW. Kippenheim: Verlag Information; 1988.
128. Birzeil, F., "Die Einwirkung von Seitenwindkraften auf den Stressenverkehr", Zeitschrift fur Verkehrssicherheit", Heft, Vol. 2, 1962.
129. Schiehlen, W., "Research Trends in Multibody System Dynamics", Journal of Multibody System Dynamics, Vol. 18, 2007, pp. 3–13.
130. Khachaturov, A. A., et al., Dynamics of Road-Tire-Automobile-Driver System. Moscow: Mashinostroenie, Publishing House; 1976 (Динамика системы дорога-шина-автомобиль-водитель, in Russian).
131. Andreev, A. F., Vantsevich, V. V., Lefarov, A. K., Differentials of Wheeled Vehicles. Moscow: Mashinostroenie Publishing House; 1987 (Дифференциалы колесных машин, in Russian).
132. Litvinov, A. S., Farobin, Y. E., Automobile. Theory of Operational Properties. Moscow: Mashinostroenie Publishing House, 1989 (Автомобиль. Теория эксплуатационных свойств, in Russian).
133. Smirnov, G. A., Theory of Motion of Wheeled Vehicles, 2nd Edition, Moscow: Mashinostroenie Publishing House; 1990 (Теория движения колесных машин, the latest edition; in Russian).
134. Grinchenko, I. V., et al., High-Terrain Mobility Wheeled Automobiles. Moscow: Mashinostroenie Publishing House; 1967 (Колесные автомобили высокой проходимости, in Russian).
135. Antonov, A. S., et al., Army Automobiles. Theory. Moscow: Department of Defense Publishing House; 1970 (Армейские автомобили. Теория, in Russian).
136. Bespalov, S. I., et al., "Theory of Motion of Battle Wheeled Vehicles", Department of Battle Vehicles and Automotive Training, Marshall R. Ya. Malinovsky Military Armored Forces Academy. Moscow: Department of Defense Publishing House; 1993 (Теория движения боевых колесных машин, in Russian).
137. Platonov, V. F., All-Wheel Drive Automobiles, 2nd Edition, Moscow: Mashinostroenie Publishing House; 1989 (Полноприводные автомобили, in Russian).

138. Platonov, V. F., et al., Multi-Purpose Track Chassis. Moscow: Mashinostroenie Publishing House; 1998 (Многоцелевые гусеничные шасси, in Russian).

139. Wehage, R. A., "Vehicle Dynamics", Journal of Terramechanics, Vol. 24, N 4, 1987, pp. 295–312.

140. McHenry, R., "The Role of Vehicle Dynamics Simulation in Highway Safety Research", The ASME Milliken Invited Lecture, ASME 16th International Conference on Advanced Vehicle Technologies. Buffalo, NY; August 17–20, 2014.

141. Amold, M., et al., "Numerical Methods in Vehicle System Dynamics: State of the Art and Current Developments", Vehicle System Dynamics, Vol. 49, N 7, 2011, pp. 1159–1207.

142. Rill, G., Simulation von Kraftfahrzeugen. Braunschweig/Wiesbaden: Friedr. Vieweg & Sohn; 1994 (in German).

143. Rill, G., Road Vehicle Dynamics. Fundamentals and Modelling. CRC Press; 2012.

144. Tomizuka, M., Hedrick, K., "Advanced Control Methods for Automotive Applications", Vehicle System Dynamics, Vol. 24, N 6-7, 1995, pp. 449–468.

145. Sharp, R. S., Peng, H., "Vehicle Dynamics Applications of Optimal Control Theory", Vehicle System Dynamics, Vol. 49, N 7, 2011, pp. 1073–1111.

146. Ulsoy, A. G., Peng, H., Çakmakci, M., Automotive Control Systems. Cambridge University Press; 2012.

147. Johansson, R., Rantzer, A., (Editors), Nonlinear and Hybrid Systems in Automotive Control. Springer; 2003.

148. van Zanten, A., et al., "The Vehicle Dynamics Control System of Bosch", SAE Paper No. 950759, Warrendale, PA, 1995.

149. van Zanten, A., et al., "Simulation for the Development of the Bosch VDC", SAE paper No. 960486, Detroit, 1996.

150. van Zanten, A., "Bosch ESP System: 5 Years of Experience", SAE Automotive Dynamics Stability Conference, Paper No. 2000-01-1633, Troy, MI; 2000.

151. Margolis, D. L., Tylee, J. L., Hrovat, D., "Heave Mode Dynamics of a Tracked Air Cushion Vehicle with Semi-Active Airbag Secondary Suspension", ASME J. Dynamic Systems, Measurement and Control, Vol. 97, N 4, 1975, pp. 399–407.

152. Ellis, J. R., Vehicle Dynamics. London: Business Book Limited; 1969.

153. Cao, D., Song, X., Ahmadian, M., "Editors' Perspectives: Road Vehicle Suspension Design, Dynamics, and Control", Vehicle System Dynamics, Vol. 49, N 1-2, 2011, pp. 3–28.

154. Ragsell, K. M., "Optimization as a Tool for Automotive Design", SAE paper 800432, 1980.

155. Siddall, J. S., Optimal Engineering Design. Marcel Dekker; 1982.

156. Arora, J. S., Introduction to Optimum Design. McGraw-Hill; 1989.

157. Mitschke, M., Wallentowitz, H., Dynamik der Kraftfahrzeuge, 4th Edition, Berlin: Springer Verlag; 2004 (in German).

158. Zomotor, A., Fahrwerktechnik: Fahrverhalten. Wurzburg: Vogel Buchverlag; 1991 (in German).

159. Edelmann, J., et al., "A Passenger Car Driver Model for Higher Lateral Accelerations", Vehicle System Dynamics, Vol. 45, N 12, 2007, pp. 1117–1129.

160. Alvarez, L., Horowitz, R., Li, P., "Traffic Flow Control in Automated Highway Systems", IFAC Journal of Control Engineering Practice, Vol. 7, N 9, Sept. 1999, pp. 1071–1078.

161. Kondo, M., "On a basic Relationship Existing between Steering and the Motion of Automobile", Transactions of JSAE, N 5, 1958, pp. 40–43 (in Japanese).

162. Yamakawa, S., "Studies on Under-and Oversteering Characteristics of an Automobile Subject to Steering Action", Transactions of JSAE, N 5, 1966, pp. 888–892 (in Japanese).

163. Ohno, T., "Steering Control on a Curved Course", Transactions of JSAE, Vol. 20, N 5, 1966, pp. 413–419 (in Japanese).

164. Yoshimoto, K., "Simulation of Man-Automotive System", Transactions of JSAE, Vol. 25, N 10, 1971, pp. 1059–1064 (in Japanese).

165. Vantsevich, V. V., "Synthesis of Driveline Systems of Multi-Wheel Drive Vehicles", ScD Dissertation (Doctor of Technical Sciences - the highest degree in the U.S.S.R.) in Automobile and Tractor Engineering, Belarusian National Technical University, awarded by the State Supreme Attestation Board, Russia, 1992 (Синтез схем привода к ведущим мостам и колесам многоприводных траспортно-тяговых машин, in Russian).

166. Vantsevich, V. V., "A New Effective Research Direction in the Field of Actuating Systems for Multi-Drive Vehicles", International Journal of Vehicle Design (UK), Vol. 15, N 3/4/5, 1994, pp. 337–347.

167. Vantsevich, V. V., Vysotski, M. S., Gileles, L. K., "A New Direction in Development of Theory of Motion of Mobile Machines", Avtomobilnaya Promishlennost (Automotive Industry), Moscow, N 3, 1998 (Новое направление в развитии теории мобильных машин, in Russian).

168. Karnopp, D., "On Inverse Equations for Vehicle Dynamics", Vehicle System Dynamics, Vol. 20, N 6, January 1991, pp. 371–379.
169. Trujillo, D. M., Busby, H. R., Practical Inverse Analysis in Engineering. CRC Press; 1997.
170. Andreasson, J., Bünte, T., "Global Chassis Control Based on Inverse Vehicle Dynamics Models", Vehicle System Dynamics, Vol. 44, N Supplement 1, January 2006, pp. 321–328.
171. Vantsevich, V. V., "Vehicle Dynamics as the Second Dynamics Problem", International Journal of Vehicle Design (UK), Vol. 25, N 3, 2001, pp. 165–169.
172. Vantsevich, V. V., "Optimization of Mass and Geometric Vehicle Parameters for Multi-Drive Wheel Trucks", International Journal of Vehicle Design (UK), Vol. 25, N 3, 2001, pp. 170–181.
173. Vantsevich, V. V., *et al.*, "An Integrated Approach to Autonomous Vehicle Systems: Theory and Implementation", International Journal of Vehicle Autonomous Systems (UK), Vol. 1, N 3/4, 2003, pp. 271–350.
174. Karnopp, D. C., Crosby, M. J., Harwood, R. A., "Vibration Control Using Semi-Active Force Generators", ASME Journal of Engineering Industry, Vol. 96, N 2, 1974, pp. 619–626.
175. Bastow, D., Howard, G., Car Suspension and Handling. London: Pentech Press; 1993.
176. Hrovat, D., "Survey of Advanced Suspension Developments and Related Optimal Control Applications", Automatica, Vol. 33, N 10, 1997, pp. 1781–1817.
177. Blundell, M. V., Harty, D., The Multibody Systems Approach to Vehicle Dynamics. Elsevier Science; ISBN, 2004 (Also published by the SAE in North America).
178. Karnopp, D., Vehicle Stability. New York: Marcel Dekker, Inc.
179. Dixon, J. C., Tires, Suspension and Handling, 2nd Edition, 1996, SAE, 621 pages.
180. Furukawa, Y., Abe, M., "Advanced Chassis Control Systems for Vehicle Handling and Active Safety", Vehicle System Dynamics, Vol. 28, 1997, pp. 59–86.
181. Abe, M., "Vehicle Dynamics and Control for Improving Handling and Active Safety: From Four-Wheel Steering to Direct Yaw Moment Control", In Proceeding of IMechE, Vol. 213, N part K, 1999, pp. 87–101.
182. Pirkovsky U. V., Shukhman, S. B., "Theory of Motion of AWD-Automobile", Elit-2000, 2001, 230 pages (Теория движения полноприводного автомобиля, in Russian).
183. Belousov, B. N., Popov, S. D., Super-Duty Wheeled Transport Equipment. MSTU Publishing House; 728 pages (Колесные транспортные средства особо большой грузоподъемности, in Russian)
184. Vantsevich, V. V., Vysotski, M. S., Gileles, L. K., Mobile Transport Machines: Interaction with Environment. Belarus, Minsk: Belaruskaya Navuka Publishing House; 1998, 303 pages (Мобильные транспортные машины: взаимодействие со средой функционирования, in Russian).
185. Karnopp, D., Margolis, D. L., Engineering Applications of Dynamics. Hoboken, New Jersey: John Wiley & Sons, Inc., 2008.
186. G. Mastinu, et al., "Integrated Controls of Lateral Vehicle Dynamics", Vehicle System Dynamics, Vol. 23, N Supplement 23, 1994, pp. 358–377.
187. S. Sato, S., et al., "Integrated Chassis Control System for Improved Vehicle Dynamics", In proceedings of the AVEC92, 1992, pp. 413–418.
188. Vlacic, L., Parent, M., Harashima, F., (Eds), Intelligent Vehicle Technologies. Theory and Applications. SAE International; 2001.
189. Pauwelussen, J. B., Pacejka, H. B., (Eds), Smart Vehicles. Lisse, the Netherlands: Swets & Zeitlinger; 1995.
190. Shladover, S., "Review of the State of Development of Advanced Vehicle Control Systems (AVCS)", Vehicle Systems Dynamics, Vol. 24, 1995, pp. 551–595.
191. Vantsevich, V. V., Vysotski, M. S., Doubovik, D. A., "Control of the Wheel Driving Forces as the Basis for Controlling Off-Road Vehicle Dynamics", SAE 2002 Transactions – Journal of Commercial Vehicles, pp. 452–459.
192. Vantsevich, V. V., "Inverse Wheel Dynamics", IMECE2006-13787, ASME Congress, Chicago, Illinois, November 5–10, 2006.
193. Vantsevich, V. V., Zakrevskij, A. D., Kharytonchyk, S. V., "Heavy-Duty Truck: Inverse Dynamics and Performance Control", IMECE2007-42659, ASME Congress, Seattle, Washington, November 10–16, 2007.
194. Vantsevich, V. V., "Multi-Wheel Drive Vehicle Energy/Fuel Efficiency and Traction Performance: Objective Function Analysis", Journal of Terramechanics, Vol. 44, N 3, July 2007, pp. 239–353.
195. Vantsevich, V. V., "Power Losses and Energy Efficiency of Multi-Wheel Drive Vehicles: A Method for Evaluation", Journal of Terramechanics, Vol. 45, N 3, 2008, pp. 89–101.
196. Vantsevich, V. V., Gray J. P., "Fuel Economy and Mobility of Multi-Wheel Drive Vehicles: Modelling and Optimization Technology", National Defense Industrial Association Ground Vehicle Systems Engineering and Technology Symposium, August 18–20, 2009.

197. Vantsevich, V. V., "Holonomic Inverse Dynamics and Performance Optimization for Autonomous Ground Vehicle Applications", USNCTAM 201-425, 16th US National Congress of Theoretical and Applied Mechanics,: State College, Pennsylvania, USA, June 27–July 2, 2010.

198. Chatillon, M. M., et al., 'Hierarchical Optimization of the Design Parameters of a Vehicle Suspension System', Vehicle System Dynamics, Vol. 44, 2006, pp. 817–839.

199. Gobbi, M., Levi, F., Mastinu, G., "Multi-Objective Stochastic Optimization of the Suspension System of Road Vehicles", Journal of Sound and Vibration, Vol. 298, 2006, pp. 1055–1072.

200. Georgiou, G., Verros, G., Natsiavas, S., "Multi-Objective Optimization of Quarter-Car Models with a Passive or Semiactive Suspension System", Vehicle System Dynamics, Vol. 45, 2007, pp. 77–92.

201. Gordon, T. M., Howell, M., Brandao, F., "Integrated Control Methodologies for Road Vehicles", Vehicle System Dynamics, Vol. 40, 2003, pp. 157–187.

202. Mokhiamar, O., Abe, M., "Simultaneous Optimal Distribution of Lateral and Longitudinal Tire Forces for the Model Following Control", Transactions of the ASME, Journal of Dynamic Systems, Measurement and Control, Vol. 126, December 2004, pp. 753–763.

203. Hirano, Y., et al., "Integrated Control System of 4WS and 4WD by H∞ Control", Proceedings of the International Symposium on Advanced Vehicle Control - AVEC'92, Paper No. 923075, Yokohama, Japan, 1992.

204. Manning, W. J., Crolla, D. A., "A Review of Yaw Rate and Sideslip Controllers for Passenger Vehicles", Transactions of the Institute of Measurement and Control, Vol. 29, 2007, pp. 117–135.

205. He, J., Crolla, D. A., Levesley, M. C., Manning, W. J., "Coordination of Active Steering, Driveline and Braking for Integrated Vehicle Dynamics Control", Proc. IMechE Part D, Journal of Automobile Engineering, Vol. 220, 2006, pp. 1401–1421.

206. Xu, L., Tseng, H. E., "Robust Model-Based Fault Detection for a Roll Stability Control System", IEEE Transaction on Control System Technology, Vol. 15, N 3, 2007, pp. 519–528.

207. Mastinu, G., Ploechl, M., (Eds.), Road and Off-Road Vehicle System Dynamics Handbook. Boca Raton, FL: CRC Press; 2014.

208. Ashok, P., Tesar, D. "The Need for Performance Map Based Decision Theory", IEEE Systems Journal, Vol. 7, NO. 4, December 2013.

209. Tesar, D., "Next Wave of Technology: Technical Development in Intelligent Machines", Updated Collection of White Papers at UTexas, Nov. 10, 2010, 140 pages.

210. Belousov, B. N., Ksenevich, T. I., Vantsevich, V. V., "Load Estimation of an Open-Link Locomotion Module for Robotic and Commercial Multi-Wheel Applications", 2013 SAE Commercial Vehicle Engineering Congress.

211. Belousov, B. N., Ksenevich, T. I., Vantsevich, V. V., "An Active long-Travel, Two Performance Loop Control Suspension of an Open-link Locomotion Module for Off-Road Applications", 2014 SAE Commercial Vehicle Engineering Congress.

212. Winterkorn, M., "VW Group has Big Plans for Low Fuel Consumption", Automotive Engineering International, June, 2013, pp. 15–16.

213. Reichart, G., Haneberg, M., "Key Drivers for a Future System Architecture in Vehicles", 2004-21-0025, Convergence Transportation Electronics Association.

214. Karnopp, D. C., Margolis, D. L., Rosenberg, R. C., System Dynamics. Modelling and Simulation of Mechatronic Systems, 4th Edition, John Wiley; 2006.

215. Jurgen, R. K., Automotive Electronics Handbook. New York: McGraw-Hill, Inc.; 1995.

216. De Silva, C. W., Modelling and Control of Engineering Systems. Boca Raton, FL: Taylor & Francis/CRC Press; 2009.

217. De Silva, C. W., Mechatronics: An Integrated Approach. Boca Raton, FL: Taylor & Francis/CRC Press; 2004.

218. Hrovat, D., Asgari, J. and Fodor, M., "Automotive Mechatronic Systems", in Mechatronic Systems, Techniques and Applications: Volume 2 – Transportation and Vehicle Systems, C. T. Leondes, Editor, Gordon and Breach Science Publishers; 2000, pp. 1–98.

219. Wiener, N., Cybernetics: Or Control and Communication in the Animal and the Machine. Paris: Hermann & Cie & Camb. Mass. (MIT Press); 1948.

220. Gray, J. P., Vantsevich, V. V., Paldan, J., "Agile Tire Slippage Dynamics for Radical Enhancement of Vehicle Mobility", Journal of Terramechanics, Vol. 65, 2016, pp. 14–37.

221. Vantsevich, V. V., Demkiv, L., Klos, S., Misko, S., Moradi, L., "An Experimental Study of Longitudinal Tire Relaxation Constants for Vehicle Traction Dynamics Modeling", DSCC2019-8994, ASME 2019 Dynamic Systems and Control Conference.

222. Vantsevich, V. V., Gorsich, D., Lozynskyy, A., Demkiv, L., Klos, S., "A Reinforcement Learning Enhanced Fuzzy Control for Real-Time off-Road Traction System", IAVSD 2017 International Symposium on Vehicle Dynamics, Paper #154. Gothenburg, Sweden, August 12–16, 2019.

223. Yi, J., et al., "On the Dynamic Stability and Agility of Aggressive Vehicle Maneuvers: A Pendulum-Turn Maneuver Example", IEEE Transactions on Control Systems Technology, Vol. 20, N 3, 2012, pp. 663–676.

224. Zhou, J., Lu, J., Peng, H, "Vehicle Stabilization in Response to Exogenous Impulsive Disturbances to the Vehicle Body", International Journal of Vehicle Autonomous Systems, Vol. 8, N 2/3/4, 2010.

225. Yamakado, M., et al, "Improvement in the Vehicle Agility and Stability by G-Vectoring Control", Vehicle Systems Dynamics, Vol. 48, Supplement, 2010, pp. 231–254.

226. Chakraborty, I., Tsiotras, P., Lu, J., "Vehicle Posture Control through Agressive Maneuvering for Mitigation of T-bone Collisions", 50th IEEE Conference on Decision and Control and European Control Conference. Orlando, FL, December 11–15, 2011.

227. Salama, M., Vantsevich, V. V., "Tire-Terrain Modeling for Agile Vehicle Dynamics Analysis", 7th American Conference of the International Society for Terrain Vehicle Systems. Tampa, Florida, November 3–7, 2013.

228. De Silva, C. W., Sensors and Actuators — Control System Instrumentation. Boca Raton, FL: Taylor & Francis/CRC Press; 2007.

229. Zeilinger, M. N., et al, "On Real-Time Robust Model Predictive Control", Automatica, Vol. 50, N 3, 2013, pp. 683–694.

230. Del Re, L., et al., (Eds.), Automotive Model Predictive Control: Models, Methods and Applications. Berlin Heilderberg: Springer Verlag; 2010.

231. Falcone, F. et al., "Linear Time Varying Model Predictive Control and Its Application to Active Steering Systems: Stability Analysis and Experimental Validation, International Journal of Robotics and Nonlinear Control, Vol. 18, N 8, 2008, pp. 862–875.

232. Borrelli, F., et al., "MPC-based Approach to Active Steering for Autonomous Vehicle Systems", International Journal for Vehicle Autonomous Systems, Vol. 3, N 2/3/4, 2005, pp. 265–291.

233. Borrelli, F., et al., "An MPC/Hybrid System Approach to Traction Control", IEEE Transaction on Control Systems Technology, Vol. 14, N 3, pp. 541–552.

234. Chang, S., Gordon, T., Model-based Predictive Control of Vehicle Dynamics", International Journal of Vehicle Autonomous Systems, Vol. 5, N 1/2, 2007, pp. 3–27.

235. Li, L., Sandu, C., Lee, J. H., Liu, Q., "Stochastic Modelling of Tire-Snow Interaction Using a Polynomial Chaos Approach", Special Issue on Snow, Journal of Terramechanics, Vol. 46, N 4, June 19, 2009, pp. 165–188 (24).

236. Canudas-de-Wit, C., et al., "Dynamic Friction Models for Road/Tire Longitudinal Interaction", Vehicle System Dynamics, Vol. 39, N 3, 2003, pp. 189–226.

237. Deur, J., et al., "A 3D Brush-type Dynamic Tire Friction Model", Vehicle System Dynamics, Vol. 42, N 3, 2004, pp. 133–173.

238. Yi, K. Hedrick, K. "Estimation of Tire-Road Friction Using Observer Based Identifiers", Vehicle System Dynamics, Vol. 31, N 4, 1999, pp. 233–261.

239. Zhu, Y., et al., "Sensitivity Analysis of Vehicle Dynamics Based on Multibody Models." Paper no. DETC2013-13212, 9 pages, Proceedings of ASME 2013 IDETC/CIE 15th International Conference on Advanced Vehicle Technologies (AVT). Portland, OR; Aug. 4–7, 2013.

240. Ahmadian, M., Blanchard, E., "Non-Dimensionalised Closed-Form Parametric Analysis of Semi-Active Vehicle Suspensions Using a Quarter-Car Model", Vehicle System Dynamics, Vol. 49, N 1–2, 2011, pp. 219–235.

241. Boggs, C., Ahmadian, M., Southward, S., "Efficient Empirical Modelling of a High- Performance Shock Absorber for Vehicle Dynamics Studies", Vehicle System Dynamics: International Journal of Vehicle Mechanics and Mobility, Vol. 48, N 4, 2010, pp. 481–505.

242. Vantsevich, V. V., Demkiv, L., Klos, S., "Analysis of Tire Relaxation Constants for Modeling Vehicle Traction Performance and Handling", ASME 2018 Dynamic Systems and Control Conference, DSCC2018-9026, September 30 – October 3, 2018.

243. Vantsevich, V. V., "Agile Dynamics Fundamentals for Tire Slippage Modeling and Control", ASME 16th International Conference on Advanced Vehicle Technologies. Buffalo, NY; August, 2014.

244. Vantsevich, V. V., Principal Investigator, Aerodynamic Intelligent Morphing System (A-IMS) for Autonomous Smart Utility Truck Safety and Productivity in Severe Environments, NSF S&AS-1849264, 2019–2023.

245. Besselink, B. C., "Development of a Vehicle to Study the Tractive Performance of Integrated Steering-Drive Systems", Journal of Terramechanics, Vol. 41, 2004, pp. 187–198.

246. Vantsevich, V. V., "Vehicle Systems: Coupled and Interactive Dynamics Analysis", Vehicle System Dynamics, Vol. 51, N 11, 2014, pp. 1489–1516.

247. Vantsevich, V. V., "AWD Vehicle Dynamics and Energy Efficiency Improvement by Means of Interaxle Driveline and Steering Active Fusion", ASME 15th International Conference on Advanced Vehicle and Tire Technologies. Portland, OR; August 4–7, 2013.

248. Patterson, M., et al., "Fusion of Driving and Braking Tire Operational Modes and Analysis of Traction Dynamics and Energy Efficiency of a 4x4 Loader", Journal of Terramechanics, Vol. 50, N 1, 2013, pp.133–152.

249. Vantsevich, V. V., Bortolin, G. "Axle Drive and Brake-based Traction Control Interaction", SAE International Journal of Commercial Vehicles, Vol. 4, N 1, 2011, pp. 49–55.

250. Paldan, J., Wei Yao, Vantsevich, V. V. "Control Fusion of a Hybrid Electric Transmission and Wheel Power Distribution System", ASME 2012 Dynamic Systems and Control Conference Ft. Lauderdale, FL; October 16–20, 2012.

251. Vantsevich, V. V., Paldan, J., Gray, J. P., "Feasibility of Hybrid-Electric Transfer Case Analysis: Lateral Vehicle Dynamics, Energy Consumption, Design Characteristics", ASME Dynamic Systems and Control Conference, San Antonio, TX; October 22–24, 2014.

252. Brooks, R. R., Yun, S. B., Deng, J., "Cyber-Physical Security in Automotive Information Technology" in Handbook on Securing Cyber-Physical Critical Infrastructure, Elsvier; 2012, pp. 655–676.

253. *Vehicle Dynamics Terminology,* SAE Standard J670, revised in 2008.

2 Essential Kinematics and Dynamics

This chapter deals with basic *kinematics* and *dynamics* of vehicle in 3D motion. Kinematics is the study of the geometry of motion, without reference to the cause of the motion. It is used to relate position and displacement, velocity and acceleration as the first- and second-order time derivatives of the displacement. Sometimes, jerk as the first-time derivative of acceleration is analyzed, too. Hence, kinematics of the vehicle motion refers to geometric and time-based properties of the motion. The study of the relationship between motion characteristics, forces and moments which cause that motion is the subject of dynamics. However, not all aspects of kinematics and dynamics will be considered through this book. The presentation is limited to only those fundamentals which relate to particular problems of vehicle dynamics. The essential kinematic and dynamic equations, which form the basis of dynamic equations of vehicle motion, are introduced in later chapters.

The vehicles studied in this book are assumed as rigid bodies. In many technical problems, a rigid body is considered as a particle with a mass, but not size if rotation of the body can be neglected. A motor vehicle is made up of many components distributed across its exterior surface. In some considerations of dynamic effects on vehicle motion, the vehicle will be considered as a lumped mass concentrated in the vehicle's center of gravity (the point at which the weight force acts) if rotational movements can be neglected. But for most elementary analyses, such as in study of vehicle kinematics, the vehicle is treated as a rigid body, of which its geometrical size and rotation are not negligible.

In general, vehicle motion has two components, translational and rotational. Translation is a motion in which a body shifts from one point to another. In rectilinear motion as a case of translational motion, a body moves along a straight line. Rotation is introduced when the vehicle rotates about a certain axis which may be either outside or intersect the vehicle. To describe physical relations between kinematic and dynamic parameters of vehicles, a coordinate system will be attached to the vehicle gravity center and then physical quantities of vehicle motions will be considered in this coordinate system. Moreover, in many vehicle dynamics technical problems, a mathematical description of motion in that coordinate system needs to be converted to another coordinate system as one wishes to express motion characteristics in thats coordinate system.

2.1 VECTOR DESCRIPTIONS AND TRANSFORMATIONS

To describe a motion of a particle or a body, space and time are essential. Apart from that, the position and displacement, velocity and acceleration are needed to characterize the motion. These variables have to be expressed in a reference system, or a *coordinate system*. The vector description and its transformation between the coordinate systems form a basic knowledge in kinematics.

VECTOR DEFINITION

Once a coordinate system is established any point in the space can be located with a 3×1 position vector.

$$\mathbf{r} = \begin{bmatrix} x \\ y \\ z \end{bmatrix} \qquad (2.1)$$

DOI: 10.1201/9781003134305-2

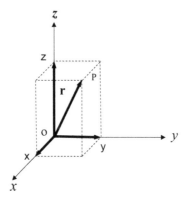

FIGURE 2.1 Vector in an orthogonal coordinate system.

Figure 2.1 shows how the coordinate system implements three mutually orthogonal vectors. The position of a particle in point P is defined at any instant by its rectangular coordinates x, y and z. Each of the coordinates can be thought of as a result of projecting the vector onto the corresponding axis. When the particle moves, the position vector describes its position in the coordinate system, and the coordinates of the particle are functions of time t. The velocity of the particle is a vector, which is defined by the derivative of the position vector with respect to t,

$$\mathbf{v} = \frac{d\mathbf{r}}{dt} = \lim_{\Delta t \to 0} \frac{\Delta \mathbf{r}(t + \Delta t) - \Delta \mathbf{r}(t)}{\Delta t} \tag{2.2}$$

Following (2.1), the velocity is given by

$$\dot{\mathbf{r}} = \begin{bmatrix} \dot{x} \\ \dot{y} \\ \dot{z} \end{bmatrix} = \begin{bmatrix} v_x \\ v_y \\ v_z \end{bmatrix} \tag{2.3}$$

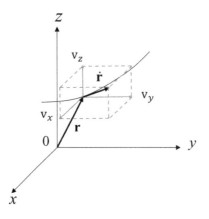

FIGURE 2.2 Velocity as a time derivative of vector **r**.

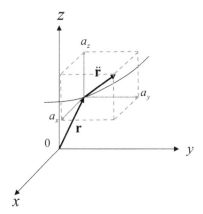

FIGURE 2.3   Acceleration as the second-time derivative of vector **r**.

Figure 2.2 illustrates the geometric relationship between the velocity and the position vector. Similarly, we obtain the acceleration by differentiating the velocity,

$$\ddot{\mathbf{r}} = \begin{bmatrix} \ddot{x} \\ \ddot{y} \\ \ddot{z} \end{bmatrix} = \begin{bmatrix} a_x \\ a_y \\ a_z \end{bmatrix} \tag{2.4}$$

which is illustrated in Figure 2.3. In such coordinate system with the rectangular components, the derivative of each motion characteristic depends only on their own components and the time variable. The position, velocity and acceleration are described particularly effective as the motion in each direction may be considered separately.

As seen from the above, vectors are used for description of the position in space, the velocity and the acceleration of bodies and particles. The same vectors are used in vehicle kinematics.

Vector Transformation in a Rotating Frame

In general case, to describe a motion of a rigid body, it is usually convenient and simpler to use several individual coordinate systems to characterize the motion in different transforms. For example, one coordinate system is fixed to the ground and another coordinate system is attached to the moving body with the origin of the attached coordinate system located at the center of the gravity of the body. Thus, velocities, accelerations and forces can be then observed and described in different coordinate systems. In this way, the mathematical handling is clear and less complex in dealing with kinematics and dynamics of the body. At the same time, it is needed to realize that the same physical quantity may be described differently in different coordinate systems. Hence, a coordinate transformation is necessary to obtain a unique reference considering the body movements.

Vehicle dynamics is concerned with expressing quantities in terms of the coordinate system attached to a moving vehicle. But also, we like to know the description of the same quantity with respect to another coordinate system, which serves as a *reference coordinate system*. Now we consider the coordinate transformation between a fixed coordinate system and a moving coordinate system. Taking as an example of two coordinate systems A_{xyz}[1] and B_{xyz} illustrated in Figure 2.4, we assume that A_{xyz} is fixed to the ground while B_{xyz} is moving with the body with the origin in its

---

[1] A presents the coordinate system while x,y,z denote the coordinate components.

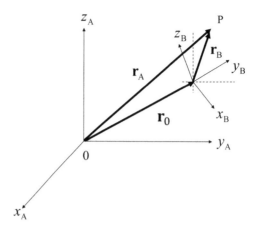

FIGURE 2.4 Vector transformation in spatial motion.

gravity center. Motion of point P on the body is described by vector $\mathbf{r}_A$ in the A_{xyz} coordinate system and vector $\mathbf{r}_B$ describes the motion of point P in B_{xyz}.

The coordinate transformation is a process to express vector $\mathbf{r}_B$, which is referenced in B_{xyz}, in the coordinate system A_{xyz} or vice versa. The effective way for the description is to study, how B_{xyz} moves relative to A_{xyz}. This relative motion of one coordinate system with regard to another system can be thought of two movements, namely translation and rotation. The translation movement means that the coordinate axes of B_{xyz} remain parallel to the corresponding axes of A_{xyz}, and the origin of B_{xyz} presents the instantaneous position of B_{xyz}. The rotation is thought for $\mathbf{r}_0$ to be constant. In particular case, $\mathbf{r}_0$ can be equal to zero, so that the origin of B_{xyz} is initially coincident with the origin of A_{xyz} and then rotates about a certain axis OD, as shown in Figure 2.5. The rotation axis indicates the direction of the *angular velocity* ω of the rotation at that instant. From the viewpoint of A_{xyz}, the rotation represents the orientation of B_{xyz} with respect to A_{xyz}. One way to describe the orientation, is to use a so-called *rotation matrix*, which has three vectors set together as the columns of a 3×3 matrix. The rotation matrix is introduced in the following. Usually in mechanics literature, a coordinate system is called a *frame* when it is considered with regard to its position and the orientation relative to another coordinate system. In the following we consider first the orientation between two frames, A_{xyz} and B_{xyz}, i.e., the relative rotation of both frames in which their origins are coincident.

In Figure 2.5 two frames are centered at 0, fixed frame A_{xyz} and frame B_{xyz}, which rotates about fixed axis *OD* with the angular velocity of ω at a given instant. Let $\mathbf{r}$ denote a vector specified and fixed in B_{xyz}. Next we shall find the mathematical description of $\mathbf{r}$ referenced to A_{xyz} by considering the position in B_{xyz} at the given instant of its rotation about *OD* (see Figure 2.5) as the result of

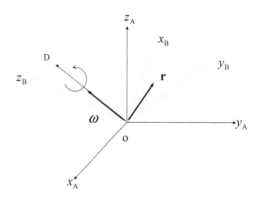

FIGURE 2.5 B_{xyz} rotated about axis OD.

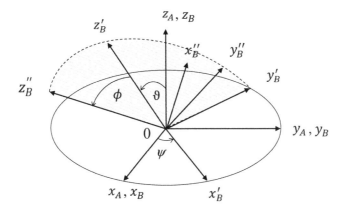

FIGURE 2.6 Euler angles.

three sequential rotations about the frames' axes, as shown in Figure 2.6. We process as follows: start with the frame B_{xyz} coincident with frame A_{xyz}. First, rotate B_{xyz} about its own z-axis by angle ψ, which moves x_B into x'_B and y_B into y'_B. Thus, x-axis and y-axis are still in the plane shown by the white ellipse in Figure 2.6. Second, rotate $B_{x'y'z}$ about y'_B-axis by angle ϑ, and thus move x'_B into x''_B and z_B into z'_B. Third and finally, rotate $B_{x''y'z'}$ about x''_B-axis by angle ϕ moving y'_B into y''_B and z'_B into z''_B. After the third rotation, y''_B-axis and z''_B-axis are in the same plane shown in thin parallel lines in Figure 2.6. Of course, axis x''_B makes 90 degrees a part with that plane. The last orientation of frame $B_{x''y'z'}$ in Figure 2.6 is identical with the position of B_{xyz} when it rotates about OD in Figure 2.5. The set of three rotation angles is called *Euler angles*. In this case, the rotation sequence follows z, y and x axes. Therefore, it is called ZYX Euler angles. Depending on the sequence of the axis rotations, different combinations of Euler angle sets can be created to mathematically describe positions of frame B_{xyz} in its rotation about OD. It is important to emphasize that the rotations of frame B_{xyz} through an Euler angle set do not represent the frame's instantaneous positions during its actual rotation.

In order to relate the position of vector **r** given in frame B_{xyz} to frame A_{xyz}, the transformation of **r** from B_{xyz} to A_{xyz} through the rotation matrices needs to be determined. For doing that, notation $\mathbf{r}_A$ denotes the original vector **r** in frame A_{xyz}, while $\mathbf{r}_B$ is vector **r** referenced in B_{xyz}. We define the following three vectors, which describe positions of vector $\mathbf{r}_B$ in the three instantaneous positions of frame B_{xyz} illustrated in Figure 2.6

$$
\mathbf{r}_\psi = \begin{bmatrix} x_\psi \\ y_\psi \\ z_\psi \end{bmatrix}, \quad
\mathbf{r}_\vartheta = \begin{bmatrix} x_\vartheta \\ y_\vartheta \\ z_\vartheta \end{bmatrix}, \quad
\mathbf{r}_\phi = \begin{bmatrix} x_\phi \\ y_\phi \\ z_\phi \end{bmatrix}
$$

i.e., these three vectors correspond to each rotation of frame B_{xyz} introduced by one of three Euler angles.

The transformation from B_{xyz} to A_{xyz} means to observe **r** from the origin of A_{xyz} when **r** describes the motion of a point in rotating frame B_{xyz}. To formulate the transformation, Figure 2.7 is used to illustrate the top view of the first rotation of frame B_{xyz} shown in Figure 2.6, in which z-axis is directed upward from plane $x_A y_A$. As seen from Figure 2.7, the coordinates of **r** in frame A_{xyz} can be calculated in the following way:

$$
\begin{aligned}
x_A &= x_\psi \cos\psi - y_\psi \sin\psi \\
y_A &= x_\psi \sin\psi + y_\psi \cos\psi \\
z_A &= z_\psi
\end{aligned}
\tag{2.5}
$$

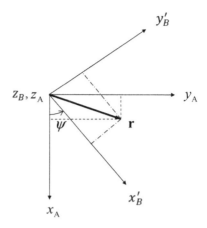

FIGURE 2.7   Coordinate transformation between A_{xyz} and $B_{x'y'z}$.

Thus, after the first rotation of frame B_{xyz}, vector $\mathbf{r}_A$, which denotes vector $\mathbf{r}$ in frame A_{xyz}, appears as follows:

$$\mathbf{r}_A = \begin{bmatrix} x_A \\ y_A \\ z_A \end{bmatrix} = \begin{bmatrix} \cos\psi & -\sin\psi & 0 \\ \sin\psi & \cos\psi & 0 \\ 0 & 0 & 1 \end{bmatrix} \begin{bmatrix} x_\psi \\ y_\psi \\ z_\psi \end{bmatrix} = T_\psi \mathbf{r}_\psi \qquad (2.6)$$

where T_ψ is the rotation matrix for the rotation of frame B_{xyz} into $B_{x'y'z}$ through angle ψ about z-axis. For the next two rotations, we use the same process graphically illustrated in Figure 2.6. For the second rotation, frame $B_{x'y'z}$ moves into frame $B_{x''y'z'}$, and vector $\mathbf{r}$ becomes

$$\mathbf{r}_\psi = \begin{bmatrix} x_\psi \\ y_\psi \\ z_\psi \end{bmatrix} = \begin{bmatrix} \cos\vartheta & 0 & \sin\vartheta \\ 0 & 1 & 0 \\ -\sin\vartheta & 0 & \cos\vartheta \end{bmatrix} \begin{bmatrix} x_\vartheta \\ y_\vartheta \\ z_\vartheta \end{bmatrix} = T_\vartheta \mathbf{r}_\vartheta \qquad (2.7)$$

where matrix T_ϑ describes the rotation of frame $B_{x'y'z}$ into $B_{x''y'z'}$. For the third rotation, the transformation of original vector $\mathbf{r}$ from frame $B_{x''y'z'}$ into frame $B_{x''y''z''}$ is operated by rotation matrix T_ϕ:

$$\mathbf{r}_\vartheta = \begin{bmatrix} x_\vartheta \\ y_\vartheta \\ z_\vartheta \end{bmatrix} = \begin{bmatrix} 1 & 0 & 0 \\ 0 & \cos\phi & -\sin\phi \\ 0 & \sin\phi & \cos\phi \end{bmatrix} \begin{bmatrix} x_\phi \\ y_\phi \\ z_\phi \end{bmatrix} = T_\phi \mathbf{r}_\phi \qquad (2.8)$$

Vector $\mathbf{r}_\phi$ actually denotes vector $\mathbf{r}_B$ in the final position of the rotational motion of frame B_{xyz}, i.e., $\mathbf{r}_B = \mathbf{r}_\phi$. The full transformation of $\mathbf{r}_B$ into $\mathbf{r}_A$ in frame A_{xyz} can be obtained by combining (2.6) through (2.8)

$$\mathbf{r}_A = T_\psi\, T_\vartheta\, T_\phi \mathbf{r}_B = T\, \mathbf{r}_B \qquad (2.9)$$

T denotes the rotation matrix from frame B_{xyz} to frame A_{xyz}. Expanding T in (2.9) provides

$$
T = \begin{bmatrix} \cos\vartheta\cos\psi & \cos\psi\sin\vartheta\sin\phi - \sin\psi\cos\phi & \cos\phi\sin\vartheta\cos\psi + \sin\phi\sin\psi \\ \sin\psi\cos\vartheta & \sin\phi\sin\vartheta\sin\psi + \cos\phi\cos\psi & \sin\psi\sin\vartheta\cos\phi - \cos\psi\sin\phi \\ -\sin\vartheta & \cos\vartheta\sin\phi & \cos\vartheta\cos\phi \end{bmatrix} \quad (2.10)
$$

In summary, the rotational transformation of a vector between two coordinate systems with a common origin, one of which is fixed to the ground and another frame rotates about an axis that passes through the origin, is represented by a 3×3 rotation matrix T given by (2.10). Another interpretation of the rotation matrix is the orientation of moving frame B_{xyz} to fixed frame A_{xyz}. In other words, the rotation matrix, T, which transforms vectors with regard to A_{xyz} and B_{xyz}, is the same as the rotation matrix, which describes positions of B_{xyz} relative to reference frame A_{xyz}.

It should be emphasized that the above-given definitions and equations work for the specified order of the rotations in Figure 2.6. Equation (2.10) is correct only for rotations performed in the defined order (ZYX Euler angles): about z-axis through angle Ψ, about y-axis through angle ϑ, and about x-axis through ϕ. The rotation matrix will change if the order of rotations changes. The above-presented example of the transformation matrix will be utilized in the following chapters of this book.

In addition, the coordinate transformation of vector $\mathbf{r}$ from A_{xyz} to B_{xyz} can be obtained in a reverse way. Thereby the inverse matrix of rotation matrix T, denoted as T^{-1}, follows from (2.10) where

$$
\mathbf{r}_B = T^{-1}\,\mathbf{r}_A \quad (2.11)
$$

$$
T^{-1} = T_\phi^{-1}\, T_\vartheta^{-1}\, T_\psi^{-1}
$$

$$
= \begin{bmatrix} \cos\vartheta\cos\psi & \cos\vartheta\sin\psi & -\sin\vartheta \\ \sin\phi\sin\vartheta\cos\psi - \cos\phi\sin\psi & \sin\phi\sin\vartheta\sin\psi + \cos\phi\cos\psi & \sin\phi\cos\vartheta \\ \cos\phi\sin\vartheta\cos\psi + \sin\phi\sin\psi & \cos\phi\sin\vartheta\sin\psi - \sin\phi\cos\psi & \cos\vartheta\cos\phi \end{bmatrix} \quad (2.12)
$$

VECTOR TRANSFORMATION IN GENERAL MOTION OF ROTATION AND TRANSLATION

We consider now the general case of the vector transformation. Here the origins of both frames, A_{xyz} and B_{xyz}, are not coincident. Previously given Figure 2.4 illustrated moving frame B_{xyz} in its rotational and translational motion relative to fixed frame A_{xyz}. Given $\mathbf{r}_B$, we can now compute vector $\mathbf{r}_A$. The position of the origin of frame B_{xyz} is presented by vector $\mathbf{r}_0$, which describes the translational movement. The rotation was described in the previous section of this chapter. It follows that vector coordinates of $\mathbf{r}_B$ projected in A_{xyz}, is obtained by addition

$$
\mathbf{r}_A = \mathbf{r}_0 + T\,\mathbf{r}_B \quad (2.13)
$$

where $\mathbf{r}_A$ denotes vector $\mathbf{r}_B$ with respect to A_{xyz}. The second term on the right side of (2.13) describes the transformation to an intermediate frame, which has the same orientation as A_{xyz}, but whose origin is coincident with B_{xyz}. Equation (2.13) can be also interpreted as a way to characterize the motion of point P in fixed frame A_{xyz} when the point's motion in frame B_{xyz} is known.

SOME PROPERTIES OF ROTATION MATRIX

In (2.9) and (2.10), the rotation matrix contains three instantaneous rotation matrices T_ψ, T_ϑ, T_ϕ that are orthonormal and their multiplicative product is the same as well, i.e.,

$$T^T = T^{-1} \tag{2.14}$$

If I_3 is denoted as a 3×3 identity matrix, (2.14) changes to (2.15)

$$T\,T^{-1} = I_3 \tag{2.15}$$

Differentiating (2.15) results in

$$\dot{T}\,T^{-1} + (\dot{T}\,T^{-1})^T = 0_3 \tag{2.16}$$

where 0_3 is a 3×3 zero matrix.
   Defining

$$S = \dot{T}\,T^{-1} \tag{2.17}$$

it results in

$$S = -S^T, \tag{2.18}$$

which means that S is a skew-symmetric matrix. Let S be assigned in the following form:

$$S = \begin{bmatrix} 0 & -\omega_z & \omega_y \\ \omega_z & 0 & -\omega_x \\ -\omega_y & \omega_x & 0 \end{bmatrix}. \tag{2.19}$$

It will show in the following that matrix S represents the angular velocity.
   Define the ω vector as

$$\omega = \begin{bmatrix} \omega_x \\ \omega_y \\ \omega_z \end{bmatrix}$$

it provides the following relation:

$$S\,\mathbf{r} = \omega \times \mathbf{r} \tag{2.20}$$

that exists for any vector $\mathbf{r}$. In this case a nine elements matrix corresponds to a 3×1 column vector through the vector cross product.

Denote matrix T as

$$\mathrm{T} = \begin{bmatrix} t_{11} & t_{12} & t_{13} \\ t_{21} & t_{22} & t_{23} \\ t_{31} & t_{32} & t_{33} \end{bmatrix},$$

Hence, $\mathrm{T}^{-1} = \begin{bmatrix} t_{11} & t_{21} & t_{31} \\ t_{12} & t_{22} & t_{32} \\ t_{13} & t_{23} & t_{33} \end{bmatrix},$

and it can be readily calculated that

$$\dot{\mathrm{T}}\mathrm{T}^{-1} = \begin{bmatrix} \dot{t}_{11}t_{11} + \dot{t}_{12}t_{12} + \dot{t}_{13}t_{13} & \dot{t}_{11}t_{21} + \dot{t}_{12}t_{22} + \dot{t}_{13}t_{23} & \dot{t}_{11}t_{31} + \dot{t}_{12}t_{32} + \dot{t}_{13}t_{33} \\ \dot{t}_{21}t_{11} + \dot{t}_{22}t_{12} + \dot{t}_{23}t_{13} & \dot{t}_{21}t_{21} + \dot{t}_{22}t_{22} + \dot{t}_{23}t_{23} & \dot{t}_{21}t_{31} + \dot{t}_{22}t_{32} + \dot{t}_{23}t_{33} \\ \dot{t}_{31}t_{11} + \dot{t}_{22}t_{12} + \dot{t}_{33}t_{13} & \dot{t}_{31}t_{21} + \dot{t}_{32}t_{22} + \dot{t}_{33}t_{23} & \dot{t}_{31}t_{31} + \dot{t}_{32}t_{32} + \dot{t}_{33}t_{33} \end{bmatrix}$$

Knowing (2.15), ω is determined as

$$\begin{bmatrix} \omega_x \\ \omega_y \\ \omega_z \end{bmatrix} = \begin{bmatrix} \dot{t}_{31}t_{21} + \dot{t}_{32}t_{22} + \dot{t}_{33}t_{23} \\ \dot{t}_{11}t_{31} + \dot{t}_{12}t_{32} + \dot{t}_{13}t_{33} \\ \dot{t}_{21}t_{11} + \dot{t}_{22}t_{12} + \dot{t}_{23}t_{13} \end{bmatrix} \tag{2.21}$$

Substituting the rotation matrix from (2.10) in (2.21) yields

$$\omega = \begin{bmatrix} \omega_x \\ \omega_y \\ \omega_z \end{bmatrix} = \begin{bmatrix} \dot{\phi}\cos\psi\cos\phi - \dot{\vartheta}\sin\psi \\ \dot{\phi}\cos\vartheta\sin\psi + \dot{\vartheta}\cos\psi \\ -\dot{\phi}\sin\vartheta + \dot{\psi} \end{bmatrix} \tag{2.22}$$

Actually ω is the *angular velocity vector* of frame B_{xyz} rotating with respect to A_{xyz}, and was described in Figure 2.5. More detail on the angular velocity is coming up in the next section.

2.2   CHANGE RATE OF VECTOR IN ROTATING FRAME

We consider the same example presented before by Figure 2.5, namely, frame A_{xyz} is fixed while frame B_{xyz} is rotating with angular velocity ω, and vector $\mathbf{r}$ is fixed in B_{xyz}, i.e., the magnitude and direction of vector $\mathbf{r}$ is constant when an observer is in the origin of B_{xyz} meaning that the observer is moving with this frame of reference. The question discussed here is on how the rate of change of vector $\mathbf{r}$ can be calculated as observed in B_{xyz}.

ANGULAR VELOCITY VECTOR

The aforementioned angular velocity, ω, describes the rotational motion of a frame in principle. In this particular case, the direction of ω indicates the instantaneous axis of rotation of frame B_{xyz} relative to frame A_{xyz}, and the magnitude of ω is the speed of rotation. Since we usually attach a moving frame to moving bodies, the angular velocity, ω, can be thought of as describing an attribute of a body attached to frame B_{xyz} and observed from frame A_{xyz}. As any vector, the angular velocity vector may be expressed in any coordinate system, including A_{xyz} and B_{xyz}. Thus, a leading super-script is useful for its reference; for example $^A\omega_B$ is the angular velocity of frame B_{xyz} relative to A_{xyz} (i.e., observed in A_{xyz}), and $^B\omega_B$ is the angular velocity of frame B_{xyz} as seen from B_{xyz} self, but for simplicity, the leading superscript of $^B\omega_B$ will be omitted so that $\omega_B = {}^B\omega_B$.

To determine ω_B, we referenced Figure 2.6, in which frame B_{xyz} executed three rotations about its principal axes. Each rotation makes a contribution to the angular velocity; hence, it can be then thought of a linear superposition of three components:

$$\omega_B = \begin{bmatrix} \omega_{x,B} \\ \omega_{y,B} \\ \omega_{z,B} \end{bmatrix} = T_\phi^{-1} T_\vartheta^{-1} \begin{bmatrix} 0 \\ 0 \\ \dot\psi \end{bmatrix} + T_\phi^{-1} \begin{bmatrix} 0 \\ \dot\vartheta \\ 0 \end{bmatrix} + \begin{bmatrix} \dot\phi \\ 0 \\ 0 \end{bmatrix} \tag{2.23}$$

The first term on the right-hand side of (2.23) is the angular velocity developed by turning through angle ψ and referenced to B_{xyz}. The second term is developed by angle ϑ and referenced to B_{xyz}. The last term is the contribution of the last rotation about x-axis through angle ϕ.

Using (2.7) and (2.8), the calculation of ω_B is straightforward, so that

$$\omega_B = \begin{bmatrix} 1 & 0 & -\sin\vartheta \\ 0 & \cos\phi & \cos\vartheta\sin\phi \\ 0 & -\sin\phi & \cos\vartheta\cos\phi \end{bmatrix} \begin{bmatrix} \dot\phi \\ \dot\vartheta \\ \dot\psi \end{bmatrix}$$

$$= \begin{bmatrix} \dot\phi - \dot\psi\sin\vartheta \\ \dot\vartheta\cos\phi + \dot\psi\cos\vartheta\sin\phi \\ -\dot\vartheta\sin\phi + \dot\psi\cos\vartheta\cos\phi \end{bmatrix} \tag{2.24}$$

For angular velocity $^A\omega_B$, which is viewed from frame A_{xyz}, we may apply again the same approach as above:

$$^A\omega_B = T_\psi T_\vartheta T_\phi \begin{bmatrix} \dot\phi \\ 0 \\ 0 \end{bmatrix} + T_\psi T_\vartheta \begin{bmatrix} 0 \\ \dot\vartheta \\ 0 \end{bmatrix} + T_\psi \begin{bmatrix} 0 \\ 0 \\ \dot\psi \end{bmatrix}$$

$$= \begin{bmatrix} \cos\psi\cos\vartheta & -\sin\psi & 0 \\ \sin\psi\cos\vartheta & \cos\psi & 0 \\ \sin\vartheta & 0 & 1 \end{bmatrix} \begin{bmatrix} \dot\phi \\ \dot\vartheta \\ \dot\psi \end{bmatrix} \tag{2.25}$$

The result confirms the calculation of (2.22). Similar to (2.9), the angular velocity transformation between the two frames can be verified and described by the same rotation matrix

$$^A\omega_B = T\,\omega_B \tag{2.26}$$

Equations (2.22) and (2.25) illustrate the relationship between the instant angular velocity, $^A\omega_B$, Euler angles ψ, ϑ, ϕ and their derivatives. The derivatives of Euler angles describe the rotational com-ponents of coordinate systems, i.e., in this case $\dot\psi$ is the original z-axis component; $\dot\vartheta$ is the rotated y-axis component, while $\dot\phi$ is the x-axis obtained after the sequential rotation of the moving frame through angles ψ and ϑ. It is important to emphasize that the angle derivatives do not describe directly the rotation of the moving frame of reference about the instantaneous axis of vector ω_B since they are not direct components of angular velocity ω_B. However, in case ϑ and ϕ are small, the angular velocity can be approximated by Euler angles' derivatives. Following (2.24), that is

$$\omega_B \approx \begin{bmatrix} \dot{\phi} \\ \dot{\vartheta} \\ \dot{\psi} \end{bmatrix} \qquad (2.27)$$

This result is utilized later in the book to analyze vehicle roll dynamics where the roll angle and the pitch angle of the vehicle are relatively small.

Example 2.1

Figure 2.8 shows two frames A_{xyz} and B_{xyz} which were initially coincident. B_{xyz} is then rotating about z-axis through angle ψ, and $\vartheta = \phi = 0$.

From (2.22) we obtain

$$\omega = \begin{bmatrix} 0 \\ 0 \\ \dot{\psi} \end{bmatrix} \qquad (2.28)$$

Using (2.10) and (2.12), we calculate S from (2.17) as follows:

$$
\begin{aligned}
S \quad &= \dot{T}\, T^{-1} \\
&= \begin{bmatrix} -\dot{\psi}\sin\psi & -\dot{\psi}\cos\psi & 0 \\ \dot{\psi}\cos\psi & -\dot{\psi}\sin\psi & 0 \\ 0 & 0 & 0 \end{bmatrix} \begin{bmatrix} \cos\psi & \sin\psi & 0 \\ -\sin\psi & \cos\psi & 0 \\ 0 & 0 & 1 \end{bmatrix} \\
&= \begin{bmatrix} 0 & -\dot{\psi} & 0 \\ \dot{\psi} & 0 & 0 \\ 0 & 0 & 0 \end{bmatrix}
\end{aligned}
\qquad (2.29)
$$

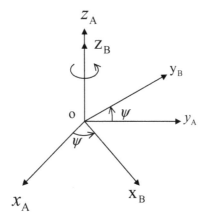

FIGURE 2.8 B_{xyz} rotates about z-axis by ψ.

which corresponds to (2.19). Given a vector $\mathbf{r}$,

$$\mathbf{r} = \begin{bmatrix} r_x \\ r_y \\ r_z \end{bmatrix},$$

we can verify the result given by (2.20)

$$S\,\mathbf{r} = \omega \times \mathbf{r} = \begin{bmatrix} -\dot{\psi}\,r_y \\ \dot{\psi}\,r_x \\ 0 \end{bmatrix} \tag{2.30}$$

❑

CHANGE RATE OF A VECTOR INVOLVING A ROTATING FRAME

The moving frame of reference, B_{xyz}, is allowed to rotate. Referring back to (2.9), the coordinate transformation between two frames in Figure 2.5 is given by

$$\mathbf{r}_A = \mathrm{T}\,\mathbf{r}_B \tag{2.31}$$

Differentiating (2.31) with respect to time, yields

$$\dot{\mathbf{r}}_A = \mathrm{T}\,\dot{\mathbf{r}}_B + \dot{\mathrm{T}}\,\mathbf{r}_B \tag{2.32}$$

Thus, the derivative of a position vector can be thought of as the linear velocity of a point in space represented by the position vector. Writing $\mathbf{r}_B$ in terms of $\mathbf{r}_A$ as given by (2.31), (2.32) extends to (2.33)

$$\dot{\mathbf{r}}_A = \mathrm{T}\,\dot{\mathbf{r}}_B + \dot{\mathrm{T}}\,\mathrm{T}^{-1}\,\mathbf{r}_A \tag{2.33}$$

which allows all terms to be referenced to A_{xyz}. Notice that the derivative of vector $\mathbf{r}_A$ in (2.33) is made of two parts. The first part is the change rate of the vector with respect to rotating frame B_{xyz} and translated into fixed frame A_{xyz}; the second part is induced by the rotation of moving frame B_{xyz} with respect to A_{xyz}.

On the other hand, the calculation can be done with respect to B_{xyz} as well. Multiplying both sides of (2.33) by inverse matrix T^{-1}, we have

$$\mathrm{T}^{-1}\dot{\mathbf{r}}_A = \dot{\mathbf{r}}_B + \mathrm{T}^{-1}\dot{\mathrm{T}}\,\mathbf{r}_B \tag{2.34}$$

On the right-hand side of (2.34), the first term is the change rate of $\mathbf{r}_B$ with respect to time within frame B_{xyz}, whereas the second term presents the change rate caused by the rotational part of motion seen in B_{xyz}. Hence, the right-hand side of (2.34) is the vector change as viewed from B_{xyz}. Indeed, (2.33) indicates the change rate as viewed from fixed frame A_{xyz}, whereas (2.34) shows its transformation into rotating frame B_{xyz}, of which origin coincides with the origin of A_{xyz}. It is the change rate of vector $\mathbf{r}_B$ in terms of the rotating frame B_{xyz}, but still relative to frame A_{xyz}. For this reason, we denote the change rate on the left side of (2.34) as $\mathbf{v}_{A/B}$, i.e.,

$$\mathbf{v}_{A/B} = \dot{\mathbf{r}}_B + \mathrm{T}^{-1}\dot{\mathrm{T}}\,\mathbf{r}_B \tag{2.35}$$

Further, $T^{-1}\dot{T}$ is the skew-symmetric matrix that presents the same angular velocity as $\dot{T}T^{-1}$ does. Based on (2.22), we have

$$T^{-1}\dot{T}\,\mathbf{r}_B = \boldsymbol{\omega}_B \times \mathbf{r}_B \tag{2.36}$$

Hence, (2.35) leads to

$$\mathbf{v}_{A/B} = \dot{\mathbf{r}}_B + \boldsymbol{\omega}_B \times \mathbf{r}_B \tag{2.37}$$

This general form shows that the absolute vector's rate of change, $\mathbf{v}_{A/B}$, with respect to a rotating frame, consists of the change rate in the length of vector $\dot{\mathbf{r}}_B$ in the rotating frame and the cross product of the angular velocity vector and the vector itself, $\mathbf{r}_B$, observed in the moving frame of reference. Rather than considering a point's velocity relative to frame A_{xyz} (as conventionally done in academic courses on dynamics), we will often consider this velocity vector referenced to frame B_{xyz}, such as $T^{-1}\mathbf{r}_A$ is $\mathbf{r}_A$ seen from B_{xyz}.

For a better understanding of the general form given by (2.37), we investigate now the second derivative of $\mathbf{r}_A$, which means the change rate of $\dot{\mathbf{r}}_A$. Differentiating (2.34), we obtain

$$\dot{T}^{-1}\dot{\mathbf{r}}_A + T^{-1}\,\ddot{\mathbf{r}}_A = \dot{\mathbf{v}}_{A/B} \tag{2.38}$$

By rearranging (2.38) as

$$T^{-1}\,\ddot{\mathbf{r}}_A = \dot{\mathbf{v}}_{A/B} - \dot{T}^{-1}\dot{\mathbf{r}}_A \tag{2.39}$$

and substituting (2.32) in (2.39), the latter can be written as follows:

$$T^{-1}\,\ddot{\mathbf{r}}_A = \dot{\mathbf{v}}_{A/B} - \dot{T}^{-1}\left(T\dot{\mathbf{r}}_B + \dot{T}\,\mathbf{r}_B\right) \tag{2.40}$$

$$= \dot{\mathbf{v}}_{A/B} - \dot{T}^{-1}\,T\left(\dot{\mathbf{r}}_B + T^{-1}\dot{T}\mathbf{r}_B\right)$$

Considering (2.16), and using (2.35) to (2.36), we have

$$T^{-1}\ddot{\mathbf{r}}_A = \dot{\mathbf{v}}_{A/B} + \boldsymbol{\omega}_B \times \mathbf{v}_{A/B} \tag{2.41}$$

which is the change rate of $\dot{\mathbf{r}}_A$ as seen in the rotating frame. As seen, (2.41) has the same form as (2.37). This confirms that the absolute rate change of $\dot{\mathbf{r}}_A$ is made of its time change in the rotating frame and its cross product with the angular velocity.

Substituting $\dot{\mathbf{v}}_{A/B}$ by using (2.37) and denoting $T^{-1}\ddot{\mathbf{r}}_A$ by $\mathbf{a}_{A/B}$, (2.41) becomes

$$\mathbf{a}_{A/B} = \ddot{\mathbf{r}}_B + \boldsymbol{\omega}_B \times \dot{\mathbf{r}}_B + \dot{\boldsymbol{\omega}}_B \times \mathbf{r}_B + \boldsymbol{\omega}_B \times \mathbf{v}_{A/B} \tag{2.42}$$

which gives the acceleration, i.e., the second time derivative of $\mathbf{r}_A$ seen from rotating frame B_{xyz} relative to A_{xyz}. Both (2.41) and (2.42) can be used for calculating the acceleration depending on given data.

CHANGE RATE OF A VECTOR IN GENERAL MOTION

In the previous section we considered the change rate of a vector described by (2.31), which is a special case of (2.13), when $\mathbf{r}_0 = 0$ (see Figure 2.4). Now we extend the results to the general case given by (2.13) for non-zero vectors $\mathbf{r}_0$.

Referring to Figure 2.4, the position of P is defined at any instant by the vector $\mathbf{r}_A$ in the fixed frame A_{xyz}, and by the vector $\mathbf{r}_B$ in the moving frame B_{xyz}. Processing the differentiation as done for (2.37) again, the velocity $\mathbf{v}_{A/B}$ of the point P relative to the fixed frame A_{xyz} is obtained by

$$\mathbf{v}_{A/B} = \mathbf{T}^{-1}\dot{\mathbf{r}}_0 + \dot{\mathbf{r}}_B + \boldsymbol{\omega}_B \times \mathbf{r}_B \tag{2.43}$$

The first term on the right-hand side of (2.43) represents the velocity of the origin of the moving frame observed from B_{xyz} to the fixed frame, and all other terms are present in (2.37). Note that the second and third terms of the right-hand side members of the above equation are referenced to frame B_{xyz}, but they are relative to the fixed frame A_{xyz}. So, $\mathbf{v}_{A/B}$ is actually the velocity relative to A_{xyz} measured by an observer moving with the frame B_{xyz}.

2.3   VELOCITIES OF POINTS ON A RIGID BODY

Figure 2.9 shows a rigid body moving with respect to an arbitrary frame A_{xyz} that is fixed to the ground. Let E and F be two points on the body, and the rigid body is rotating with ω about the axis through point E. Furthermore, $\mathbf{v}_E$ and $\mathbf{v}_F$ are velocities of points E and F, respectively. Assume that these two variables are observed from the rotating body, and we compute the velocity of point F, $\mathbf{v}_F$.

To describe the motion of a point that moves with the rigid body, it is convenient to use a frame that moves with the rigid body, and such a frame is body-fixed. In Figure 2.9, we attach a frame B_{xyz} to the body, whose origin is coincident with point E. A_{xyz} is the reference frame relative to which the motion of rigid body is described. One can see that Figure 2.9 is the same as the Figure 2.4 if the position vector of point F with regard to point E is denoted as $\mathbf{r}_{FE}$ (which is vector $\mathbf{r}_B$ in Figure 2.4). Since E and F are fixed in the body, the distance between them remains unchanged, i.e., $\dot{\mathbf{r}}_{FE} = 0$. Following (2.43), we have

$$\mathbf{v}_F = \mathbf{v}_E + \boldsymbol{\omega} \times \mathbf{r}_{FE} \tag{2.44}$$

This concludes that the velocity of point F is equal to velocity of E plus the rotational component caused by the rotation of the body about an axis through E.

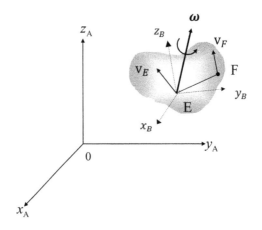

FIGURE 2.9   Velocity of point F on the rigid body.

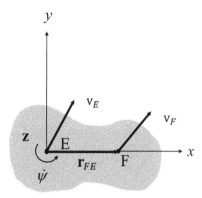

FIGURE 2.10   Velocity of point F on the slab.

Example 2.2

Figure 2.10 shows a slab placed in a frame whose z-axis is directed out of the page plane. Point E is located at the origin of the frame. Point F is moving when the body rotates with angular velocity $\dot{\psi}$ about z-axis at that instant. The attached frame is rotating with the slab. It is to assume that at this instant the velocity of point E and the position of point F are given by

$$\mathbf{v}_E = \begin{bmatrix} v_{EX} \\ v_{EY} \\ 0 \end{bmatrix} \text{ and } \mathbf{r}_{FE} = \begin{bmatrix} r_{FE} \\ 0 \\ 0 \end{bmatrix}$$

respectively.

To determine the velocity of point F, we apply (2.44):

$$\mathbf{v}_F = \begin{bmatrix} v_{EX} \\ v_{EY} \\ 0 \end{bmatrix} + \begin{bmatrix} 0 \\ 0 \\ \dot{\psi} \end{bmatrix} \times \begin{bmatrix} r_{FE} \\ 0 \\ 0 \end{bmatrix}$$

$$= \begin{bmatrix} v_{EX} \\ v_{EY} + \dot{\psi}\, r_{FE} \\ 0 \end{bmatrix} \tag{2.45}$$

❏

2.4   VEHICLE VELOCITIES AND ACCELERATIONS

The basic kinematic relationships of vectors and associated rates of change with regard to various coordinate systems were introduced in the previous sections. In the following, the results will be applied to vehicle curvilinear motion to determine vehicle velocities and accelerations.

The vehicle is considered a rigid body moving on a horizontal surface. The vehicle has an attached coordinate system V_{xyz} with its origin at the center of gravity as shown in Figure 2.11, where we use the following notations:

a   vehicle CG acceleration vector

v   vehicle CG velocity vector

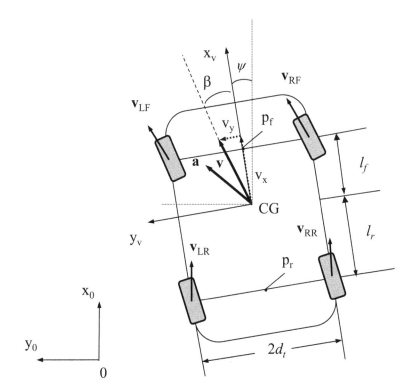

FIGURE 2.11   Vehicle kinematic diagram.

ψ vehicle yaw angle
β vehicle sideslip angle
d_t half wheel track
l_f distance from CG to the front axle
l_r distance from CG to the rear axle
p_f midpoint of the front axle
p_r midpoint of the rear axle
$\mathbf{v}_{\mathrm{LF}}$ velocity of the left front wheel
$\mathbf{v}_{\mathrm{RF}}$ velocity of the right front wheel
$\mathbf{v}_{\mathrm{LR}}$ velocity of the left rear wheel
$\mathbf{v}_{\mathrm{RR}}$ velocity of the right rear wheel

VEHICLE VELOCITY AND YAW RATE

Referring to the above graph, the vehicle is moving with velocity $\mathbf{v} = \left(v_x, v_y\right)^T$ in the earth-fixed initial coordinate system 0_{xyz} while simultaneously rotating about z-axis with an angular velocity $\dot{\psi}$, which is also called vehicle *yaw rate*. The yaw angle ψ is angle between the vehicle x_v-axis of the vehicle-fixed coordinate system and the x_0-axis of the initial coordinate system, 0_{xyz}. The x_v-axis is aligned with the vehicle heading direction. On the other hand, we can express the vehicle velocity components in terms of the angle between vehicle velocity and vehicle heading direction, i.e.,

$$v_x = v cos\beta$$
$$v_y = v sin\beta$$

(2.46)

Angle β is called *vehicle sideslip angle*, which is the angle between the vehicle x_v-axis of the vehicle-fixed coordinate system and the vehicle CG velocity vector. Both yaw rate and sideslip angle will be detailed later in Chapter 8.

VELOCITIES OF WHEELS

The vehicle is undergoing a planar motion; therefore, the linear velocity of a wheel center can be seen as the velocity of a point on the vehicle body. The vehicle coordinate system V_{xyz} is a rotating frame attached to the vehicle CG. Vehicle velocity $\mathbf{v}$ is considered to be viewed in V_{xyz} relative to 0_{xyz}. Using (2.44), the velocity of the left front wheel is given by

$$
\mathbf{v}_{LF} = \mathbf{v} + \begin{bmatrix} 0 \\ 0 \\ \dot{\psi} \end{bmatrix} \times \begin{bmatrix} l_f \\ d_t \\ 0 \end{bmatrix}
$$

$$
= \begin{bmatrix} v_x - d_t\,\dot{\psi} \\ v_y + l_f\,\dot{\psi} \\ 0 \end{bmatrix}
\tag{2.47}
$$

$\mathbf{v}_{LF}$ is the left-front wheel velocity observed in V_{xyz} relative to frame 0_{xyz}, its components are along the coordinate axes of the frame V_{xyz}, respectively. Accordingly, the velocities of other three wheels are

$$
\mathbf{v}_{RF} = \begin{bmatrix} v_x + d_t\dot{\psi} \\ v_y + l_f\dot{\psi} \\ 0 \end{bmatrix}
\tag{2.48}
$$

$$
\mathbf{v}_{LR} = \begin{bmatrix} v_x - d_t\dot{\psi} \\ v_y - l_r\dot{\psi} \\ 0 \end{bmatrix}
\tag{2.49}
$$

$$
\mathbf{v}_{RR} = \begin{bmatrix} v_x + d_t\dot{\psi} \\ v_y - l_r\dot{\psi} \\ 0 \end{bmatrix}
\tag{2.50}
$$

VEHICLE ACCELERATION

The vehicle CG acceleration is the first derivative of vehicle velocity $\mathbf{v}$. Hence, we apply (2.41) and obtain

$$
\boldsymbol{a} = \begin{bmatrix} a_x \\ a_y \\ 0 \end{bmatrix} = \begin{bmatrix} \dot{v}_x \\ \dot{v}_y \\ 0 \end{bmatrix} + \begin{bmatrix} 0 \\ 0 \\ \dot{\psi} \end{bmatrix} \times \begin{bmatrix} v_x \\ v_y \\ 0 \end{bmatrix}
$$

$$
= \begin{bmatrix} \dot{v}_x - v_y\dot{\psi} \\ \dot{v}_y + v_x\dot{\psi} \\ 0 \end{bmatrix}
\tag{2.51}
$$

The interpretation is similar to the velocity. Acceleration $\boldsymbol{a}$ is the CG acceleration viewed in V_{xyz} relative to 0_{xyz}, and its components are along the axes of the vehicle coordinate system V_{xyz}.

Substituting (2.46) in (2.51), we have

$$\boldsymbol{a} = \begin{bmatrix} \dot{v}\cos\beta - v\sin\beta\left(\dot{\beta}+\dot{\psi}\right) \\ \dot{v}\sin\beta + v\cos\beta\left(\dot{\beta}+\dot{\psi}\right) \\ 0 \end{bmatrix} \tag{2.52}$$

In the case that angle is small, $\beta \approx 0$, the acceleration can be approximated as

$$\boldsymbol{a} \approx \begin{bmatrix} \dot{v} \\ v\left(\dot{\beta}+\dot{\psi}\right) \\ 0 \end{bmatrix} \tag{2.53}$$

Example 2.3

The front and rear axle centers are denoted as P_f, P_r respectively and given in Figure 2.11. Determine the velocities and accelerations of these two centers. Using (2.44), the velocities are

$$\boldsymbol{v}_f = \begin{bmatrix} v_x \\ v_y + l_f\dot{\psi} \\ 0 \end{bmatrix} \tag{2.54}$$

$$\boldsymbol{v}_r = \begin{bmatrix} v_x \\ v_y - l_r\dot{\psi} \\ 0 \end{bmatrix} \tag{2.55}$$

and the accelerations follow from the above equations:

$$\boldsymbol{a}_f = \dot{\boldsymbol{v}}_f + \begin{bmatrix} 0 \\ 0 \\ \dot{\psi} \end{bmatrix} \times \boldsymbol{v}_f = \begin{bmatrix} \dot{v}_x - v_y\dot{\psi} - l_f\dot{\psi}^2 \\ \dot{v}_y - l_f\ddot{\psi} + v_x\dot{\psi} \\ 0 \end{bmatrix} \tag{2.56}$$

$$\boldsymbol{a}_r = \dot{\boldsymbol{v}}_r + \begin{bmatrix} 0 \\ 0 \\ \dot{\psi} \end{bmatrix} \times \boldsymbol{v}_r = \begin{bmatrix} \dot{v}_x - v_y\dot{\psi} + l_r\dot{\psi}^2 \\ \dot{v}_y + l_r\ddot{\psi} + v_x\dot{\psi} \\ 0 \end{bmatrix} \tag{2.57}$$

❏

Example 2.4

Figure 2.12 shows a plane motion of a vehicle-trailer combination in a fixed universe frame 0_{xyz}. The vehicle-trailer combination is illustrated by an articulated two-rod combination. Rod l_1, which

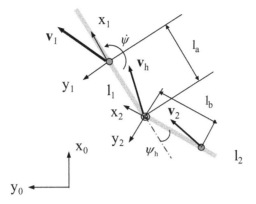

FIGURE 2.12   Diagram for trailer velocity calculation.

represents the vehicle, has frame x_1y_1 attached to the vehicle's center of gravity, while rod l_2, which represents the trailer, has frame x_2y_2 attached at the joint point (i.e., the trailer hitch). The vehicle, rod l_1, is turning with velocity $\mathbf{v}_1$ and yaw rate $\dot{\psi}$ observed from the frame attached to l_1. Assuming that the trailer, rod l_2, has a small articulation angle ψ_h, what is the acceleration of the CG of the trailer observed from the frame x_1y_1 of l_1?

With respect to frame x_1y_1 of l_1, we first determine the velocity at the trailer hitch joint:

$$\mathbf{v}_h = \begin{bmatrix} v_{1x} \\ v_{1y} - l_a\dot{\psi} \\ 0 \end{bmatrix} \tag{2.58}$$

The angular velocity of l_2 rod observed in frame x_2y_2, denoted as $\dot{\psi}'_2$, is angular velocity in frame x_1y_1 plus the new component caused by angular velocity in frame x_2y_2, and thus

$$\dot{\psi}'_2 = T_2^{-1}\begin{bmatrix} 0 \\ 0 \\ \dot{\psi} \end{bmatrix} + \begin{bmatrix} 0 \\ 0 \\ \dot{\psi}_h \end{bmatrix} = \begin{bmatrix} 0 \\ 0 \\ \dot{\psi} + \dot{\psi}_h \end{bmatrix}$$

We can then calculate the velocity $\mathbf{v}'_2$ at the CG of l_2 rod, which imitates the trailer, by applying (2.44)

$$\mathbf{v}'_2 = T_2^{-1}\mathbf{v}_h + \begin{bmatrix} 0 \\ -l_b\left(\dot{\psi} + \dot{\psi}_h\right) \\ 0 \end{bmatrix} \tag{2.59}$$

where $\mathbf{v}'_2$ is referenced to the frame of l_2. The second term on the right-hand side of (2.59) is the component caused by the rotation of l_2, and T_2 is defined

$$T_2 = \begin{bmatrix} \cos\psi_h & -\sin\psi_h & 0 \\ \sin\psi_h & \cos\psi_h & 0 \\ 0 & 0 & 1 \end{bmatrix} \tag{2.60}$$

The acceleration at the CG of l_2 is the derivative of $\mathbf{v}'_2$. With respect to frame of l_2, it is determined by applying (2.41):

$$\mathbf{a}'_2 \quad = \dot{\mathbf{v}}'_2 + \begin{bmatrix} 0 \\ 0 \\ \dot{\psi} + \dot{\psi}_h \end{bmatrix} \times \mathbf{v}'_2$$

$$= \begin{bmatrix} \dot{v}_{xh} \cos\psi_h + \dot{v}_{yh}\sin\psi_h + \left(v_{xh}\sin\psi_h - v_{yh}\cos\psi_h\right)\dot{\psi} + l_b\left(\dot{\psi} + \dot{\psi}_h\right)^2 \\ -\dot{v}_{xh}\sin\psi_h + \dot{v}_{yh}\cos\psi_h + \left(v_{yh}\sin\psi_h + v_{xh}\cos\psi_h\right)\dot{\psi} - l_b\left(\ddot{\psi} + \ddot{\psi}_h\right) \\ 0 \end{bmatrix} \quad (2.61)$$

Then the result can be translated back to l_1 frame through T_2, under the condition that ψ_h is small
so that $\sin\psi_h \approx 0$ and $\cos\psi_h \approx 1$, it yields:

$$\mathbf{a}_2 \quad = T_2\mathbf{a}'_2$$

$$= \begin{bmatrix} \dot{v}_{1x} - v_{1y}\dot{\psi} + l_a\dot{\psi}^2 + l_b\left(\dot{\psi} + \dot{\psi}_h\right)^2 \\ \dot{v}_{1y} + v_{1x}\dot{\psi} - \left(l_a + l_b\right)\ddot{\psi} - l_b\ddot{\psi}_h \\ 0 \end{bmatrix} \quad (2.62)$$

which presents the acceleration of l_2 observed from the x_1y_1 frame of l_1.

❏

Referring to the vehicle in Figure 2.13, we analyze the vehicle acceleration in the case when the
vehicle is rotating first about the z-axis with angular velocity $\dot{\psi}$ and then about a new position of
the x-axis with angular velocity $\dot{\phi}$, i.e., the frame V_{xyz} moves into $V_{x''y''z''}$. There is no rotation about
y-axis, and angle ϕ is assumed small. The reader can see the original position of the x-axis, y-axis
and z-axis in Figure 1.6.

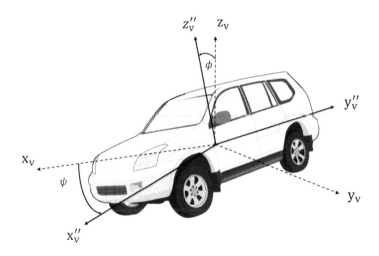

FIGURE 2.13   Vehicle body in rotation.

Comparing to result (2.51), the acceleration is extended by a component along the z-axis:

$$
\boldsymbol{a} = \begin{bmatrix} a_x \\ a_y \\ 0 \end{bmatrix} = \begin{bmatrix} \dot{v}_x \\ \dot{v}_y \\ 0 \end{bmatrix} + \begin{bmatrix} \dot{\phi} \\ 0 \\ \dot{\psi} \end{bmatrix} \times \begin{bmatrix} v_x \\ v_y \\ 0 \end{bmatrix}
$$

$$
= \begin{bmatrix} \dot{v}_x - v_y \dot{\psi} \\ \dot{v}_y + v_x \dot{\psi} \\ v_y \dot{\phi} \end{bmatrix} \tag{2.63}
$$

2.5   NEWTON'S AND EULER'S EQUATIONS

The discussion so far has focused on kinematics, i.e., positions, velocities and accelerations. We have not yet considered the relationships between forces and the motions produced by the forces, i.e., essential dynamics. Dynamics fundamentals are used to predict the motion caused by given forces or to determine the forces required to produce a given motion, which is inverse dynamics problem formulation.

Dynamics of vehicles constitutes a theme of a very high complexity, which cannot be covered here in the completeness it deserves. However, the modeling of vehicle dynamics in this book is purposed for the control design, and thus models can be derived in simplified and explicit forms, which are accurate enough for describing actual vehicle motions. The fundamental formulations are well suited to those applications. In particular, Newton's equation and Euler's equation can be directly applied when the vehicle body is considered as a lumped mass with a symmetrical rigid body rotating about the center of gravity. For many analyses, vehicles are considered as multi-body systems, i.e., the vehicle body and the wheels. In our vehicle applications, as indicated in the previous sections, we use a reference frame attached on the moving body, with the origin located at the center of the gravity of the body. The frame of reference is selected in the way that the vehicle body is symmetrical with respect to the rotation axes shown in Figure 1.6.

NEWTON'S EQUATION

Figure 2.14 shows a rigid body in translational spatial motion acted upon by several external forces F_1, F_2, F_3, etc. The vector sum of the external forces acting on the body is equal to

$$
\sum F = m\,\boldsymbol{a} \tag{2.64}
$$

where m is the body mass and $\boldsymbol{a}$ is the acceleration of the gravity center CG. It implies that acceleration vector $\boldsymbol{a}$ must have the same line of action as the resultant external force $\sum F$. Equation (2.64) is also referred to as Newton's second law of motion of a rigid body.

Equivalently, (2.64) can be written in the component form using a coordinate system, which is most convenient for the problem at hand, i.e.,

$$
\sum F_x = m\,a_x, \ \sum F_y = m\,a_y, \ \sum F_z = m\,a_z \tag{2.65}
$$

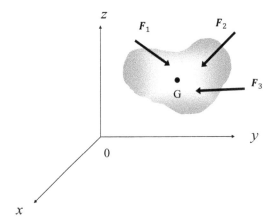

FIGURE 2.14   Forces on a rigid body in translational spatial motion.

ANGULAR MOMENTUM

Figure 2.15 shows a rigid body in rotational and translational spatial motion with fixed frame A_{xyz}. A second frame B_{xyz} is attached to the body with origin at the gravity center, which is in point G. The angular velocity of the rigid body is $\boldsymbol{\omega}$ referenced to the fixed frame A_{xyz}. The angular momentum H_G can be observed differently from both frames. With respect to B_{xyz}, $\boldsymbol{H}_G$ is determined as:

$$
\boldsymbol{H}_G = \begin{bmatrix} I_{xx} & -I_{xy} & -I_{xz} \\ -I_{yx} & I_{yy} & -I_{yz} \\ -I_{zx} & -I_{zy} & I_{zz} \end{bmatrix} \begin{bmatrix} \omega_x \\ \omega_y \\ \omega_z \end{bmatrix} \tag{2.66}
$$

$$
= \boldsymbol{I}_B \boldsymbol{\omega}
$$

where $\boldsymbol{I}_B$ is called *inertia matrix* or *inertia tensor*. Scalar elements I_{xx}, I_{yy} and I_{zz} are the *mass moments of inertia* while elements I_{xy}, I_{xz} and I_{yz} are known as the *mass products of inertia*. They are given by

$$
I_{xx} = \int\int\int_V \left(y^2 + z^2\right)\rho\,dxdydz
$$

$$
I_{yy} = \int\int\int_V \left(x^2 + z^2\right)\rho\,dxdydz
$$

$$
I_{zz} = \int\int\int_V (x^2 + y^2)\rho\,dxdydz
$$

$$
I_{xy} = \int\int\int_V xy\rho\,dxdydz \tag{2.67}
$$

$$
I_{xz} = \int\int\int_V xz\rho\,dxdydz
$$

$$
I_{yz} = \int\int\int_V yz\rho\,dxdydz
$$

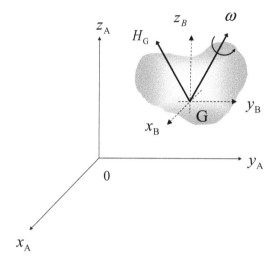

FIGURE 2.15 Angular momentum of a rigid body in space motion.

where the mass of the rigid body is distributed as differential volume elements with uniform density ρ. The inertia matrix is a function of the location and the orientation of the reference frame. As the position and the orientation of the axes are changed relative to the body, the moments and products of inertia will also change. There is one unique orientation of axes for a given origin for which the products inertia become to be zero, i.e.,

$$
I_B = \begin{bmatrix} I_{xx} & 0 & 0 \\ 0 & I_{yy} & 0 \\ 0 & 0 & I_{zz} \end{bmatrix}
\tag{2.68}
$$

where the moments of inertia take on stationary values [1]. The axes of the frame when so aligned are known as *principal axes* and the corresponding mass moments are the *principal moments of inertia*.

Example 2.5

Figure 2.16 shows a solid rectangular parallelepiped of mass m with dimensions of $3a \times 2b \times 3c$. The center of gravity is at the geometric center of the body. A coordinate system is attached to the body. The position of the origin of the xyz –coordinate system is located at point 0 that is distanced by dimensions a, b, c from three faces of the parallelepiped. Thus, the CG coordinates, x_G, y_G, z_G, are $(-0.5a, 0, 0.5c)$. Determine the moments of inertia of the parallelepiped in the xyz-system.

The moments of inertia of the parallelepiped about x', y', z axes that go through the center of the gravity (not shown in Figure 2.16) in parallel with x, y, z axes are calculated as follows:

$$
I_{x'x'} = \frac{1}{3}m\left[b^2 + (1.5c)^2\right]
$$

$$
I_{y'y'} = \frac{1}{3}m\left[(1.5a)^2 + (1.5c)^2\right]
$$

$$
I_{z'z'} = \frac{1}{3}m\left[(1.5a)^2 + (b)^2\right]
$$

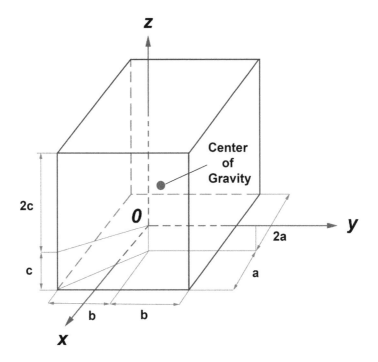

FIGURE 2.16   A rectangular parallelepiped.

The moments of inertia about x, y, z axes are determined using the parallel-axis theorem:

$$I_{xx} = I_{x'x'} + m\left(y_G^2 + z_G^2\right)$$

$$I_{yy} = I_{y'y'} + m\left(x_G^2 + z_G^2\right)$$

$$I_{zz} = I_{z'z'} + m\left(x_G^2 + y_G^2\right)$$

By substituting the given data in the above equations, the final expressions for the moments of inertia about x, y, z axes come as follows:

$$I_{xx} = \left(\frac{b^2}{3} + c^2\right)m$$

$$I_{yy} = \left(a^2 + c^2\right)m$$

$$I_{zz} = \left(a^2 + \frac{b^2}{3}\right)m$$

❑

EULER'S EQUATION

In general, for a rigid body rotating about gravity center CG with angular velocity
$\boldsymbol{\omega} = \begin{bmatrix} \omega_x & \omega_y & \omega_z \end{bmatrix}^T$, the sum of external moments, $\sum \boldsymbol{T}_G$, required to be acting on the body to cause the motion is given by

$$\sum \boldsymbol{T}_G = \dot{\boldsymbol{H}}_G \tag{2.69}$$

The angular momentum, $\boldsymbol{H}_G$, is a space vector, and its time derivative can be calculated by applying (2.66) and results in

$$\sum \boldsymbol{T}_G = \boldsymbol{I}_B\, \dot{\boldsymbol{\omega}} + \boldsymbol{\omega} \times \boldsymbol{I}_B \boldsymbol{\omega} \tag{2.70}$$

If the *principal axes* are chosen as the axes of the coordinate system, the factors I_{xy}, I_{xz}, I_{yz} will be zero, and (2.70) becomes

$$\sum T_{Gx} = I_{xx}\dot{\omega}_x - \left(I_{yy} - I_{zz}\right)\omega_y\omega_z$$
$$\sum T_{Gy} = I_{yy}\dot{\omega}_y - \left(I_{zz} - I_{xx}\right)\omega_z\omega_x \tag{2.71}$$
$$\sum T_{Gz} = I_{zz}\dot{\omega}_z - \left(I_{xx} - I_{yy}\right)\omega_x\omega_y$$

These equations are known as *Euler's equations*.

2.6 POWER AND EFFICIENCY

In vehicle engineering, often mechanical power transfer as well as its efficiency are considered. *Power* is defined as the time rate at which work is done. For an electric motor or an internal combustion engine, power is a much more important criterion than the actual amount of work to be performed. For a rigid body in rotation with angular velocity ω and under torque T, the power transmitted to the rigid body, denoted as P, is the product of ω and T

$$P = T\omega \tag{2.72}$$

Typically, mechanical power transfer happens through *gears*. The gears are used to multiply the torque and at the same time to reduce the speed at the output of a gear set. Examples in vehicles are given such as the transmission and the final drive, which provides the power transfer from the transmission to the differential. Consider a simple schematic of a transmission in Figure 2.17 with an input and output shafts,

The velocity ratio is defined in terms of the ratio of the output and input speeds, i.e.,

$$u = \frac{\omega_{out}}{\omega_{in}} \tag{2.73}$$

The power transmitted by torque T_{in} applying to the shaft rotating with ω_{in} is given by P_{in}. In this process both the torque and the speed are changed at the output shaft, denoted as T_{out} and ω_{out} respectively. The ratio of the output power and the input power is defined as the *efficiency* of the transmission, which is

$$\eta = \frac{P_{out}}{P_{in}} = \frac{T_{out}\omega_{out}}{T_{in}\omega_{in}} \tag{2.74}$$

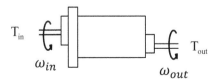

FIGURE 2.17   Schematic of a transmission.

In an ideal transmission, the motion is frictionless and the power supplied at the input shaft is equal to the power available at the output shaft, i.e., $\eta = 1$, then we have the i-ratio of the torques related to the velocity ratio as follows:

$$i = \frac{T_{out}}{T_{in}} = \frac{\omega_{in}}{\omega_{out}} = \frac{1}{u} \tag{2.75}$$

Ratio i of the torques in (2.75) is known as the gear ratio, i.e., the ratio of the products of the tooth numbers of the driven gears, z_{driven}, to the products of the tooth numbers of the drive gears, z_{drive} (this is for non-planetary gear sets)

$$i = \frac{\Pi z_{driven}}{\Pi z_{drive}} \tag{2.76}$$

It follows that the torque is increased while the speed is reduced and vice versa. In a real transmission, however, the power is lost through friction and inertia, and consequently the output power is always smaller than input power, i.e., efficiency $\eta < 1$. In this case, (2.75) can be rewritten in

$$T_{out} = \eta i T_{in} \tag{2.77}$$

REFERENCES

1. A. Bedford and W. Fowler, Engineering Mechanics: Dynamics, Prentice Hall, 2002.
2. J. J. Craig, Introduction to Robotics, 4th ed., New Jersey: Pearson Education, 2017.

3 Vehicle Longitudinal Dynamics

In modern highway vehicles, dynamic behavior is determined by the driving torques applied to the wheels and forces acting on the vehicle from the tire-surface interactions, the gravity components and the aerodynamic forces. The vehicle motion equations presented in this chapter describe vehicle longitudinal dynamics, as well as wheel-road dynamics. The term of vehicle longitudinal dynamics is concerned with both a straight line motion and a curvilinear motion of a vehicle. In the latter case, vehicle characteristics of motion are studied in the direction of the longitudinal axis of the vehicle or in the vehicle heading direction. The modeling of longitudinal dynamics provides insight to the following vehicle characteristics:

- Accelerating and braking performance on the level ground and grade.
- Driveline torque distribution among the front and rear axles, effect of driving and braking torque.
- Influence of aerodynamic and rolling resistance forces on vehicle motion.

Also, wheel dynamics is an essential aspect of vehicle longitudinal dynamics. The dynamic movement of a vehicle, with exception of the aerodynamic forces, is provided through the interaction of the wheels with roads, i.e., the forces that determine on how the vehicle turns, brakes and accelerates are originated at the tire-road patches. Hence, as a component of the vehicle, the wheel needs to be examined to determine what dynamics characteristics will be produced. In this context, the influence of lateral forces on the tires cannot be disregarded from vehicle longitudinal dynamics analysis. Both longitudinal and lateral tire forces are introduced in this chapter.

In this chapter, we begin with essential equations of wheel dynamics and tire-road forces. Likewise, elemental equations of vehicle forces and moments are then derived – the way, in which the longitudinal motion of vehicle arises.

In the description of longitudinal dynamics, we assume that the normal and pitch movements are negligible and the right and left wheels have the same dynamic behavior, so that the forces and torques can be considered for each axle (i.e., a vehicle bicycle model is used in this analysis). Road grade, i.e., the longitudinal gravity component will directly contribute to the driving and braking force, either in a positive (uphill) or negative (downhill) sense.

To find the relation of the forces, a free-body analysis is performed first. Thereby the entire vehicle is considered to consist of three components – the front and rear wheels and the vehicle body. Figure 3.1 shows free-body diagrams for both the vehicle body and the wheels, where the vehicle is modeled as a bicycle in one-dimensional movement. First, the vehicle body and wheels are understood in an ideal situation regardless of any disturbances, and then some additional effects are added to the wheel dynamics consideration. The basic connection between the vehicle body and the wheels is the axles and the suspensions. Hence, the forces acting on the axles in the vehicle body are reflected as reaction forces in the free-body diagram of the wheels. Physically, the axles are viewed to be associated with the wheels as both components are connected and rotate simultaneously. Therefore, we consider an equivalent mass of the rotating components for each wheel, which include the mass of the wheel and the mass of the axle reduced to the axis of the wheel. The vehicle body is usually represented by the *sprung mass* while the wheels, brakes

DOI: 10.1201/9781003134305-3

and axles represent the *unsprung mass*. The steering is usually split between the sprung and the unsprung mass. The *unsprung mass* that is also known as an *equivalent mass* since it includes additions, formed by reduced masses of several other vehicle components. For the description in Figure 3.1, the following notations are used:

M gross mass of the entire vehicle

m_v mass of the vehicle body (sprung mass)

m_w equivalent mass of the rotating wheel (unsprung mass)

a vehicle acceleration/deceleration in the longitudinal direction

g acceleration of gravity

h_s CG height of the vehicle

h_v CG height of the sprung mass

γ_z road grade angle

F_D resultant longitudinal force at the wheel axle

F_x resultant longitudinal force from the road to the tires

F_N normal force/reaction from the road to the tires

F_Z load on the wheel axle

F_{Ad} aerodynamic drag force

F_{Li} lift force on the vehicle

T_a driving torque at the axle

T_b braking torque at the axle

f subscript for the front wheel

r subscript for the rear wheel

Ideally, the physical parameters and geometric dimensions illustrated in Figure 3.1 represent the real system, i.e., those kinematic parameters remain constant without any variation. In practice, the tire/wheel position will experience variations as it rolls due to forces and moments at the contact with the ground. Also in real vehicles, some mechanical constrictions in the body structure are necessary to adjust due to weight imbalance and dimensional and stiffness variations of the vehicle body. The combination of all these effects may result in elasticity in the suspension and in slight dimensional changes of the vehicle kinematics [1]. In certain situations, those stiffness and dimensional variations can be assumed to be constant, for instance, in straight-line driving. Hence, the vehicle is considered here as a rigid body system.

From the equation of motion for the center of gravity, the mass of the total vehicle, spring mass and unspring mass are interrelated in case $\gamma_z = 0$:

$$M = m_{wf} + m_{wr} + m_v$$

(3.1)

$$Mh_s = m_v h_v + \left(m_{wr} + m_{wf}\right) r_w$$
$$Ml_r = m_{wf}\left(l_r + l_f\right) + m_v l_r$$

(3.2)

The above equations basically explain the definition and determination of the gravity center of the vehicle. Based on the free-body diagrams, dynamic equations of the wheel and the vehicle will be derived in the following sections. The reader can notice that the effect of tire rolling resistance on the wheel moment balance is not explicitly shown in Figure 3.1. This will be done later in this

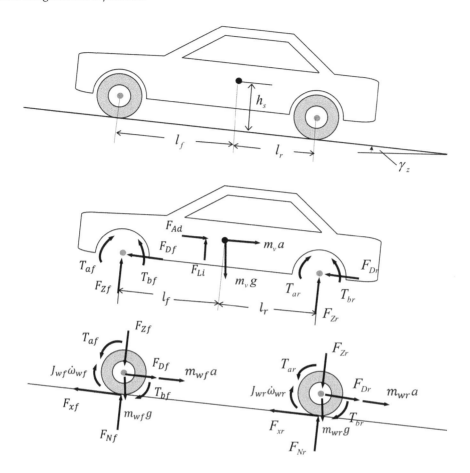

FIGURE 3.1 Free-body diagrams of vehicle body and wheels.

chapter. Also, the resultant forces in the front and rear tire-road patches result from the circumferential forces generated by the driving and braking torques, the wheel rolling resistance forces, and the gravity components of the wheels. Force F_x is also known as the net tractive force [2].

3.1 DYNAMICS OF WHEEL AND TIRE

3.1.1 BASIC EQUATIONS

For deriving dynamic equations of wheel movement, the wheel coordinate system W_{xyz} defined in Chapter 1 is referenced. The index w is omitted in the following description. Figure 3.2 shows a three dimensional free-body diagram of the wheel with corresponding coordinate system, where axis x is forward, and with axis z directed upward in order to form a right-hand set. Axis x is pointed to the heading direction of the wheel being along the wheel plane. The force exerted on the tire by the road, denoted by F_x, is the longitudinal force (the net tractive force). The force in the road plane that is parallel to y, is denoted as F_y. This is the lateral force in the tire patch. In the z direction, force F_N is the wheel normal reaction, which is normally exerted on the road plane. Force F_D opposed to the heading direction is the reaction force from the wheel axle.

The wheel has two basic movements – driving and braking. In the driving mode, torque T_a is applied through the axle shaft to accelerate the wheel forward while braking torque T_b will reduce the wheel speed. Usually, the wheel is in either one of two modes since a human driver does not

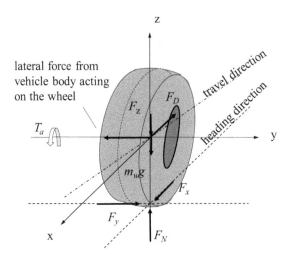

FIGURE 3.2 Forces and torque on a driven wheel.

apply the throttle and the brake pedal at the same time. However, both the driving and the braking torque can act concurrently if an automatic control system is in use for controlling the wheels. Of course, the resultant torque at the tire is dependent on the above-mentioned torques. According to the SAE definition, a positive applied torque is defined as the torque that causes a positive longitudinal force (see the force's positive direction in Figure 3.2). With the equivalent mass m_w of the wheel and the acceleration a, longitudinal force F_x is given by Newton's equation (2.65),

$$F_x - F_D = m_w a \tag{3.3}$$

The above equation can also be derived with the use of Figure 3.1 when a wheel moves on a horizontal road. Further, we apply Euler's equation presented through (2.69) to the moment balance about the wheel axis, and obtain

$$T_a - T_b - F_x r_w = J_w \dot{\omega}_w \tag{3.4}$$

where J_w denotes the rotational inertia of the wheel with the equivalent mass of m_w, and $\dot{\omega}_w$ is the rotational acceleration of the wheel. Radius r_w is the effective rolling radius of the wheel in the driven mode (at zero torque [2]). In some literatures [3], there are also other wheel radii defined, such as the loaded and the unloaded radius. Equation (3.4) will be further extended in Section 3.1.2 with an explicit representation of the rolling resistance moment.

So far, the wheel dynamics has been understood in an ideal situation regardless of any aero drag disturbances, etc. In the following sections, we will continuously investigate several effects, which explain the wheel dynamics in various driving situations influenced some additional forces and moments. At the end of the sections, we will see the total moment balance on the wheel.

Effective Rolling Radius

The effective rolling radius of the tire is the ratio of the linear wheel speed in the heading direction to the rotational/angular speed of the wheel loaded with zero torque and moved by a force applied to the wheel axis in the direction of motion (the reader can refer to the front wheel of a bike to visualize this mode of the wheel rolling). The linear wheel speed is equal to the effective wheel circumference multiplied by rotational speed n_w measured in revolution per second

$$v_w = 2\pi r_w n_w \tag{3.5}$$

which is equivalent to

$$v_w = r_w \omega_w \tag{3.6}$$

here, ω_w is the rotational speed of the wheel measured in radian per second. Velocity v_w is named the wheel theoretical velocity since it is determined by the effective rolling radius at zero wheel torque, i.e., in the absence of tire slippage (more details on slippage and wheel velocities can be found in Chapter 4).

If distance l_w, which the wheel travels, and corresponding wheel angle ϕ_w are known in the same time period t, the linear wheel theoretical speed and wheel angular speed can be expressed as

$$v_w = \frac{l_w}{t}; \; \omega_w = \frac{\phi_w}{t} \tag{3.7}$$

The effective rolling radius can be calculated by (3.6):

$$r_w = \frac{l_w}{\phi_w}$$

The latter expression is useful when the effective rolling radius is determined in experimental tests.

There are several factors impacting on the effective rolling radius such as the normal reaction in the tire patch, the inflation pressure and the vehicle speed. For the control design purpose in this book, it is assumed that the effective rolling radius is equal to the unloaded radius of the tire.

3.1.2 Rolling Resistance

There have been different approaches developed for defining and determining the wheel rolling resistance force [3]. Below, the commonly used interpretation of the rolling resistance is given. When a wheel is moved forward by a longitudinal force applied to the axis of the wheel at zero torque, the tire deflects under the normal load. The distribution of the normal load along the length of the tire patch is not symmetrical about the line that is normal to the road and runs through the axis of rotation. Instead, the distribution is asymmetrical due to the hysteresis of the tire material and thus the resultant of the distributed load, which is the normal reaction, F_N, does not go through the axis of rotation, but is shifted toward the direction of the wheel's motion. Such off-set, a_R, of the normal reaction creates a moment, T_R, that acts opposite to the direction of rotation, i.e., the moment resists to the wheel motion and is named as the moment of resistance to rolling

$$T_R = F_N a_R \tag{3.8}$$

In wheel dynamics analysis, the rolling resistance force, F_R, is also introduced as a force in the tire patch that is opposite to the direction of motion. In a steady motion of a wheel at zero torque, the rolling resistance force is overcome by the force F_D that is opposite to F_R and acts from the vehicle frame on the axis of the wheel. Based on experiment results, in the first approximation the rolling resistance force is considered proportional to the normal reaction in a linear matter

$$F_R = f_R F_N \tag{3.9}$$

The rolling resistance coefficient f_R is a dimensionless factor that expresses the effects of the complicated and interdependent physical properties of tire and the road, and it usually ranges from 0.01 to 0.05 for cars [4].

The physical meaning of the rolling resistance coefficient can be seen by analyzing equations for the moment of resistance to rolling and the rolling resistance force. By substituting $F_R r_w$ in the equation for the moment and changing F_R for $f_R F_N$, one can come with the following expression for the rolling resistance coefficient: $f_R = \dfrac{a_R}{r_w}$. At low speed f_R is under 0.01 while at higher speeds it rises approximately linearly with speed because of increased flexing work and vibration in the tire body. In addition, the rolling resistance can also be affected by tire temperature, ground material as well as the tire inflation pressure. Nevertheless, the rolling resistance coefficient is often assumed to be constant for calculation and control design.

For a vehicle, which is subjected to lateral forces (e.g., to a centrifugal force when the vehicle is making a turn), the rolling resistance coefficient f_{RC} of a tire is made up of coefficient f_R for a straight-line motion and an increase Δf_R [5],

$$f_{RC} = f_R + \Delta f_R \tag{3.10}$$

where the increase Δf_R given by

$$\Delta f_R \approx \mu_y \sin\alpha \tag{3.11}$$

is determined through the lateral friction coefficient μ_y and tire slip angle α. The friction coefficient is considered in the next section. The slip angle is the angle between the heading and the travel direction of the wheel, and will be detailed more in Chapter 4.

3.2   TIRE FORCE PROPERTIES

The tire is the connection element between the vehicle and the road surface. The area between the tire and the road ultimately influences vehicle dynamics. Therefore, the force relations at the tire-road patch have a decisive impact on the driving and braking behavior of the vehicle. In this section, the forces at the tire-road patch are introduced. The friction force plays a central role for vehicle traction/acceleration and braking performance. The use of the term "friction" is pretty conditional here since the tire-road interaction is very complex and the use of the "gripping force" is much more appropriate to address the interaction between the tire and the road. However, the term "friction" will be further used in the book to make it aligned with many existing publications on vehicle dynamics.

3.2.1   LONGITUDINAL TIRE FORCE

The longitudinal tire force is the force exerted on the tire in the heading direction of the wheel. As Figure 3.2 shows, the force is presented by F_x. The magnitude of F_x is considered proportional to the normal reaction

$$F_x = \mu_x F_N \tag{3.12}$$

μ_x is called the *normalized tractive force* or the *current friction coefficient* that is dependent on material characteristics and environment conditions of the tire-road contact. The word "current" means that this friction coefficient does not characterize the maximum friction properties on the tire-road contact and, thus the maximum tractive force. Instead, the current friction coefficient corresponds to a magnitude of the tractive force that is developed by the wheel at "current" time moment. Thus, the current friction coefficient is a fracture of the maximum/peak friction coefficient that is determined by the Coulomb model. A detailed description of the relationship between the current friction coefficient and the tire slip will consider tractive and brake case separately and be presented in the Chapter 4.

Normal Force Dependence

The Coulomb model describes the longitudinal tire force in the tire-road surface as long as the variation of the normal force and remains in a relatively small range. At high values of the normal force, for instance at a range of more than 5000 N at a 165/80R13 tire, the tire force will not increase linearly with the normal force increase, but rather tends to decrease [6]. This effect might be traced back to the elastic deformation of the tire structure. Consequently, it is strongly related to tire materials. The extended relation can be then approximated as

$$F_x = \mu_{x0} F_N \left(1 - c_{FN} F_N^2\right) \tag{3.13}$$

where c_{FN} is tire structure coefficient, usually between 0 and 0.04. μ_{x0} is constant and depending on tire type.

3.2.2 LATERAL TIRE FORCE

Generating lateral force is one of the important functions of a tire to control directions of the vehicle motion, such as in turns or for lane changes, etc. By steering vehicle wheels, the tire advances at the slip angle in its travel direction, so the lateral elastic deformation of the tire develops by this process. The tread elements become sideways deflected as well as other elements of the tire, and the lateral force emerges throughout the contact patch.

The term "lateral force" is used differently in literatures. Some of publications mean the force perpendicular to the vehicle longitudinal coordinate system while others mean the force perpendicular to the wheel plane of rotation. The basic reason for different definitions is that the resultant tire force can be split in different component directions. Certainly, the definition of the lateral force and its direction will finally have an influence on determining vehicle lateral dynamics. Choosing a wheel-fixed coordinate system, it is convenient to consider the lateral force aligned with the y-axis of the coordinate system.

Similar to the longitudinal direction, the lateral force acting on the tire depends on the normal force and the current friction coefficient as

$$F_y = \mu_y F_N, \tag{3.14}$$

where μ_y denotes the *lateral current friction coefficient*. The y-direction was defined before as the direction perpendicular to the wheel plane. The travel direction of the wheel and the wheel plane build the slip angle.

Similar to the F_x case, the lateral tire force can be extended to

$$F_y = \mu_{y0} F_N \left(1 - c_{FN} F_N^2\right) \tag{3.15}$$

to consider the normal force dependency. Here, μ_{y0} is constant depending on tire type.

Aligning Moment

Actually, the lateral force does not act at the center of the tire-road contact, but at the rear of the contact patch by a distance known as the pneumatic trail. Figure 3.3 displays the buildup of the lateral force. The left picture shows the tire rolling straight ahead, where no lateral force appears. The right picture illustrates the tire profile being subjected to an external lateral force that generates a lateral force/reaction in the tire patch, which deflects the tire. Thereby an elastic deflection arises. The integration of the forces over the contact patch, which are distributed asymmetrically, yields the lateral force, F_y. Therefore, the lateral force at this position causes a moment about the z-axis called as the aligning moment

$$T_l = n_{al} F_y \tag{3.16}$$

where n_{al} denotes the pneumatic trail.

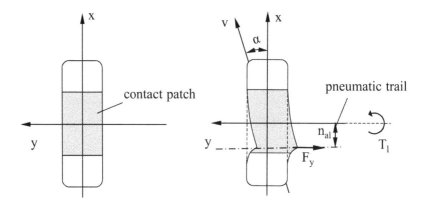

FIGURE 3.3 Tire top view with lateral force.

The pneumatic trial is usually not constant when the slip angle increases. When an increase of the slip angle occurs, there will be a sliding of the tire elements at the rear part of the contact patch. The lateral force is no longer proportional to the slip angle, and the centroid moves back toward the centerline so that the pneumatic trial goes to zero as the slip angle approaches $90°$ (the tire is skidding aside). Hence, the aligning moment may reduce even though the lateral force increases. This effect produces a maximum of the aligning moment at a slip angle, which is usually around $3°$ [7].

Relation to Wheel Slip Angle

Since the lateral force is one of the key tire properties in vehicle dynamics and the slip angle characterizes actual movement of the tire, the relationship between the lateral force and the slip angle is of a fundamental importance to the directional control and stability of vehicle.

Theoretically, the relationship can be formulated as $\mu_y = f(\lambda_y)$. λ_y is the so-called wheel slip. Later in Chapter 4, slip λ_y will be introduced and its dependency on α will be described. Slip angle of $\alpha = 90°$ can be possible when a tire is in full lateral skid. However, the measurement of the force is indeterminate for large α value since the vehicle is only controllable at lower slip angles, usually smaller than $10°$. For this reason, a diagram of the lateral force over the slip angle is preferred to demonstrate their relationship within the range that makes practical meaning. Figure 3.4 shows an example, where the lateral force is plotted against slip angle with varying normal forces. The curve is in accordance with experimental data and ends up mostly with slip angle less than $10°$. We can see a rough linearity when $\alpha \leq 5°$.

On the other side, from Figure 3.3 we see that the lateral force comes from the integration of the forces distributed over the contact patch. The larger the lateral force, the bigger the slip angle is. This effect assumes a proportionality between them, at least for small slip angles. Consequently, from the above-given discussion we recognize the linearity between the lateral force and small slip angle values, i.e.,

$$F_y = C_\alpha \alpha \tag{3.17}$$

C_α is the proportional factor, called *cornering stiffness*.

Coming from the same background, lateral force F_y demonstrates the same behavior as the aligning moment when the slip angle becomes larger. The force reaches its maximum at small α values. Experimental measurements show a close linear approximation for $\alpha \leq 3°$, which can be observed at a lateral acceleration less than 4 m/s². Within this range of the lateral acceleration the cornering stiffness and the pneumatic trial can be considered constant. Usually, vehicles controlled by average drivers behave also in this linear range of the lateral force and the slip angle.

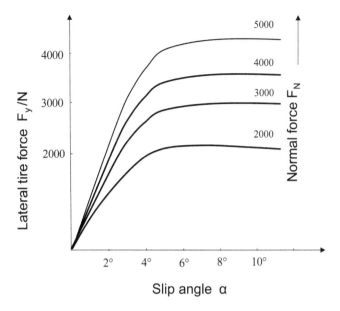

FIGURE 3.4 Lateral force over slip angle for various loads.

3.2.3 CAMBER ANGLE AND CAMBER FORCE

In the previous sections, we considered that the rolling wheel stands upright on the road, i.e., the wheel plane is normal to the road plane. In the case that the wheel is inclined to the road, it causes an additional force so that the lateral force and the aligning moment change as well. Camber is the inclination of the wheel away from the normal plane to the road and will produce a lateral force, called the camber force, sometimes the camber thrust. Figure 3.5 shows the camber angle.

The camber force is a function of the tire stiffness properties and the camber angle. The camber angle is formed by the inclination of centerline of the wheel. It can be influenced by the vehicle body roll and the jounce of the wheels. The force acts in line with the direction that the tire is inclined, and is therefore a force exerted by the road into the tire. For small camber angles, the camber force can be approximated as a product of camber stiffness C_γ and camber angle γ_c,

$$F_\gamma = C_\gamma \gamma_c \tag{3.18}$$

Then, from (3.17) the wheel lateral force is extended to the following algebraic summation:

$$F_y = C_\alpha \alpha + C_\gamma \gamma_c \tag{3.19}$$

The typical value of the camber stiffness is 35N/degree for a modern radial tire. Since the camber angle is fairly small, its lateral effect is generally secondary. The camber angle produces a much less lateral force than the slip angle. About 20% of the camber angle is required for the slip angle to get an equal lateral force.

3.2.4 KAMM CIRCLE

For vehicle dynamics, especially at high lateral acceleration it is essential to know the relation of the longitudinal and lateral forces, the aligning moment and the slip angle.

The maximum of the longitudinal friction force is determined by (3.12) when the lateral force equals to zero. In a driving situation, where both the longitudinal and the lateral forces occur, the

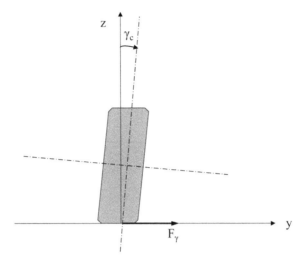

FIGURE 3.5 Camber force.

resultant force will not be greater than the maximum friction force accomplished with peak friction coefficient μ_H. Mathematically it means

$$\sqrt{F_x^2 + F_y^2} \leq \mu_H F_N \tag{3.20}$$

If increasing F_x or F_y so that the resulting force determined by the square root in (3.20) reaches maximum friction force $\mu_H F_N$, the tire will be in complete slide. Equation (3.20) can be interpreted via a so-called Kamm circle (friction circle) as shown in Figure 3.6. It concludes that the available friction force, which is the maximum force that tire patch can develop, must be shared in the longitudinal and lateral directions. Increasing the longitudinal force will reduce the potential for the lateral one. The available friction force in the longitudinal direction is lower if the lateral force exists. For a given lateral force, the slip angle will increase when the longitudinal force becomes larger. This can be seen by recording the lateral force vs. the slip angle of the tire by varying the

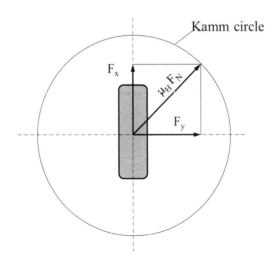

FIGURE 3.6 Kamm friction circle.

longitudinal force. The aligning moment keeps the same shape over the slip angle, but will become lower when the longitudinal force increases. More details can be seen in [7, 8].

The Kamm circle assumes isotropic properties of the tire-road contact in the longitudinal and lateral directions. Due to some actual anisotropy, the circle can transform to an ellipse.

3.3 TOTAL FORCE AND MOMENT LOADS ON WHEELS

At this moment of consideration, we still handle the forces in the longitudinal and lateral directions independently. As long as an individual wheel is considered in a dynamics control application, the independent treatment is more convenient since control inputs come from different sources. The longitudinal tire force is created by applying torques about the wheel spin axis while the lateral force can be achieved by steering the wheel. The longitudinal wheel dynamics is usually modeled for designing a wheel speed control by applying the engine torque or the brake torque. This means, the engine or the braking torque is regulated to reach a wheel's linear speed target in the heading direction. The lateral force will be of decisive importance when the lateral dynamics of the vehicle is considered, and it is usually used to calculate the wheel speed target rather than to control the wheel speed (the reader can find more on lateral dynamics in Chapter 8).

We have considered the rolling resistance force, the camber angle, as well as the aligning moment, all of which contribute additional rotation moments at the wheel and summarized in Figure 3.7. Due to the camber angle, the wheel axle builds an angle of γ_c to the y-axis. The algebraic summation of the all moment effects constitutes the total moment of the wheel about the wheel axle. Extending (3.4), we arrive at the total moment balance equation:

$$T_a - T_b - F_x r_w - T_R \cos \gamma_c - T_l \sin \gamma_c = J_w \dot{\omega}_w \tag{3.21}$$

Note that the rolling resistance moment, T_R, is tire moment vector tending to rotate the tire about the y-axis. The total wheel rolling resistance force is determined by dividing the last two components of the left-hand side of (3.21) by the effective rolling radius (the same definitions is also accepted by SAE)

$$F_R = \frac{T_R \cos\gamma_c + T_l \sin\gamma_c}{r_w} \tag{3.22}$$

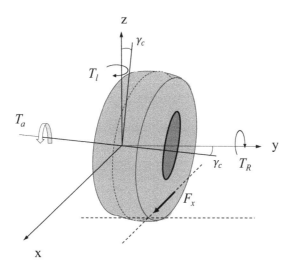

FIGURE 3.7 Torques and force on a driven wheel.

Now, the force and moment balances can be stated for the front and rear axles. When considering each axle, both left and right wheels are assumed to have the same dynamic behavior. Considering the free-body diagram in Figure 3.1 and using (3.3), we have the force balances in both x- and z-directions

$$F_{xf} - F_{Df} - m_{wf} g \sin\gamma_z = m_{wf} a \tag{3.23}$$

$$F_{Nf} - F_{Zf} - m_{wf} g \cos\gamma_z = 0 \tag{3.24}$$

where subscript f stands for the front. Substituting the rolling resistance and the aligning moment given in (3.22) into (3.21), the moment balance of the front axle is

$$T_{af} - T_{bf} - F_{xf} r_w - F_{Rf} r_w = J_{wf} \dot{\omega}_{wf} \tag{3.25}$$

The equations for rear axle are similarly derived

$$F_{xr} - F_{Dr} - m_{wr} g \sin\gamma_z = m_{wr} a \tag{3.26}$$

$$F_{Nr} - F_{Zr} - m_{wr} g \cos\gamma_z = 0 \tag{3.27}$$

$$T_{ar} - T_{br} - F_{xr} r_w - F_{Rr} r_w = J_{wr} \dot{\omega}_{wr} \tag{3.28}$$

Note that here the parameters in the equations refer to the entire axle. So, m_{wf}, m_{wr}, J_{wf} and J_{wr} present the equivalent wheel masses and the equivalent moments of inertia.

3.4   EQUATIONS OF VEHICLE MOTION

3.4.1   Vehicle Forces and Moments

From the free-body diagram of the vehicle body in Figure 3.1, the force balance in the *x*-direction can be expressed as follows:

$$F_{Df} + F_{Dr} - F_{Ad} - m_v g \sin\gamma_z = m_v a \tag{3.29}$$

where the aerodynamic drag force, denoted as F_{Ad}, is described in the next section. Assuming that there is no acceleration in *z*-direction, the normal forces are in balance with aerodynamic lift force, F_{Li}, and the component of gravity:

$$F_{Zf} + F_{Zr} + F_{Li} - m_v g \cos\gamma_z = 0 \tag{3.30}$$

The lift force will be described in the next section. At this moment, the rotation about the vehicle *y*-axis, which is the pitch of the vehicle, is not considered. The resultant moment about the center of gravity of the sprung mass is equal to zero:

$$F_{Zr} l_r - \left(T_{ar} - T_{br} + T_{af} - T_{bf} \right) - F_{Zf} l_f - \left(h_v - r_w \right)\left(F_{Dr} + F_{Df} \right) - F_{Li} l_{Al} = 0 \tag{3.31}$$

with an assumption of the use of the effective rolling radius.

Summing (3.23), (3.26) and (3.29), the total longitudinal tractive force on the vehicle is

$$F_{xf} + F_{xr} = Ma + F_{Ad} + Mg\sin\gamma_z \tag{3.32}$$

The sum of the longitudinal tractive forces is equal to three force resistance components. The first component is the inertia force acting at the center of gravity opposite to the direction of the acceleration. The second component is the aerodynamic force acting on the body of the vehicle, and the last component is the gravity force component that is parallel to the road. Longitudinal forces F_{xf}, F_{xr} result in different directions for braking and acceleration. In braking, they oppose the moving direction. If the vehicle is accelerating, F_{xf} and F_{xr} act as tractive forces. In order to make positive tractive forces, the engine torque must overcome all of the resistance forces specified in the right-hand side of (3.32).

By summing (3.25) and (3.28) we arrive at the vehicle moment balance equation:

$$T_{af} + T_{ar} - T_{bf} - T_{br} - \left(F_{xf} + F_{xr}\right)r_w - \left(F_{Rf} + F_{Rr}\right)r_w = J_{wf}\dot{\omega}_{wf} + J_{wr}\dot{\omega}_{wr} \tag{3.33}$$

Recognizing that

$$\dot{\omega}_{wf} = \dot{\omega}_{wr} = \frac{a}{r_w}, \tag{3.34}$$

combining (3.32) with (3.33), provides

$$T_{af} + T_{ar} = T_{bf} + T_{br} + \left(F_{Ad} + Mg\sin\gamma_z\right)r_w + \left(F_{Rf} + F_{Rr}\right)r_w + Mar_w + \frac{J_{wf}a + J_{wr}a}{r_w} \tag{3.35}$$

To interpret (3.35), we see that the driving torque is the effort available to overcome the braking toques, the aerodynamic force as well as the rolling resistance, allowing acceleration of the vehicle body and the wheels.

3.4.2 Aerodynamic Forces

Aerodynamic forces interact with the vehicle causing drag, lift and lateral forces. Since air flow over a vehicle is complex, approximations are made in developing empirical models for the aerodynamic forces, which are produced mainly from the pressure drag. To give an idea about the pressure distribution, Figure 3.8 shows roughly the air flow around a moving vehicle. As the air hits obstacles such as the front of the hood and the windshield, the pressure increases as the flow turns upward. As the air flow passes the surfaces such as the front hood, the roof, the taillight and the trunk, the pressure decreases because the air flow rises turning and following the contour.

From the vehicle dynamics point of view, the center of gravity of the vehicle is an important reference point for forces and moments. The aerodynamic forces are not dependent on the location of the center of gravity and the weight distribution of the vehicle between the wheels, but rather the shape of the vehicle. The aerodynamic force is assumed to act on the body at the center of pressure (CP), which normally located ahead of the CG of the vehicle. The assumption of the CP location is due to the fact that the vehicle is turned away when it is subject to a side wind. As shown in Figure 3.8, CP is assumed at the same height of the CG.

Drag Force

The aerodynamic drag is the largest aerodynamic force encountered by cars at highway speeds. The overall drag on a vehicle comes from many sources, different car parts and components. Approximately 65% of the drag results from the body [9]. In general, the direction of the drag

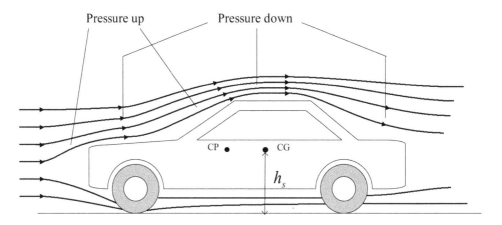

FIGURE 3.8 Air flow over a vehicle body.

force is determined by the driving direction and the wind gust. The resultant wind velocity v_{ad} with respect to vehicle motion is the vector sum of the vehicle velocity v and the wind velocity v_{aw}, as shown in Figure 3.9. The angle τ_w between the driving direction and the wind gust is called *wind angle*, which can be computed as

$$cos\, \tau_w = \frac{v_{ad}^2 + v^2 - v_{aw}^2}{2 v_{ad} v} \tag{3.36}$$

To determine the magnitude of the drag forces, empirical models were developed based on experimental verifications [4]. In case of a headwind or tailwind, i.e., $\tau_w = 0$, the drag force is defined:

$$F_{Ad} = \frac{\rho}{2} c_{Ad} A v_{ad}^2 \tag{3.37}$$

where ρ is the air density and c_{Ad} is aerodynamic drag coefficient. A denotes the front area of the vehicle. In practice, v_{ad} is often assumed to be equal to the vehicle speed since the wind velocity is mostly not available. The air density is a function of temperature, pressure and humidity. At standard conditions (15°C, 1.02bar) it yields $\rho = 1.22\,\frac{kg}{m^3}$. Drag coefficient c_{Ad} is usually determined experimentally from wind tunnel tests. It can range from 0.2 to 0.9 depending on the car shapes [9].

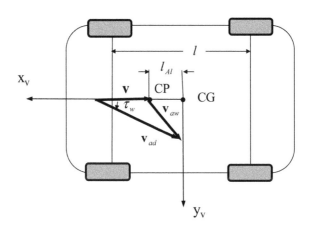

FIGURE 3.9 Influence of the total wind velocity.

Of course, these are empirical data measured under steady wind conditions while the wind behavior changes continuously during normal driving practice.

Lateral Force

In crosswinds, the lateral wind components impose a side force on the vehicle. The aerodynamic lateral force is characterized by

$$F_{Al} = \frac{\rho}{2} c_{Al} A v_{ad}^2 \tag{3.38}$$

where c_{Al} is the aerodynamic lateral force coefficient. Wind tunnel tests show that the lateral force coefficient is a function of wind angle τ_w. Approximately it can be considered linear:

$$c_{Al} = c_{Al}^* \tau_w \tag{3.39}$$

where c_{Al}^* denotes the linear ratio between the aerodynamic lateral force coefficient and the wind angle [8]. The aerodynamic lateral force is assumed to act on the body at the center of pressure (CP). The air also exerts a yaw moment on the vehicle. In practice, wheelbase l is preferably used, so that the yawing moment can be given:

$$M_{Al} = \frac{\rho}{2} c_{Am} l \, A v_{ad}^2 \tag{3.40}$$

where c_{Am} is the wind yawing moment coefficient, and l is the wheel base. Both c_{Am} and c_{Al} are determined in wind tunnel measurements.

From another point of view, the moment arises due to the distance between center of pressure CP and CG. Therefore, M_{Al} can be built as a product of the lateral force and distance l_{Al} (see Figure 3.9):

$$M_{Al} = \frac{\rho}{2} c_{Al} l_{Al} A v_{ad}^2 \tag{3.41}$$

Equating (3.40) and (3.41), the location of CP can be determined by

$$l_{Al} = \frac{c_{Am}}{c_{Al}} l \tag{3.42}$$

Later in Chapter 8, we will see how the aerodynamic effects are considered in the vehicle dynamics models.

Lift Force

As shown in Figure 3.8 above, some low pressure regions exist along the contour on the top the vehicle. The lift force is caused by this pressure differential from the top to the bottom. Since the source of the wind is the same, a similar characteristic for the lift force can be quantified as

$$F_{Li} = \frac{\rho}{2} c_{Li} A v_{ae}^2 \tag{3.43}$$

where c_{Li} is the aerodynamic lift coefficient, which is also dependent on the overall shape of the vehicle, normally ranging from 0.3 to 0.5 in modern passenger cars [9]. It acts in the normal direction. Therefore, the lift force may consequently have an impact on handling through the control forces available at tires in the normal direction.

To simplify the application, the lift force will be divided into two forces applied to the front and the rear axle, respectively. The front axle force is modeled by

$$F_{Li\_f} = \frac{\rho}{2} c_{Li\_f} A v_{ae}^2, \tag{3.44}$$

and the rear force comes from

$$F_{\text{Li\_r}} = \frac{\rho}{2} c_{\text{Li\_r}} A v_{ae}^2 \tag{3.45}$$

$c_{\text{Li\_f}}$ and $c_{\text{Li\_r}}$ are lift coefficients of the front and rear force, respectively.

3.4.3   DYNAMIC AXLE LOADS

In the previous section we derived the vehicle force and moment equations. From there, the axle loads can be calculated, which show the relationship between the normal force acting on an axle and other vehicle dynamic parameters. Quantifying the axle loading is an important measure in analysis of acceleration and braking performance because the axle loads influence the tractive effort obtainable at each axle, which affects the acceleration, gradeability and the maximum speed.

The tire will experience a normal force from the road, denoted by F_{Nf} and F_{Nr} for front and rear respectively. Substituting (3.29) and (3.35) in (3.31), and combining the result with (3.30), we obtain

$$\begin{aligned}
F_{Zf} l = & - F_{Li}\left(l_r + l_{Al}\right) - h_v F_{Ad} - \left(F_{Rf} + F_{Rr}\right)r_w - J_{wf}\dot{\omega}_{wf} - J_{wr}\dot{\omega}_{wr} \\
& + m_v g l_r \cos\gamma_z - h_s\left(m_v a + m_v g \sin\gamma_z\right) \\
& - \left(Ma + Mg\sin\gamma_z - m_v a - m_v g \sin\gamma_z\right)r_w
\end{aligned} \tag{3.46}$$

Considering (3.24) leads to

$$\begin{aligned}
F_{Nf} l = & \left(m_{wf} l + m_v l_r\right)g\cos\gamma_z - \left(m_v h_v + m_{wf} r_w + m_{wr} r_w\right)g\sin\gamma_z \\
& - \left(m_v h_v + m_{wf} r_w + m_{wr} r_w\right)a - \left(J_{wf}\dot{\omega}_{wf} + J_{wr}\dot{\omega}_{wr}\right) \\
& - F_{Li}\left(l_r + l_{Al}\right) - h_v F_{Ad} - \left(F_{Rf} + F_{Rr}\right)r_w
\end{aligned} \tag{3.47}$$

Then we make use of the relationship between the vehicle and wheel masses, and solve for F_{Nf} as

$$\begin{aligned}
F_{Nf} = & Mg\left(\frac{l_r}{l}\cos\gamma_z - \frac{h_s}{l}\sin\gamma_z\right) - a\left(M\frac{h_s}{l} + \frac{J_{wf}}{r_w l} + \frac{J_{wr}}{r_w l}\right) \\
& - \left\{F_{Li}\frac{\left(l_r + l_{Al}\right)}{l} + F_{Ad}\frac{h_v}{l} + \left(F_{Rf} + F_{Rr}\right)\frac{r_w}{l}\right\}
\end{aligned} \tag{3.48}$$

Basically, the loads consist of the static load

$$Mg\left(\frac{l_r}{l}\cos\gamma_z - \frac{h_s}{l}\sin\gamma_z\right)$$

and the dynamic load

$$a\left(M\frac{h_s}{l} + \frac{J_{wf}}{r_w l} + \frac{J_{wr}}{r_w l}\right),$$

plus the load through the lift force, the air drag and the rolling resistance represented by

$$F_{Li} \frac{(l_r + l_{Al})}{l} + F_{Ad} \frac{h_v}{l} + \left(F_{Rf} + F_{Rr}\right)\frac{r_w}{l}.$$

Knowing that

$$F_{Nf} + F_{Nr} + F_{Li} - Mg\cos\gamma_z = 0, \tag{3.49}$$

the rear axle load equation can be reduced to

$$F_{Nr} = Mg\left(\frac{l_f}{l}\cos\gamma_z + \frac{h_s}{l}\sin\gamma_z\right) + a\left(M\frac{h_s}{l} + \frac{J_{wf}}{r_w l} + \frac{J_{wr}}{r_w l}\right)$$
$$+ \left(F_{Li}\frac{(l_r + l_{Al})}{l} + F_{Ad}\frac{h_v}{l} + \left(F_{Rf} + F_{Rr}\right)\frac{r_w}{l}\right) \tag{3.50}$$

To conclude, the axle loads on both the front and the rear contain the static loads, which are dominant when vehicle is not in acceleration or deceleration. A road grade will increase the load on the rear axle and lower the load on the front. The lift force will reduce the loads while the air drag and the rolling resistance loads decline on the front and raise on the rear. When the vehicle is accelerating, the front dynamic loads decrease and the rear dynamic loads build up.

REFERENCES

1. J. Reimpell, Fahrwerktechnik: Grundlagen, Würzburg: Vogel Verlag, 1995.
2. A. F. Andreev, V. I. Kabanau, V. V. Vantsevich, Driveline Systems of Ground Vehicles: Theory and Design, Boca Raton, London, New York: CRC Press, 2010.
3. J. C. Dixon, Tires, Suspension and Handling, Warrendale: SAE, 1996.
4. M. Mitschke, Dynamik der Kraftfahrzeuge, Band A: Antrieb und Bremsung, Berlin: Springer-Verlag, 1982.
5. H. P. Willumeit, Modelle und Modellierungsverfahren in der Fahrzeugdynamik, Stuttgart: B. G. Teubner, 1998.
6. M. Burckhardt, Radschlupf-Regelsysteme, Würzburg: Vogel Verlag, 1993.
7. M. Mitschke, Dynamik der Kraftfahrzeuge, Band C: Fahrverhalten, Berlin: Springer-Verlag, 1990.
8. A. Zomotor, Fahrwerktechnik: Fahrverhalten, Würzburg: Vogel Verlag, 1991.
9. T. D. Gillespie, Fundamentals of Vehicle Dynamics, Warrendale: SAE, 1992.

4 Tire and Wheel Characteristics

The tire is the connection element between the vehicle and the road surface. The contact area between the tire and the road determines the dynamics of vehicles ultimately. Therefore, the tire-road interaction plays an important role regarding the behavior of vehicle driving and braking. The tire characteristics are resulted from the dynamical variables (inputs/outputs), tire structure design and tire material. In this regard, the tire can be considered as both a sensor and the actuator. The tire physical parameters and characteristics can be interpreted in a form of characteristic curves, data fields as well as lookup tables relating to vehicle dynamic variables such as the forces and moments/torques, the tire inflation pressure and the tire/road contact pressure characteristics, speeds and accelerations/decelerations. Based on the tire technical data, it is usually sufficient for a vehicle manufacturer to meet the standards on vehicle safety and vehicle performance requirements. However, those technical data are insufficient or unsuitable for vehicle dynamics control systems, which manage the tire forces and control the wheel heading direction.

For tire dynamics control purposes, a simple mathematical approach is necessary to quantify essential tire properties, and thus to manage tire dynamics. This kind of modeling of the tire-road interaction relies on empirical equations that are built on experimental tire data in different conditions of the road surface, rather than on a physics-based model approach, which is time consuming for the online operation. The transparency of the empirical models, which is characterized by clear and simple mathematical handling, is preferred to the high accuracy of the physics-based models. This chapter presents essential tire-road characteristics, including tire longitudinal slippage, tire friction and lateral force – slip angle characteristics that are based on empirical approach.

As long as the wheel in rolling, the applied torque generates a longitudinal force in the tire patch that overcomes forces of resistance to motion, including the rolling resistance force, aero drag, the gravity component on a grade, inertia, etc. Due to the tire-road sliding adhesion and tire elastic deflections, the wheel moves forward under the action of the driving torque [1]. When the tire deflects in the circumferential direction and the driving torque increases, its rolling radius decreases [2]. As a result, the linear velocity of the center of the wheel is less as compared to the linear velocity of the wheel moving under zero torque when the wheel is moved by a force applied to the wheel axis. The bigger the driving torque, the smaller the rolling radius and so is the wheel linear velocity. This velocity reduction is estimated by tire slip or tire slippage, which is caused by the circumferential deflection of the tire and, at large torques, by a motion of some or all points in the tire patch relative to the road. In the brake mode, when the braking torque goes up, the rolling radius increases. At big brake moments, the tire can be a longitudinal skid relative to the road.

For the tire slip description, it is sufficient to mathematically characterize the phenomenon of the sliding adhesion and the circumferential deflection without an exact physical consideration in the background. There have been many slip definitions in use. For example, the slip definitions are comprehensively described in reference [3]. The wheel travel direction in its translational motion, namely the direction of wheel linear velocity, is chosen as the longitudinal slip direction. If a wheel is moving with a side-slip angle, the longitudinal slip direction is aligned with the x-axis of the wheel coordinate system. It also makes easier to consider the lateral force in vehicle dynamics analysis.

Here, for the slip definition, the tire coordinate system T_{xyz} defined in Chapter 1 is referenced. The index T of each axis in the coordinate system is omitted in the following description. As shown in Figure 4.1, the positive direction of the x-axis is forward, and with z upward in order to form a

DOI: 10.1201/9781003134305-4

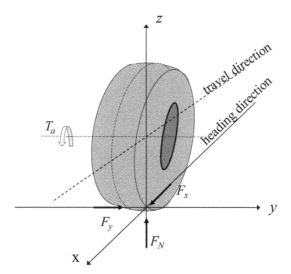

FIGURE 4.1   Tire coordinate system.

right-hand set. Axis x is pointed to the heading direction of the wheel being along the wheel plane. The resultant force exerted on the tire by the road in the longitudinal direction, denoted by F_x, is the longitudinal force. A force in the road plane along with the y-axis, denoted by F_y, is the lateral force. In the z-direction force F_N is the normal force, which is normal to the road plane. For the sake of simplicity, the forces at the axis of the wheel and moments are not shown (except the wheel torque).

4.1   BRAKE SLIP

When a vehicle undergoes its planar motion, its braking movement needs to be considered in a two-dimensional space, i.e., within xy-coordinate system. As shown in Figure 4.2, the coordinate system previously described in Figure 4.1 is used for the brake slip definition, where the x-direction is the wheel heading direction.

Since the wheel is not driven straight ahead consistently, so that it can build an angle α – wheel slip angle – between the wheel travel direction and the heading direction. Denoting v as the wheel's center actual velocity and v_w as the wheel theoretical longitudinal velocity at zero torque, vector $v_{sum,B}$ in Figure 4.2 illustrates the difference between the previously mentioned velocities. The velocity difference in the x-direction is given by (4.1)

$$\Delta v_{x,B} = v\cos\alpha - v_w \tag{4.1}$$

with

$$v_w = \omega_w r_w \tag{4.2}$$

where ω_w is the angular velocity of the wheel and r_w denotes the tire rolling radius at zero torque. The difference of the velocities given by (4.1) is the measure of the velocity change in the longitudinal direction caused by a non-zero brake torque, which is known as the tire brake slip in the x-direction, i.e., in the heading direction.

$$\lambda_{x,B} = \frac{v\cos\alpha - v_w}{v} \tag{4.3}$$

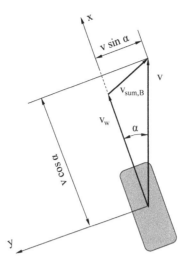

FIGURE 4.2 Tire top view: brake slip definition.

Using the slide velocity in the y-direction shown in Figure 4.2 as $v\sin\alpha$, the brake slip in the y-direction can be specified as

$$\lambda_{y,B} = \frac{v\sin\alpha}{v} = \sin\alpha \tag{4.4}$$

For $\alpha = 0$, when the slip in the y-direction is zero, the longitudinal brake slip becomes

$$\lambda_B = \frac{v - v_w}{v} \tag{4.5}$$

where λ_B is called brake slip at zero slip angle. Note that this brake slip will be used further in this chapter as a basic component in various equations to relate the tire slips to the forces in the tire patch. This is because slip λ_B can be easily measured and related to the tire forces and, thus, be utilized to characterize the tire motion in a simple mathematical form.

4.2 TRACTIVE SLIP

In the traction operational mode, e.g., the vehicle is accelerated, the magnitude of v_w is larger than v, as shown in Figure 4.3. The tire slip value according to the definition (4.5) will be negative and tends to infinity if the vehicle speed moves to zero. For this reason, the wheel theoretical longitudinal velocity at zero driving torque is used as a reference velocity. Hence, the tractive longitudinal slip is defined as:

$$\lambda_{x,T} = \frac{v_w - v\cos\alpha}{v_w} \tag{4.6}$$

The longitudinal tractive slip characterizes the change in the longitudinal velocity of the wheel caused by the applied driving torque in the heading direction of axis x.

The side slip is

$$\lambda_{y,T} = \frac{v\sin\alpha}{v_w} \tag{4.7}$$

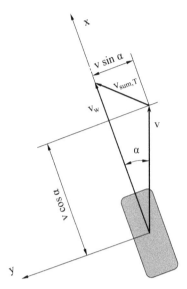

FIGURE 4.3 Tractive slip definition.

For $\alpha = 0$ and zero side tractive slip, the longitudinal tractive slip becomes

$$\lambda_T = \frac{v_w - v}{v_w} \tag{4.8}$$

where λ_T is called tractive slip at zero slip angle. Similar to λ_B in (4.5), λ_T will be further used in different slip equations as a basic term.

4.3 TIRE FRICTION PROPERTIES

The tire current friction coefficient was introduced in Chapter 3. The coefficient corresponds to a magnitude of the tractive force that is developed by the wheel at "current" time moment of motion, not the maximum friction force in the tire-road contact patch. Here, we present the current friction coefficient in a form of a μ-curve varying with the tire slip. Actually, the current friction coefficient is a measure for the ability of a tire to develop its braking and tractive force. The higher the force is, the larger the current friction coefficient will be (at a given magnitude of the wheel normal reaction). More technical detail is presented below beginning with the consideration of the brake mode of operation, and the results for the traction mode are then summarized straightaway following the brake case.

TIRE FRICTION IN BRAKING

From (3.12), the longitudinal current friction coefficient (the word "longitudinal" is further omitted in this sub-section) is defined as the ratio of the braking force, which is the resultant force in the tire patch, and the normal force shown in Figure 3.1:

$$\mu_{x,B} = \frac{F_{x,B}}{F_N} \tag{4.9}$$

The current friction coefficient in (4.9) depends on the tire-road surface materials, tire tread, weather conditions and also some other forces influence relationship (4.9). In terms of its physics, the current friction coefficient illustrates the percentage of the normal force that is currently (i.e., in a particular

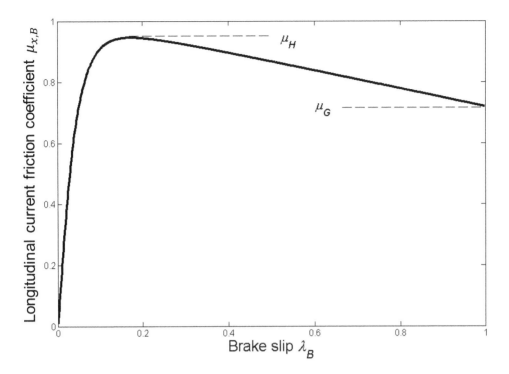

FIGURE 4.4 A typical μ-curve at zero slip angle.

moment of motion) is converted into the longitudinal force of the wheel. As known, the longitudinal slip, $\lambda_{x,B}$, depends on the braking torque and, accordingly, the braking force, $F_{x,B}$. Thus, it is easy to realize that there exists a relationship between $\mu_{x,B}$ and $\lambda_{x,B}$. According to (4.3) and (4.5), $\lambda_{x,B}$ is a function of the slip at zero slip angle, λ_B and slip angle α. Indeed, using (4.3) and (4.5) one can come up with

$$\lambda_{x,B} = \cos\alpha - \left(1 - \lambda_B\right) \tag{4.10}$$

so that the relation between $\mu_{x,B}$ and $\lambda_{x,B}$ can be formulated as follows:

$$\mu_{x,B} = f\left(\lambda_{x,B}\right) = f\left(\lambda_B, \alpha\right) \tag{4.11}$$

implying that $\mu_{x,B}$ is a function of both brake slip λ_B at zero slip angle and the slip angle itself. In case of $\alpha = 0$, $\mu_{x,B}$ is a function of λ_B only. Function (4.11) is highly nonlinear and can be obtained and verified by measurements. Figure 4.4 shows an example of such curve at $\alpha = 0$.

The maximum of friction coefficient in Figure 4.4 is given by μ_H, called the *peak friction coefficient*, which corresponds to the maximum braking force:

$$F_{B,max} = \mu_H \, F_N \tag{4.12}$$

The peak friction coefficient presents the maximal percentage of the normal force that can be converted into the longitudinal force on a particular road surface. Usually, μ_H is established in the slip range given by $0.1 < \lambda_B < 0.3$. At the moment $\lambda_B = 0$, the wheel is in the operational mode at zero torque. The current friction coefficient, $\mu_{x,B}$, responds to increasing braking torque and slip rate with a radical rise followed by a sharp descent when μ_H is reached. It stretches then into the unstable range down to *sliding friction coefficient* μ_G. The wheel will be locked in the unstable range. In a braking situation, μ_H corresponds to the highest braking force which can be generated and is only

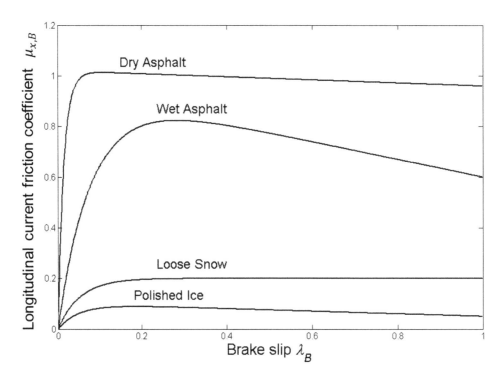

FIGURE 4.5   μ-curves on different road surfaces.

theoretically possible to achieve because the operation of the brake system is unstable at this point. During traction, the wheel will accelerate and spin off. The vast majority of braking and traction processes operate at the slip rates within the stable area, when increasing the slip will be followed by a corresponding rise in available longitudinal force.

As mentioned before, the current friction coefficient varies to reflect changes of different factors. Figure 4.5 presents the μ-curves for different road surfaces. The coefficient is higher under the dry road condition since a larger contribution of the adhesive friction. The adhesion diminishes when the road surface is covered with water.

The wheel slip angle is one of the factors that have influence on the friction coefficient as the above-presented equations indicated.

The current friction coefficient in the lateral direction (y-direction) is defined as follows:

$$F_{y,B} = \mu_{y,B}\, F_N \qquad (4.13)$$

$$\lambda_{y,B} = \sin \alpha \qquad (4.14)$$

The resultant slip of $\lambda_{x,B}$ and $\lambda_{y,B}$ takes the form of the vector sum

$$\begin{aligned}\lambda_{\mathrm{sum},B} &= \sqrt{\lambda^2_{x,B} + \lambda^2_{y,B}} \\ &= \sqrt{2(1-\lambda_B)(1-\cos\alpha) + \lambda_B^2}\end{aligned} \qquad (4.15)$$

which can be related to a resultant current friction coefficient, $\mu_{\mathrm{sum},B}$, which, in its turn, is linked to $\mu_{x,B}$ and $\mu_{y,B}$.

In practical engineering, the μ-curve is just known through experimental data. A mathematical approximation of the experimental data is useful for analyzing and understanding this characteristic. Such an approximation will be also useful for real-time control design as well as for offline calculations in dynamics simulations to constitute the tire-road relation. It is convenient to point out, that the μ-curve approximation is merely a mathematical approach heuristically based on the experimental data, and very often it does not include any physical background considering the tire-road friction mechanism. The following Examples 4.1 and 4.2 show the mathematical approximations for the brake and tractive slip, respectively. Furthermore, the components of the current friction coefficients in relation to their slips are highlighted in a vector geometry explanation. In the end, the reader is able to simulate and display the dependency of the current friction coefficients to the slip angles, which is frequently used in the control of tire dynamics.

Example 4.1

In LabVIEW® simulation, use the following mathematical model of the brake slip to approximate the μ-curve that is given in reference [3] by the following equation:

$$\mu_{x,B} = C_1\left(1 - e^{-C_2\lambda_B}\right) - C_3\lambda_B, \tag{4.16}$$

where $C_1 = 1.02$, $C_2 = 67.9$, $C_3 = 0.06$. The simulation is to demonstrate the influence of the slip angle, α, in the $\mu_{x,B}$ function of slip λ_B.

The resultant force $F_{sum,B}$ in the tire patch, which is the vector sum of $F_{x,B}$ and $F_{y,B}$, depends on $\mu_{sum,B}$, hence on $\lambda_{sum,B}$. Equation (4.16) can be used to relate both parameters

$$\mu_{sum,B} = C_1\left(1 - e^{-C_2\lambda_{sum,B}}\right) - C_3\lambda_{sum,B} \tag{4.17}$$

Note that (4.16) is special case of (4.17). Based on the vector geometry shown in Figure 4.2, the ratio between the forces is valid for the velocity components as well when the tire-road interaction is considered isotropic in the longitudinal and lateral directions, i.e.,

$$\frac{F_{sum,B}}{F_{x,B}} = \frac{v_{sum,B}}{\Delta v_{x,B}} \tag{4.18}$$

which can lead to the following relations:

$$\mu_{x,B} = \frac{\mu_{sum,B}}{\lambda_{sum,B}}\lambda_{x,B} \tag{4.19}$$

$$\mu_{y,B} = \frac{\lambda_{y,B}}{\lambda_{x,B}}\mu_{x,B}, \tag{4.20}$$

where $\lambda_{sum,B}$ is determined by (4.15). Substituting $\lambda_{sum,B}$ into (4.16), $\mu_{sum,B}$ is calculated as the function of $\lambda_{sum,B}$. Equations (4.10) and (4.14) are used for the calculation of $\lambda_{x,B}$ and $\lambda_{y,B}$, respectively, where the input data are slip angle α and λ_B.

The LabVIEW simulation implementation for this example is presented in Appendix A.1. The simulation results are illustrated in Figure 4.6. The $\mu_{x,B}$-function shows that increasing the slip angle reduces the maximum μ_H and shifts it to large λ_B values. The maximum of the lateral current friction coefficient $\mu_{y,B}$ increases when the slip angle goes up. Unlike the longitudinal current friction coefficient, $\mu_{y,B}$ reaches its maximum at $\lambda_B = 0$. Exactly speaking, the tire obtains the higher side force when longitudinal slip is small, and the side force vanishes as slip tends to 100%. Also, at a constant longitudinal force, the tire slippage is higher in the presence of a lateral load, i.e., the higher the lateral force, the higher tire slip while the longitudinal force is constant.

The μ-curve for different surfaces, such as wet asphalt, snow, ice, can be approximated by different sets of parameters C1, C2 and C3.

❏

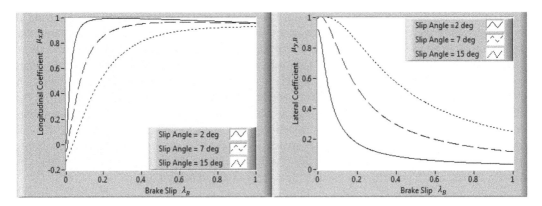

FIGURE 4.6 $\mu_{x,B}$-curve simulation with slip angle $\alpha = 2°, 7°, 15°$.

TIRE FRICTION IN TRACTION

From (4.6) and (4.8), the longitudinal tractive slip can be expressed in

$$\lambda_{x,T} = 1 - (1 - \lambda_T)\cos\alpha \tag{4.21}$$

and the side slip as

$$\lambda_{y,T} = (1 - \lambda_T)\sin\alpha \tag{4.22}$$

The resultant tractive slip is built as

$$\lambda_{\text{sum},T} = \sqrt{2(1 - \lambda_T)(1 - \cos\alpha) + \lambda^2_T} \tag{4.23}$$

Similar to the braking force, the tractive force in the longitudinal direction and the lateral force in the side direction are determined by the normal force and the current friction coefficients:

$$F_{x,T} = \mu_{x,T} F_N \tag{4.24}$$

$$F_{y,T} = \mu_{y,T} F_N \tag{4.25}$$

Below, Example 4.2 presents an analysis of the tire forces and slips in the traction mode similar to the above-given Example 4.1 for the braking force and slip.

Example 4.2

In LabVIEW simulation, use the following mathematical model given in reference [3] by the following equation

$$\mu_{x,T} = C_1(1 - e^{-C_2\lambda_T}) - C_3\lambda_T, \tag{4.26}$$

to approximate the $\mu_{x,T}$-curve. The simulation is to demonstrate the influence of slip angle, α, in the $\mu_{x,T}$ function of slip λ_T. As shown in the previous example, the μ -curve for different surfaces, such as wet asphalt, snow, ice, is approximated by different sets of parameters C_1, C_2 and C_3.

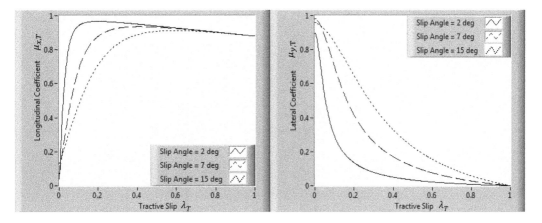

FIGURE 4.7 $\mu_{x,T}$-curve simulation with slip angle $\alpha = 2°, 7°, 15°$.

Similar to (4.17), we have

$$\mu_{sum,T} = C_1\left(1 - e^{-C_2\lambda_{sum,T}}\right) - C_3\lambda_{sum,T},\tag{4.27}$$

Based on the vector geometry shown in Figure 4.3, the ratio between the components of the current friction coefficient is valid for the slip components as well, hence

$$\mu_{x,T} = \frac{\mu_{sum,T}}{\lambda_{sum,T}}\lambda_{x,T}\tag{4.28}$$

$$\mu_{y,T} = \frac{\lambda_{y,T}}{\lambda_{x,T}}\mu_{x,T}\tag{4.29}$$

where $\lambda_{sum,T}$ is determined by (4.23). Substituting $\lambda_{sum,T}$ into (4.26), $\mu_{sum,T}$ is calculated as the function of $\lambda_{sum,T}$. Using slip angle α and λ_B as the input data, (4.21) and (4.22) are used for the calculation of $\lambda_{x,T}$ and $\lambda_{y,T}$, respectively.

The LabVIEW simulation implementation for this example is presented in Appendix A.2. The simulation results are illustrated in Figure 4.7. The tractive μ-curves will form similar shapes as in the braking mode.

❑

For the conclusion, it should be emphasized that the tire-road friction characteristics in the longitudinal and lateral directions are essential for driving and braking performance of tires and then vehicles. In this regard, the μ-curve characteristics as functions of the tire slips are useful to rationalize the vehicle behavior under different road conditions.

The next example demonstrates how the maximum braking force may change on different road surfaces.

Example 4.3

Consider a vehicle moving on a level ground, and the weight loads are the same on the right and left wheels. The road has a *split-μ* surface so that the left wheels are on a surface with the peak coefficient of $\mu_{Hl} = 0.3$ while the right wheels have $\mu_{Hr} = 0.8$ (Figure 4.8). How may the vehicle behave when the driver makes a spike brake apply? We assume that the vehicle is not equipped with any braking force control/distribution systems.

When a brake mechanism is applied, the braking forces at all four wheels are emerged. The braking forces on the high μ-side are higher than those on the low μ-side. This is simply because

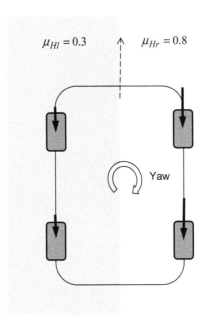

FIGURE 4.8 Vehicle braking on *split-μ* road surface.

of the different μ_H values and equal weighting left to right, see (4.9), in which the current friction coefficient, $\mu_{x,B}$, becomes equal to μ_H. The unbalanced forces result in a clockwise yaw movement. The vehicle may end up spinning off the path without a steering input. The wheels on the low μ-side can lose its capability to take lateral forces because the increasing slip decreases the lateral current friction as shown in Figure 4.6.

❑

REFERENCES

1. T. D. Gillespie, Fundamentals of Vehicle Dynamics, Warrendale: SAE, 1992.
2. A. F. Andreev, V. I. Kabanau, V. V. Vantsevich, Driveline Systems of Ground Vehicles: Theory and Design, Boca Raton, London, New York: CRC Press, 2010.
3. M. Burckhardt, Radschlupf-Regelsysteme, Würzburg: Vogel Verlag, 1993.

5 Acceleration Analysis

The performance in longitudinal acceleration of a motor vehicle is determined by engine power and traction capability. The engine power is transmitted to the wheels through the powertrain that contains several mechanical systems such as engine, transmission and driveline. The traction capability is characterized by the torque transformation from the driven wheels to the forces in the tire-road patches, and thus the traction capability depends on the tire/road characteristics.

The driveline as a part of the powertrain includes all the components from the output shaft of the transmission to the drive wheels. As shown in this chapter, the driveline configuration very much determines the engine power transmission to the wheels, and thus has a significant impact on vehicle longitudinal dynamics and acceleration performance. Several major layouts of different driveline system configurations are discussed (Front-wheel drive, Rear-wheel drive, Four-wheel drive and All-wheel drive) and then basic modeling techniques of torque transfer and distribution through the drivelines are presented to be used in controller designs. Control techniques of torque vectoring are also introduced in the book. For sure, the driveline is a vibrating system because of different masses and stiffness as well as damping of the mechanical elements. Considering the vibration will unnecessarily increase the complexity of the vehicle dynamics modeling and complicate the control design. Therefore, we focus on steady engine toque distribution and the tire longitudinal traction properties.

5.1 DRIVELINE TORQUE DISTRIBUTION

In Section 3.4 we studied the vehicle dynamic behavior driven by the torques at the axles. The driving torques are delivered by the engine through the drivetrain, which includes the main clutch or the torque converter, the transmission and the driveline system, to the drive wheels. This section concerns on how the engine power is transmitted and delivered to the drive wheels. We show the different driveline configurations first, and then the basic modeling of the torque transfer through the entire drivetrain.

5.1.1 DRIVELINE CONFIGURATION

Vehicle original engineering manufacturers have different layouts for delivering engine power to the drive wheels. For light trucks and passenger cars, there are basically four driveline layouts or configurations: *rear-wheel drive*, *front-wheel drive*, *all-wheel drive* and *four-wheel drive*. In all layouts, the engines may be located differently depending on vehicle designs, more configurations and details were given in references [1] and [2]. In the following subsections, several typical examples are presented.

Rear-Wheel Drive

Figure 5.1 shows the most typical case of a rear-wheel-drive (RWD) vehicle. The engine is placed in the front, and the torque is transferred to the rear drive axle. Beginning with engine, it provides an output power into a main clutch or a torque converter. Unlike the main clutch, the torque converter can increase the engine torque. The main clutch is used in vehicles with manual transmissions while the torque converter is used for vehicles with automatic transmissions. Usually, the main clutch is a separate unit. It transmits the engine power and disengages the engine and the transmission when shifting gears. From the transmission, the power flow passes the driveshaft and comes to the final drive and the differential. The final drive turns the power flow 90° and the differential splits the

DOI: 10.1201/9781003134305-5

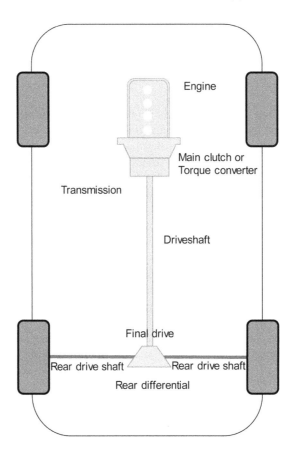

FIGURE 5.1   Rear-wheel-drive configuration.

power between the right and left wheels of the rear drive axle allowing both wheels to rotate at different speeds and have the same torques (with no friction assumed in the differential). As the last one in the powertrain, the rear axle shafts transmit the torque to the rear wheels.

Front-Wheel Drive

In the front-wheel-drive system (FWD), the engine unit including the main clutch and transmission can be mounted longitudinally or transverse at the front drive axle. The transverse mounting is a common case as shown in Figure 5.2, where the engine is along with the axle and the transmission is connected by the final drive set to the differential. Since the engine and drive wheels are on the same side of the vehicle, there is no need to accommodate a driveshaft between the engine and the wheels. In the common automotive language both FWD and RWD are also called *two-wheel drive* (2WD).

Four-Wheel Drive

Four-wheel drive (4WD), sometimes also referred to as Four by Four (4 × 4), is a kind of combination of FWD and RWD, and able to transmit and deliver power to the front and rear drive axles. A specialized gearbox, called a *transfer case*, is designed to split the power among the front and the rear drive axles, and to realize the function of varying gear ratios. Figure 5.3 presents an example of such system. The transfer case is connected to the output shaft of the transmission. The power received from the transmission is distributed by the transfer case among the front and the rear drive axles. The driver can activate 4WD system with buttons or switches. Typical four-wheel-drive vehicles also have two-wheel-drive mode (2WD), in which it sends power only to the rear. The transfer case can be electrically controlled to switch back and forth between two-wheel and four-wheel-drive modes.

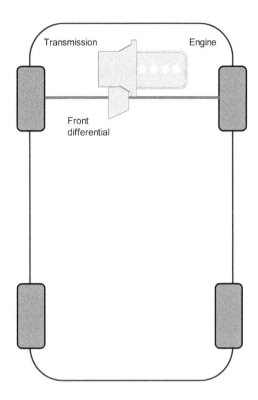

FIGURE 5.2   Front-wheel-drive configuration.

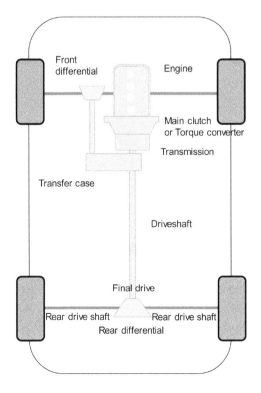

FIGURE 5.3   Four-wheel-drive configuration.

Normally, 4WD vehicles have a selection of two gear ranges – High range and Low range. They are called 4-Hi and 4-Lo mode, respectively. As the transfer case accepts input from the transmission, the torque will be first increased according to the *gear ratio* of the transfer case and then distributed between front and rear axles. For example, the 4-Hi range can be set to 1 gear ratio while 4-Lo has 5.2 gear ratio. Gear ratio of 1 means no torque multiplication between the input shaft of the transfer case and the shaft that brings the torque to the power-dividing unit that splits that torque between two output shafts of the transfer case. Gear ratio of 5.2 says that the torque at the transfer case's power-dividing unit is 5.2 times higher than the input torque, but the rotation speed is 5.2 times slower.

In general, Low range gears are extra reduction gears, which slow down the vehicle and increase the torque available at the drive axles. They are used during slow-speed or extreme off-road maneuver, such as rock-crawling, or when pulling a heavy load.

All-Wheel Drive

The all-wheel drive (AWD) is a type of the four-wheel drive. In AWD vehicles, the permanently driven axle is the front-drive axle, and the rear axle is automatically engaged and disengaged based on different control inputs. Figure 5.4 shows this type of configuration. For mechanic AWD, the power-transmitting unit can contain, for example, a viscous coupling to allow a speed difference between the front and rear wheels that is needed for an automatic lock of the viscous coupling. Some manufactures prefer to use an electronic control unit to control the torque split between the front and rear based on traction or wheel speed conditions. All-wheel-drive systems increase the traction

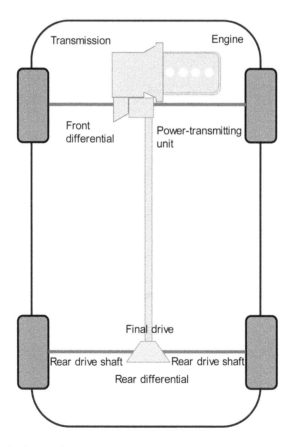

FIGURE 5.4   All-wheel-drive configuration.

capability of vehicles and are useful for all road conditions while four-wheel-drive vehicles perform strong in extreme low-traction conditions.

The following example shows a driving situation, in which a vehicle with different type of driveline performs differently.

Example 5.1

A vehicle is turning into a curve on a low-μ road. While the vehicle is accelerating, the tire slip is increasing and the drive wheels begin to spin. How may the vehicle behave if it has a) a front-wheel drive, b) a rear-wheel drive or c) an all-wheel drive system?

a. Figure 5.5 shows a FWD car turning left into the curve. While the front wheels are slipping, they may lose the ability to withstand lateral forces. Due to the centrifugal force action, the lateral forces at the rear tires act toward the inside of the curve, which causes a clockwise yaw moment; F_{LR}, F_{RR} are lateral forces of the left rear and the right rear tires, respectively. At this moment the front end tends to drift out of the curve, and the driver has to increase the steering input – a phenomenon known as *understeer*.
b. Figure 5.6 shows a RWD car turning left into the curve. While the rear wheels are spinning, they may lose the ability to take lateral forces. The lateral forces at the front tires act toward the inside of the curve, which causes a counterclockwise yaw moment. F_{LF}, F_{RF} are the lateral forces of the left front and the right front tires, respectively. The rear end of the car tends to swing around and pass the front end – a phenomenon known as *oversteer*. Both under-steer and oversteer will be introduced and analyzed later in Chapter 9.
c. For an AWD vehicle, its behavior depends on how the torque is distributed among two drive axles. Since the total torque is distributed on two drive axles and assuming the drive-line provides positive torques at both axles, there could be more lateral forces generated at the all four tires, and they can be more balanced on the front and the rear. Therefore, the car may act more stable than FWD and RWD. More on stability of AWD vehicles can be found in references [2].

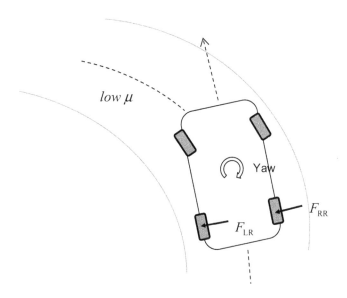

FIGURE 5.5   FWD vehicle turning into a curve on low-μ surface.

❑

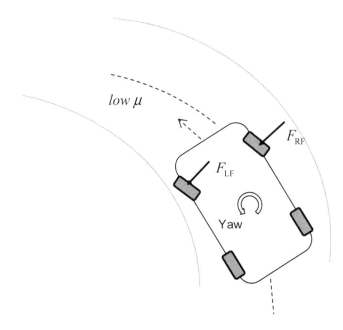

FIGURE 5.6   RWD vehicle turning into a curve low-μ surface.

Differential

A *differential,* or an *open differential,* is a mechanical device with a planetary gear train, and transmits/splits power from a single power source to two output shafts of the differential. In all-wheel-drive vehicles with a differential driveline system, the *center differential* couples the front and rear axles, and the front and the rear differentials connect the left and right drive wheels. The differential is designed to drive a pair of wheels while allowing them to rotate at different speeds. When cornering, the inner wheel rotates slower than the outer wheel, which is needed to cove different travels during the same time period. Without a differential, a vehicle experiences difficulties with its turnability, i.e., with its ability to make turns.

The differential has a fix ratio between its output shafts that is unity in symmetrical differentials. For this reason, the open symmetrical differential always applies the same amount of torque to the wheels that the differential connects. Figure 5.7 shows the physical interpretation of the torque split with regard to the wheel dynamics when one wheel is on asphalt and the other wheel is on an icy road, where "*l*" and "*r*" indicate left and right, respectively; the normal reactions at the wheels are the same, F_N; so do the rolling radius in the driven mode, r_w. Both surfaces are characterized

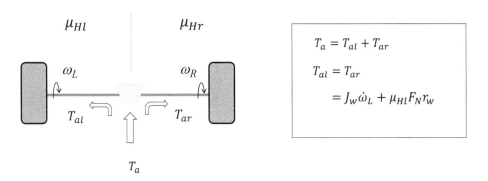

$$T_a = T_{al} + T_{ar}$$

$$T_{al} = T_{ar}$$

$$= J_w \dot{\omega}_L + \mu_{Hl} F_N r_w$$

FIGURE 5.7   Torque transfer by open differential.

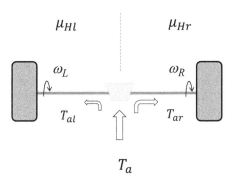

FIGURE 5.8 Torque transfer by locking differential.

by different peak friction coefficients of low and high values, $\mu_{Hl} < \mu_{Hr}$ (for example, 01 and 0.8). For such road conditions, the maximum traction capability of the drive axle can be analyzed in the following way. If the current friction coefficient at the left wheel reaches the peak friction coefficient, i.e., $\mu_l = \mu_{Hl}$, the torques at the output shafts of the differential are the same and equal to the torque developed at the wheel with the lower peak friction coefficient. This is because the gear ratio between the output shafts, which is the gear ratio of the differential, is unity. Such low torque is not enough to move the right wheel that is on asphalt, and the wheel stops, $\omega_R = 0$. The left wheel on the icy road begins to spin with the doubled angular velocity of the differential, $\omega_L = 2\omega_a$. The sum of the output torques is the torque applied to the differential (no friction between the differential elements is assumed).

If one wheel comes off the ground, it will spin in the air because both wheels keep receiving the same amount of torque, which may not be able to move the other wheel that is on the ground.

Another type of differentials is a *locking differential*. This differential has the same parts as an open differential, but adds an electric, pneumatic or hydraulic mechanism to lock two elements together. They are either the housing of the differential and one of its output shafts or, which is much rare, two output shafts. When the differential is unlocked, it operates as the open differential. The lock mechanism is activated by a switch, and when activated, both wheels will rotate at the same speed. Figure 5.8 shows the physical interpretation for the locking differential in the case when the left wheel is on a *low-μ* surface, i.e., $\mu_{Hl} < \mu_{Hr}$. The locking differential always distributes a bigger torque to the wheel that has a higher peak friction coefficient. Therefore, the maximum traction capability of the axle can be observed in case both wheels reach the peak friction coefficients as shown in equations included in Figure 5.8. Comparing the components of the equations in Figure 5.7 and Figure 5.8, one can notice the advantage of the usage of the locking differential to improve traction performance of vehicles.

When driving on surfaces with different grip properties and the vehicle traction load (caused by the grade, trailer, etc.) is less than the maximum traction capability, the torques of the wheels connected by a locking differential are also different as described by (5.1)

$$T_{ar} - T_{al} = \left(\mu_r - \mu_l\right) F_N r_w \tag{5.1}$$

Knowing $T_{al} + T_{ar} = T_a$, the two equations above combine to give individual torques

$$T_{ar} = \frac{T_a}{2} + \frac{1}{2}\left(\mu_r - \mu_l\right) F_N r_w$$

$$T_{al} = \frac{T_a}{2} - \frac{1}{2}\left(\mu_r - \mu_l\right) F_N r_w \tag{5.2}$$

Thus, half of the torque difference is reduced from the wheel with lower friction properties but is added to the wheel with higher μ_H. This simple mathematical example concludes that the locking differential provides better traction by transferring the torque from the wheel in a poor friction condition to the wheel that has a better friction and can develop a bigger traction effort. If one wheel ends up off the ground, the other wheel won't be influenced. Both wheels will continue to rotate at the same speed as if nothing had changed. All torque from the case of the differential will go to the wheel that maintains its contact with the ground. That's the reason why the locking differential improve vehicle traction and mobility when driving off-road. Unlike the open differentials, the locking differentials prevent wheel spin by transmitting the excessive driving torque to the wheels with higher peak fiction coefficient. However, it may cause disturbing moments in steering and vehicle stabilization due to unequal torque distribution, for example in cornering or in straight motion where the tires have different frictions. Therefore, this mechanical method to increase traction has a switchable option for the driver to manage the traction capability on various road surfaces.

An advanced technology is electronically-controlled differential. An electronic control unit uses inputs from multiple sensors, including the wheel rotational speed and the steering input angle, etc., and adjusts the distribution of the torque among the wheels. In general, this kind of power delivery is called *torque vectoring*. The method can also be used for electrical vehicles, on which the drive axles or wheels can be controlled individually by the electric motors. The individual actuation of the electric drivelines allows a continuous generation of the wheel torques required to achieve the maximum acceleration as well as improve stability in critical driving conditions.

Thus, torque vectoring performs basic differential tasks while also distributing the torque among the wheels to improve turnability, stability and traction of vehicles on a need basis. Torque vectoring can be employed on all-wheel drive vehicles to vary torque between the front and the rear wheels, but also on the front or the rear-wheel drive to vary the torque between two wheels. More detail on modeling open and locking differentials, self-locking differentials and torque vectoring can be found in references [2] and [3]. In Section 5.2.6, an example of a torque vectoring method is presented.

The following example shows how the driving torque affects the wheels on different road surface.

Example 5.2

A front-wheel-drive vehicle is accelerating on a level ground. When, the road becomes a *split-μ* surface, the left wheel is on surface with peak coefficient $\mu_{Hl} = 0.3$ while the right wheel has $\mu_{Hr} = 0.8$, see Figure 5.9. The total mass weight $M_{gf} = 12571$ N on the front axle with an open differential is distributed equally between the left and right wheels. The torque applied to the front axle is given by T_{af}, and the tire effective rolling radius $r_w = 0.35$ m. How will be the torque amount distributed to the left and right wheels if a) $T_{af} = 1100$ Nm, b) $T_{af} = 3300$ Nm? How the torque distribution may affect the movement of the vehicle? Disregard the rotation inertia of the wheels and the powertrain.

    a. The torque is transmitted to both left and right wheel equally, that means

$$T_{al} = T_{ar} = \frac{1}{2}T_{af} = 550 \text{ Nm}$$

The maximum friction torque on the left wheel that can be supported by the grip with the road is

$$T_{al\_max} = 0.5\mu_{Hl}M_{gf}r_w = 660 \text{ Nm}$$

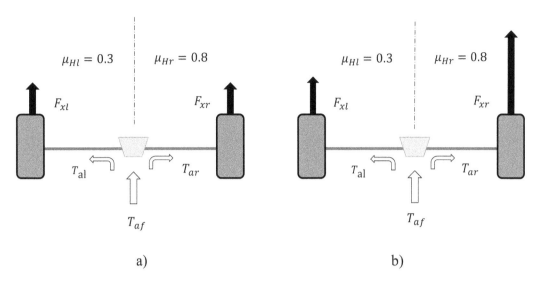

FIGURE 5.9 Friction limitation on the tractive forces.

As seen, $T_{al} < T_{al\_max}$, i.e., torque T_{al} can be taken by the road. The same occurs with the right wheel torque, T_{ar}, that is much less than the wheel's maximum friction torque determined by the peak friction coefficient of 0.8.

If the torque supplied to both drive wheels, $T_{af} = 1100$ Nm, can overcome the rolling resistance of all four wheels and also provide a tractive force, the vehicle will continue to accelerate. The reader can see this from (3.25) at $J_{wf} = 0$ and $T_{bf} = 0$.

b. The torque is transmitted to both left and right wheel equally, that means

$$T_{al} = T_{ar} = \frac{1}{2}T_{af} = 1650 \text{ Nm}$$

Here, the driving torque that the driveline is trying to supply to the left wheel, $T_{al} = 1650$ Nm, is much higher than the maximum friction torque that the wheel is able to support in its grip with the road, $T_{al\_max} = 660$ Nm. The maximum tractive force of the left wheel is limited by the peak friction coefficient

$$F_{xl\_max} = 0.5\mu_{Hl}M_{gf} = 1886 \text{ N}$$

When the wheel is approaching the maximum tractive force, its slip will be drastically increasing and the wheel begins to spin. Thereby the lost torque $T_{al} - T_{al\_max} = 990$ Nm is worn out to increase the rotational speed of the left wheel.

The torque at the right wheel is equal to the torque at the left wheel, i.e., $T_{ar} = 660$ Nm, which is less than the maximum friction torque of $T_{ar\_max} = 0.5\mu_{Hr}M_{gf}r_w = 1760$ Nm. Thus, the right wheel can take torque T_{ar} of 660 Nm. If the sum of the torques (660 Nm + 660 Nm) can overcome the rolling resistance of the four wheels and provide a tractive force [see (3.25) and (3.28)], the vehicle will continue to accelerate. Otherwise, the vehicle will stop. ❏

5.1.2 POWER DELIVERY THROUGH POWERTRAIN

As shown in the different driveline configurations above, the power transfer over the entire powertrain follows similar physical paths, i.e., from the engine to the main clutch or the torque converter, to the transmission and then through a driveline system to the drive wheels. From the control system point of view, the power delivery can be considered as an open loop system, in which the power is the product of the torque and the rotational speed at the input and output of each component of the powertrain as illustrated in Figure 5.10 that is based on a RWD system. Thereby, the main

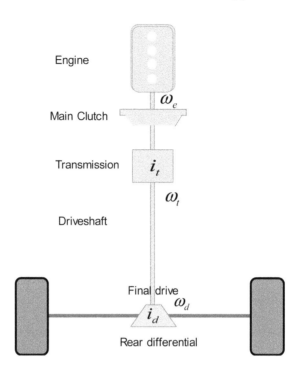

Engine

ω_e

Main Clutch

Transmission i_t

ω_t

Driveshaft

Final drive
ω_d
i_d

Rear differential

FIGURE 5.10 Power transfer through the powertrain.

clutch, the transmission, the drive shaft as well as the drive axle are assumed to be stiff, nonlinearity of the clutch and no flexibility are is included.

The engine torque, denoted as T_e, is usually measured at steady speeds on a dynamometer. The output torque, T_t, delivered by the main clutch, which is considered completely engaged, as the input to the transmission is reduced by the amount required to accelerate the total rotation inertia of the engine and the main clutch and internal friction $T_{e,fr}$ in the engine. The friction in the main clutch is not included since the clutch is locked and friction in bearings is disregarded. This yields the equation

$$T_e - T_t - T_{e,fr} = J_{e-cl}\dot{\omega}_e \tag{5.3}$$

where J_{e-cl} is the total inertia of the engine and the main clutch rotation, and ω_e is the output speed, which is the speed of the engine since the main clutch is fully locked.

The purpose of the transmission is to match the engine torque and the speed to vehicle driving conditions. The transmission provides a high gear ratio (i.e., a high torque) at low vehicle speed and a low gear ratio (i.e., a low torque) at high speed. Having several gear ratios, it allows efficient operations in a wide speed range. Because of the power loss due to the internal friction and rotating inertia of the transmission, (2.75) is applied here. At the output shaft of the transmission, the rotational speed is decreased

$$\omega_t = \omega_e / i_t \tag{5.4}$$

while the torque is amplified by transmission gear ratio i_t:

$$T_t i_t - T_d - d_t \omega_t = J_t \dot{\omega}_t \tag{5.5}$$

where J_t is the transmission rotational inertia. d_t is the viscous damping coefficient of the transmission assuming the internal friction to be described by viscous damping. T_d is the torque at the output shaft of the transmission that is assumed to be the same at the driveshaft (no rotational inertia or internal friction in the driveshaft assembly is considered).

The final drive and the differential have the aim to transmit the power from the driveshaft to the drive axles. At this point of the driveline, the power flow changes from the longitudinal plane of the vehicle to the transverse, i.e., to the wheels. Both wheels are assumed to have the same grip condition. By design, the rotational speed after the final drive is reduced by gear ratio i_d, hence the torque is increased. So similar to the transmission, the speed ω_d and torque T_w at the differential's two outputs can be characterized by rotating inertia J_d of the differential and the final drive and their viscous damping, d_d:

$$\omega_t = i_d \omega_d \tag{5.6}$$

$$T_d i_d - T_w - d_d \omega_d = J_d \dot{\omega}_d \tag{5.7}$$

Combining (5.3), (5.5) and (5.7), we get a lumped model of all powertrain components, in which the components are reduced to the drive wheels:

$$T_e i_t i_d - T_w - T_{e,fr} i_t i_d - d_t i_d^2 \omega_d - d_d \omega_d = \left(J_e i_t^2 i_d^2 + J_t i_d^2 + J_d \right) \dot{\omega}_d \tag{5.8}$$

The first term on the left-hand side is the total effective engine torque delivered from the crankshaft to the drivetrain as seen from the drive axle, i.e., reduced to the drive axle. To get onto the drive axle, the torque has to overcome the internal frictions in the engine, the transmission, the final drive and the differential, and to accelerate the inertias of their rotating components. In the industrial design, an efficiency factor, η_e, is preferred instead of the internal frictions. For this reason, (5.8) can be formulated as

$$T_e i_t i_d \eta_e - T_w = \left(J_e i_t^2 i_d^2 + J_t i_d^2 + J_d \right) \dot{\omega}_d \tag{5.9}$$

Note that η_e, as a product of individual efficiencies, considers all the efficiencies of the drivetrain components. It varies widely with engine level, typically in the range of 80%– 90%. T_w is the torque on the drive axle, i.e., on both wheels.

More accurate modelling for the drivetrain components needs to include nonlinearity of the main clutch and flexibility of the driveshaft. The model parameter estimation as well as the experimental data verification is shown in references [2] and [4].

5.2   LONGITUDINAL ACCELERATION

Based on the results presented in Chapter 3, we can continue to investigate the torque distribution among the drive wheels during the vehicle acceleration. The torque distribution among the front and rear axles very much depends on the driveline power-dividing unit, which is employed in the transfer case, and the tire-road conditions characterized, in particular, by the peak friction coefficient. Since the maximum traction surfaces in low transmission gear and at low vehicle speeds, the inertia of rotating masses as well as the aerodynamic force can be neglected. Considering Figure 3.1 without the road grade, the inertia of rotating masses and the aerodynamic drag, the arising forces on the vehicle in straight-line acceleration, including the longitudinal forces, F_{Tf} and F_{Tr}, are illustrated in Figure 5.11.

where the following notations are used:

M gross mass of the vehicle
a vehicle longitudinal acceleration
g acceleration of gravity

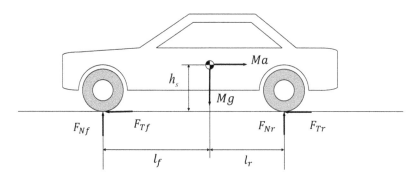

FIGURE 5.11   Forces acting on the vehicle in acceleration.

F_{Tf} tractive force from road to the front tires
F_{Tr} tractive force from road to the rear tires
F_{Nf} vertical force on the front wheels
F_{Nr} vertical force on the rear wheels
h_s height of CG (center of gravity)
l_r longitudinal distance between rear axle and CG
l_f longitudinal distance between front axle and CG

The longitudinal forces are presented by the tractive forces, i.e., $F_{xf} = F_{Tf}$ and $F_{xr} = F_{Tr}$. From (3.32), we have

$$F_{Tf} + F_{Tr} = Ma \tag{5.10}$$

Basically, the entire vehicle accelerates as a unit, which can be represented as one lumped mass located at its center of CG with the gross mass of the vehicle. Using the reference to gravitational acceleration g as

$$a_g = \frac{a}{g} \text{ and } M_g = Mg \tag{5.11}$$

equation (5.10) can be reformulated to

$$F_{Tf} + F_{Tr} = M_g a_g \tag{5.12}$$

Now a_g represents the *normalized longitudinal acceleration* being normalized to the gravity acceleration.

5.2.1   Driving Force Distribution

According to SAE J670, the *driving force* of a wheel is the longitudinal force resulting from the torque application [5]. In a wheel free body diagram, the driving force is a reaction in the tire patch acting from the road on the tire in the heading direction. The driving force overcomes the rolling resistance and generates a positive longitudinal force in the tire patch, F_x, i.e., the force in the heading direction needed to overcome force F_D, which acts from the vehicle body on the driven wheel and is applied at the wheel axis of rotation (see Figure 3.1). Longitudinal force F_x, which is the difference between the driving force and the rolling resistance, is the resultant force from the road on the tire that actually moves the wheel (and the vehicle) forward as shown in Figure 3.1. Thus, longitudinal force F_x is the tractive force (also defined as the net tractive force in Chapter 3).

In 4WD and AWD vehicles, the wheel torque split, i.e., the driving force distribution between the front and rear driven wheels, can significantly impact vehicle longitudinal and lateral dynamics, ability to make turns, and energy efficiency. To provide an analytical foundation to study such impact, the driving force distribution can be conveniently characterized by the tractive force distribution between the driven wheels. Indeed, as it is usually conventional in road vehicle dynamics and presented in Chapter 4, the tractive force of a wheel is functionally linked to tire slippage that influences the actual speed of the wheel. As a component of the driving force, the tractive force can characterize the wheel (and then the vehicle) traction potential to accelerate and overcome external resistance of road grades and aero drag forces.

If knowing the tractive forces at the front and rear drive wheels in 4WD and AWD vehicles, the driving forces can be easily computed by adding the rolling resistance to the tractive forces and, then, the front and rear wheel driving torques follow from the driving forces and rotational inertia of the wheels. This approach to characterize the driving force distribution is accepted and considered in the following.

Now we consider first the tractive forces at both axles relating the forces to the acceleration represented by a_g.

$$F_{Tf} = \mu_f F_{Nf} \tag{5.13}$$

$$F_{Tr} = \mu_r F_{Nr} \tag{5.14}$$

where μ_f and μ_r are the longitudinal current friction coefficients of front and rear tires respectively. Substituting (5.13) and (5.14) into (5.12), we obtain

$$\mu_f F_{Nf} + \mu_r F_{Nr} = M_g a_g \tag{5.15}$$

The above equation presents the force relation in the longitudinal direction of vehicle motion.

As the second step, we formulate the moment balance equations with regard to the front and rear axle respectively (Figure 5.11).

$$F_{Nf}\left(l_f + l_r\right) = M_g l_r - a_g M_g h_s \tag{5.16}$$

$$F_{Nr}\left(l_f + l_r\right) = M_g l_f + a_g M_g h_s \tag{5.17}$$

The force $a_g M_g$, acting at the vehicle's center of gravity, causes a pitching moment. A dynamic load transfer between the front and rear axles takes place during the acceleration, i.e., the load is transferred from the front axle to the rear axle in proportion to the acceleration and the ratio of the CG height to the wheel base. By defining

$$\chi = \frac{h_s}{l_f + l_r} \tag{5.18}$$

$$\zeta = \frac{l_f}{l_f + l_r} \tag{5.19}$$

both (5.16) and (5.17) can be rewritten in the form of the normal reactions at the front and rear wheels

$$F_{Nf} = M_g\left(1 - \zeta - a_g \chi\right) \tag{5.20}$$

$$F_{Nr} = M_g\left(\zeta + a_g \chi\right) \tag{5.21}$$

In order to determine a relationship between tractive forces F_{Tf} and F_{Tr} as the measure of the actual acceleration a_g and the weight distribution, we substitute (5.13) and (5.14) into (5.20) and (5.21) respectively, which lead to

$$\frac{F_{Tf}}{M_g} = \mu_f \left(1 - \zeta - a_g \chi\right) \tag{5.22}$$

$$\frac{F_{Tr}}{M_g} = \mu_r \left(\zeta + a_g \chi\right) \tag{5.23}$$

Further, along with the expression for a_g obtained from (5.12), (5.22) becomes

$$\frac{F_{Tf}}{M_g} = \mu_f \left(1 - \zeta - \frac{F_{Tf} + F_{Tr}}{M_g} \chi\right) \tag{5.24}$$

This equation shows the relationship between $\dfrac{F_{Tf}}{M_g}$ and $\dfrac{F_{Tr}}{M_g}$, which is reflected in Figure 5.12. If μ_f is kept constant, (5.24) is a straight-line equation with intersection points $\left(\dfrac{\mu_f \left(1 - \zeta\right)}{1 + \mu_f \chi}, 0\right)$ and $\left(0, \dfrac{1 - \zeta}{\chi}\right)$ on $\dfrac{F_{Tf}}{M_g}$ and $\dfrac{F_{Tr}}{M_g}$ – axis, respectively. In the same way, (5.23) becomes

$$\frac{F_{Tr}}{M_g} = \mu_r \left(\zeta + \frac{F_{Tf} + F_{Tr}}{M_g} \chi\right) \tag{5.25}$$

intersecting with $\dfrac{F_{Tf}}{M_g}$ and $\dfrac{F_{Tr}}{M_g}$ - axis at $\left(-\dfrac{\zeta}{\chi}, 0\right)$ and $\left(0, \dfrac{\mu_r \zeta}{1 - \mu_r \chi}\right)$ respectively when μ_r is kept constant. Considering (5.12), the intersection point of the two lines, which is the joint solution of (5.24) and (5.25), determines the actual acceleration of the vehicle.

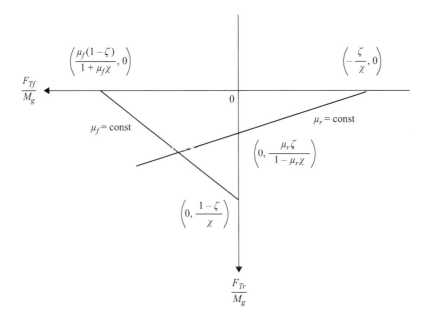

FIGURE 5.12 Friction limitation on the tractive forces.

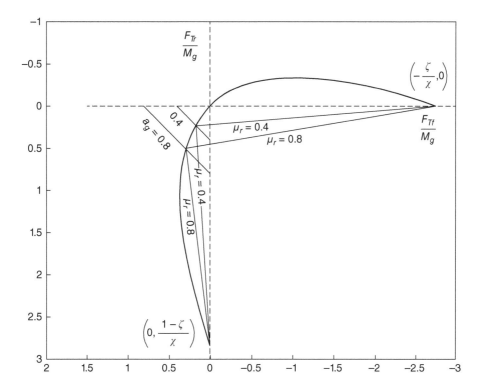

FIGURE 5.13 Ideal driving force distribution.

Apparently, if both μ_f and μ_r are varying parameters in different road conditions, (5.24) and (5.25) will result in a set of straight-lines. As shown in Figure 5.13, (5.24) is graphically visualized by a set of straight-lines going through the intersection coordinate $\left(0, \dfrac{1-\zeta}{\chi}\right)$, while (5.25) is visualized by a set of straight-lines going through intersection coordinate $\left(-\dfrac{\zeta}{\chi}, 0\right)$.

5.2.2 IDEAL DRIVING FORCE DISTRIBUTION

As (5.15) describes, the current friction coefficients, taken together with the normal reactions of the front and rear wheels, influence the acceleration that a vehicle can attain. Equation (5.15) can be reformulated as

$$\left(\mu_f - a_g\right)F_{Nf} + \left(\mu_r - a_g\right)F_{Nr} = 0 \tag{5.26}$$

because the sum of the vertical force at each axle is equal to vehicle weight M_g. For a given maximal acceleration, the wheels of an axle or both axles will be over-driven and spin up by increasing the driving torque of that axle or both axles. However, if reducing the torque, vehicle traction capability is not utilized. Consequently, we have an ideal case, in which the maximum tractive force can be achieved on both the front and the rear tires, that is

$$\mu_f = \mu_r = a_g \tag{5.27}$$

Condition (5.27) indicates an optimum utilization of the potential acceleration capacity of the vehicle, and tractive force distribution among the drive axles under this condition is called "ideal". In the ideal case, the vehicle can achieve the maximum acceleration performance (in g's) equivalent to the current friction coefficients between the tires and the road surface.

For the ideal tractive force distribution, (5.22) and (5.23) become by substituting a_g for μ_f and μ_r

$$\frac{F_{Tf}}{M_g} = a_g \left(1 - \zeta - a_g \chi\right) \tag{5.28}$$

$$\frac{F_{Tr}}{M_g} = a_g \left(\zeta + a_g \chi\right) \tag{5.29}$$

Substituting (5.12) into (5.29), we obtain the ideal relationship between the tractive forces on the front and rear wheels at the maximum of acceleration

$$\left(\frac{F_{Tf}}{M_g} + \frac{F_{Tr}}{M_g}\right)^2 \chi + \left(\frac{F_{Tf}}{M_g} + \frac{F_{Tr}}{M_g}\right)\zeta - \frac{F_{Tr}}{M_g} = 0 \tag{5.30}$$

It shows that the relationship between $\dfrac{F_{Tr}}{M_g}$ and $\dfrac{F_{Tf}}{M_g}$ depends on χ and ζ, which are merely related

to the geometrical parameters of the vehicle. This relationship can also be interpreted as the joint solution of (5.24) and (5.25). Figure 5.13 gives a graphical illustration of (5.30), which is a parabola built by intersection points of the straight lines, which meet the condition given by (5.27). The straight-lines with a_g value equal to 0.4 and 0.8 result from (5.12). The parabola displays the ideal relationship between the front and rear tractive forces that should be generated by the wheel torques to provide the maximum acceleration of the vehicle. Both the front and the rear forces as well as the acceleration always have the same direction, hence the same coordinate sign, so that the ideal distribution curve in Figure 5.13 is merely considered in the positive coordinate area.

As long as the tractive force distribution follows the ideal curve, i.e. the current friction coefficients are equal to that one required for the maximum acceleration and the optimal tractive capability. In Figure 5.13, the negative coordinate area in the first quadrant layouts the ideal curve for the braking force distribution that will be studied in the next chapter. Basically, both acceleration and braking are a dynamic duo operating in oppose force directions and tied together in terms of mathematical formulations. We will consider this form again with regard to the braking force distribution and show their common characteristics in Chapter 6.

Fixed Force Distribution

In some AWD vehicles, the ratio of the tractive force distributed to the front and rear axles is not very much adjustable. At the same time, we have already seen that the individual tractive forces at the drive axles must be controlled in order to reach the maximum acceleration performance. The basic issue here is to understand the influence of the friction coefficient on the tractive force distribution and on the achievable acceleration the vehicle. In Section 5.2.4 an example shows the effect of the tractive force distribution with a fixed ratio.

5.2.3   TRACTION CAPABILITY AT DIFFERENT DRIVELINE CONFIGURATIONS

One of the most important questions that an engineer wants to answer is what the maximum acceleration can be achieved in a particular given tire-road condition. That answer expresses the traction capability of a vehicle since the acceleration relates to the tractive forces as seen from (5.15). With an adequate engine torque, the maximum acceleration is determined by the current friction coefficients between the tires and roads. As discussed before, driveline configuration impacts the torque split between the drive axles. Therefore, the driveline configurations play a significant role for traction capability assessment.

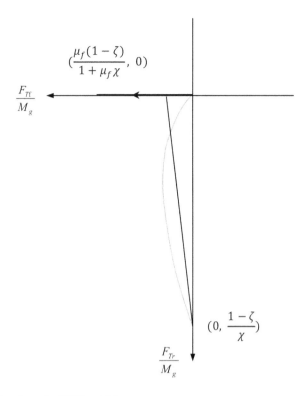

FIGURE 5.14   Tractive force for FWD vehicle.

Front-Wheel Drive

In the front-wheel-drive vehicle, the engine torque is only delivered to the front axle, i.e., $F_{Tr} = 0$. We can obtain the range for the tractive force of the front wheels by referring to (5.24), which defines a set of straight-lines going through points with coordinates of $\left(0, \dfrac{1-\zeta}{\chi}\right)$ and $\left(\dfrac{\mu_f(1-\zeta)}{1+\mu_f\chi}, 0\right)$ in the force distribution graph of $\dfrac{F_{Tf}}{M_g}$ vs. $\dfrac{F_{Tr}}{M_g}$ as shown in Figure 5.14. Since the power comes to the front wheels only, the range of the available front tractive force is determined by the coordinate of $\dfrac{\mu_f(1-\zeta)}{1+\mu_f\chi}$ on the horizontal axis.

Thus, as seen from Figure 5.14, the available tractive force for a FWD vehicle is determined on the $\dfrac{F_{Tf}}{M_g}$ - axis by varying parameter μ_f and follows from the equation

$$\frac{F_{Tf}}{M_g} = \frac{\mu_f(1-\zeta)}{1+\mu_f\chi} \tag{5.31}$$

The maximum acceleration is given when the current friction coefficient has its maximum μ_H, that is

$$a_{g,FWD} = \frac{1-\zeta}{\dfrac{1}{\mu_H}+\chi} \tag{5.32}$$

where μ_H is the peak coefficient of the friction. For the purpose of characterizing the maximum acceleration in traction, it is common to refer to the peak friction coefficient as it was done in (5.32).

However, in practice, the achievable maximum acceleration is considered when the current friction coefficient approaches but still less than the peak friction coefficient because the friction force begins to drop of beyond the peak point and tire slip drastically increases.

Rear-Wheel Drive

Similarly, for the rear-wheel-drive vehicle, the available tractive force is moving on $\dfrac{F_{Tr}}{M_g}$-axis with

$$\frac{F_{Tr}}{M_g} = \frac{\mu_r \zeta}{1 - \mu_r \chi} \tag{5.33}$$

depending on parameter μ_r as shown in Figure 5.15. The maximum acceleration in turn can be calculated as

$$a_{g,RWD} = \frac{\zeta}{\dfrac{1}{\mu_H} - \chi} \tag{5.34}$$

All-Wheel Drive

For *all-wheel drive* (AWD) and *four-wheel drive* (4WD) vehicles, the torque is delivered to the front and rear axles in a defined ratio, that can be introduced here as the torque bias

$$\Phi_T = \frac{T_{ar}}{T_{af} + T_{ar}} \tag{5.35}$$

Φ_T can be a fixed value or a variable one. As an example, open differential has a fixed torque bias that provide a constant torque split among two drive wheels or among drive axles. According to reference [3], earlier models of Audi Quattro and BMW 325iX equipped with central differentials had $\Phi_T = 0.5$ and $\Phi_T = 0.63$, respectively. The numerical value of 0.5 means that the torques at the front and rear wheels are the same. The magnitude of 0.63 explain that 63% on the input torque is delivered to the rear axle and 37% is supplied to the front axle. Some self-locking differentials vary their torque biases in different grip conditions, such as asphalt road, gravel, soft soil, etc. In such mechanisms, for instance, the torque bias can be a function of the input torque, which is different in

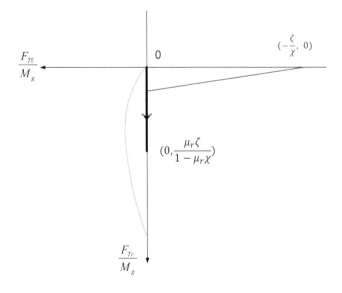

FIGURE 5.15   Driving torque for RWD vehicle.

various conditions. In torque vectoring systems, the torque bias can be actively controlled as discussed in this chapter. Additional information on various power-dividing units and their torque biases can be found in reference [2].

Under the same assumption made in this chapter earlier for (5.10), we get by solving (5.10) and (5.35) for the tractive forces

$$F_{Tf} = (1 - \Phi_T) a_g M_g \tag{5.36}$$

$$F_{Tr} = \Phi_T a_g M_g \tag{5.37}$$

Substituting both equations above into (5.22) and (5.23) respectively, the current friction coefficients that correspond to a given torque bias are calculated as

$$\mu_f = \frac{(1 - \Phi_T) a_g}{1 - \zeta - a_g \chi} \tag{5.38}$$

$$\mu_r = \frac{\Phi_T a_g}{\zeta + a_g \chi} \tag{5.39}$$

At the same time, the torques to be transmitted to the front and rear wheels are limited by the peak friction coefficient between the tire and the road. If more torque is provided by the drivetrain than the friction can support, the wheel will spin up. The wheel spinning has been considered so far as a boundary on the vehicle traction capability. Whether the front or rear wheels start to spin, it is subject to the torque distribution for a given coefficient of the peak friction coefficient. To identify the beginning of the wheel spin, the numerical values of μ_f and μ_r, which comes from (5.38) and (5.39), should be compared with μ_H. If $\mu_f \geq \mu_H$, the front wheels are in spin. The rear wheels spin if $\mu_r \geq \mu_H$.

If the current friction coefficient of the front wheels is higher than that at the rear wheels, i.e. $\mu_f > \mu_r$, the front wheel can start spinning first and ((5.38) determines the limit for the acceleration, i.e., (5.40) is solved for the acceleration maximum when $\mu_f = \mu_H$

$$a_{g, max} = \frac{1 - \zeta}{\dfrac{1 - \Phi_T}{\mu_H} + \chi} \tag{5.40}$$

For a higher current friction coefficient of the rear wheels, $\mu_f < \mu_r$, the acceleration limit is resulted from (5.39)

$$a_{g,max} = \frac{\zeta}{\dfrac{\Phi_T}{\mu_H} - \chi} \tag{5.41}$$

If Φ_T can be controlled variably through the driveline system, so that both current friction coefficients can be made equal, i.e. $\mu_f = \mu_r$. In this case, the acceleration limit becomes

$$a_{g,ideal} = \mu_H \tag{5.42}$$

which corresponds to the ideal force distribution between the two axles. Torque bias Φ_T, which corresponds to the ideal torque distribution between the front and rear wheels, is computed through (5.38) and (5.39)

$$\Phi_T = \zeta + a_g \chi \tag{5.43}$$

It should be emphasized again that (5.40), (5.41) and (5.42) are only mathematically right. As it was commented on (5.32) before, the actual achievable maximum acceleration is considered when the current friction coefficient approaches but still less than the peak friction coefficient.

If the torque between front and rear is distributed according to the ratio given by (5.43), F_{Tf} and F_{Tr} will operate right on the ideal parabola curve so that the vehicle can achieve the best acceleration performance (with an assumption that the effective rolling radii of the front and rear tires are the same). Note that the ideal torque distribution only applies to the straight-line acceleration of vehicles. Indeed, the lateral force impacts the wheel torque (so do the tractive force balance), especially when the vehicle tends to yaw.

One design option to get the analytical result presented by (5.43) is using a mechanical solution to lock the center differential. When the center differential lock function is enabled, the front and rear drives are rigidly coupled, which corresponds to the condition of $\omega_{wf} = \omega_{wr}$. Under this condition and provided that the tire slips of both drive axles are identical, then the current friction coefficients equalize at both axles, and the torque is transferred over the locked differential in proportion to the normal forces on the axles. More details on influence of the effective rolling radii of the front and rear wheels on the power split between the drive axles can be found in reference [2].

Example 5.3

At the 2nd gear of a rear-wheel-drive car, the engine outputs a torque of 335Nm, assuming the main clutch is locked. The vehicle has geometric parameters $l_r = 2.42$ m, $l_f = 1.74$ m, $h_s = 0.72$ m and mass $M = 2866$ kg, as well as drivetrain parameters $i_t = 2.34$, $i_d = 3.31$, $\eta_e = 0.84$. The effective rolling radius is $r_w = 0.38$ m. Calculate the current friction coefficient μ between the rear tires and the road in the *quasi-static* case, in which the inertial terms as well as the angular accelerations of the wheels are neglected.

Using (5.9), the torque transmitted to the rear axle is

$$T_w = T_e i_t i_d \eta_e = 2179 \, \text{Nm}$$

Considering that the rear wheels are not slipping, the available tractive force should be equal to the tractive force at the rear, so the current friction coefficient can be calculated by using (5.33):

$$\frac{T_w}{r_w M g} = \frac{\mu_r \zeta}{1 - \mu_r \chi}$$

so that

$$\mu_r = 0.45$$

❑

5.2.4   Vehicle Stability in Driving Mode of Operation

As aforementioned, the tractive force distribution impacts not only acceleration performance but also vehicle stability during the driving mode of operation, which means yaw stability as well as the ability of keeping a driving course. The vehicle stability will be more and deeply investigated in Chapter 9. Here, stability is explained in a plausible way with regard to the tractive force distribution.

The above-mentioned stability is dependent on spin-up of the drive wheels. A wheel spin-up occurs when more torque is tended to apply to the wheel than the friction between the tire and road can support. Once a wheel spins up it loses the ability to take an external lateral force and to generate a lateral tire force needed to keep the vehicle oriented on the road. There are three cases that will be further considered: front wheels spin up, rear wheels spin up, or both front and rear wheels spin up.

Spin-up of the front wheels causes loss of ability to steer the vehicle, but it will stay in a stable condition as seen in Figure 5.16a. Once a yaw disturbance acts on the vehicle, which can be potentially caused by unwanted lateral forces (wind, etc.) or road lateral grade, the resultant force acting on the vehicle is no longer along with longitudinal direction of the vehicle. Consequently, the lateral

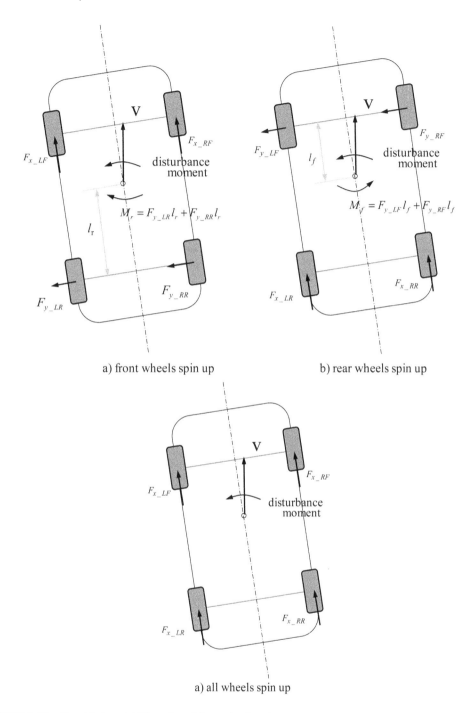

a) front wheels spin up

b) rear wheels spin up

a) all wheels spin up

FIGURE 5.16 Considering stability when driven wheels spin up.

forces (which are the reactions from the road) will be generated on the tires to keep the force balance on the vehicle. As the front wheels are spinning up, they are unable to generate the lateral tire forces. However, the yaw moment resulted through rear wheels is oriented against the yaw disturbance. In this case, the vehicle will try to continue its straight motion despite any steering inputs. Thus, the front spin-up affect can be mitigated in a straight-line acceleration. If the vehicle is in cornering, the vehicle may drift to the side out of its trajectory path, that will be the understeer case previously shown in Example 5.1.

Further, Figure 5.16b displays the situation where the rear drive wheels spin up. Any yaw disturbance can initiate a rotation of the vehicle as it will be supported rather than compensated by the yaw moment developed through the front wheels. The vehicle can get into an unstable condition since it may be no longer controllable. If the vehicle is in cornering, that will be the oversteer case discussed before in Example 5.1.

In Figure 5.16c the spin-up of both the front and the rear wheels is illustrated. All wheels lose the lateral force, and the vehicle can drift to the side of the road. Even the vehicle may have a rotation caused by the yaw disturbance, this situation is less critical than an alone spin-up of the rear wheels since a correction is still possible. If the driver is able to keep the steering wheel straight and reduce the driving torques at the wheels, the steering direction can be corrected.

Here, we use an example to present the relation between the tractive force distribution and the acceleration with respect to vehicle stability.

Example 5.4

A vehicle with two drive axles has the following geometric parameters: l_r = 1.55 m, l_f =1.24 m, h_s= 0.65 m. It is driven straight on a road with μ_H =0.6. Discuss possible tractive force distributions that correspond to three different fixed torque biases between front and rear. The three torque biases are denoted as TB-A, TB-B and TB-C in Figure 5.17.

From (5.18) and (5.19), we obtain $\chi = 0.233$ and $\zeta = 0.444$. A constant torque bias implies a constant proportioning of the torque split between the front and rear axles. That in turn presents a straight line on the graph starting at the origin in Figure 5.17 and extending downward and to the left. The slope of the straight line is the ratio of the tractive forces at the front and the rear axles. Each ratio is characterized by a torque bias introduced by (5.35). TB-A, TB-B and TB-C are illustrated corresponding to each straight-line respectively.

TB-A is designed to get the maximum acceleration when $\mu_H = 0.8$ since it intersects with the ideal distribution parabola at $a_g = 0.8$. Driving on the road with $\mu_H = 0.6$, the rear axle reaches the maximum acceleration at the intersection point, which shows $a_g = 0.52$. At this point the rear wheels spin up first and cannot take side forces, and the vehicle will be unstable.

For TB-B, on the same road the front wheels spin up first at the intersection point between $\mu_{Hf} = 0.6$ and the distribution line of TB-B. If the slip can be limited on the front wheels only, the tractive force on the rear axle can increase to achieve the intersection of line $\mu_{Hf} = 0.6$ and the ideal curve. However, with TB-B, which provides a bigger tractive force to the rear wheels, the rear axle will still spin up on a road with $\mu_{Hr} = 0.3$ since the distribution line in this region is below the ideal distribution curve.

The preference for fixed ratio is usually a tangent on the ideal distribution curve at the origin, presented as TB-C. This is because, in this case the front wheels will spin up first and the vehicle will be more stable as discussed before.

❑

In general, the biasing of the torque to the front wheels constitutes the preferred torque split for stability. However, the traction potential cannot be fully utilized if the available maximum torque on the rear remains low so that the current friction coefficient at the rear tires is always far away from the peak friction coefficient. In an extreme case of no torque on the rear wheels, we have then a FWD vehicle. Therefore, a compromise is necessary between traction capability and vehicle stability.

At the other end, RWD obviously is at disadvantage from the stability point of view since a rear-wheel-drive vehicle becomes unstable when the rear wheels spin. A rear-wheel-drive vehicle also has a more equitable balance of the vehicle's weight front-to-rear. During the acceleration the vehicle weight will be shifted to the rear, which leads to an improved traction.

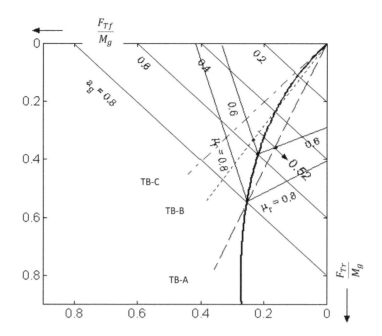

FIGURE 5.17 An example of determining tractive forces.

Under all conditions, spinning wheels will result in a loss of either steerability or stability. This suggests that, ideally, the torque distribution should be adjustable to ensure optimum or close to optimum traction performance while keeping up with stability under various operating conditions.

5.2.5 Design Implementation of Ideal Torque Distribution

In today's vehicle design, the sophistication of all-wheel-drive technology is reaching the point where the driving torque to each wheel can be independently controlled. The *torque vectoring* technology, i.e., controlling the torque split between the front and rear wheels and left and right wheels in real time, has become popular. In electric vehicles, electric motors allow for very much flexible torque management between drive axles or individually between drive wheels. This offers vehicle stability enhancements similar to those provided by dynamic stability control, in which primarily the braking torque is controlled at spinning wheels. In those vehicle dynamic stability control systems, presented by references [6] and [7], the engine torque reduction only serves as a secondary source to stabilize the vehicle when excessive engine torque is applied by the driver. Independent control of the driving torque distribution in vehicles with four drive wheels would therefore be especially beneficial during acceleration close to the limit of stability.

From (5.27), we know that the vehicle can achieve its optimal acceleration performance when the tire-road current friction coefficients on both the front and the rear wheels are equal; this means

$$\frac{F_{xf}}{F_{Nf}} = \frac{F_{xr}}{F_{Nr}} \tag{5.44}$$

Otherwise, the torque either excessively accelerates wheels above the vehicle speed or provides much less tractive force for the wheels than the tire-road friction can support. Since the torque bias, Φ_T, directly affects the balance between F_{xf} and F_{xr}, it can be determined how Φ_T relates to all other vehicle parameters.

Considering only the driving torques in (3.35), i.e., $T_{bf} = T_{br} = 0$, and combining it with (5.35), we can solve for torque T_{ar} at the rear wheels

$$T_{ar} = \Phi_T \left[\left(F_{Ad} + Mg \sin \gamma_z \right) r_w + \left(F_{Rf} + F_{Rr} \right) r_w + Ma r_w + \frac{J_{wf} a + J_{wr} a}{r_w} \right] \tag{5.45}$$

Then, solve F_{xr} using (3.28) and (5.45) by eliminating T_{ar}, and substituting F_{xr} into (3.32), the tractive force on the front wheels results in

$$F_{xf} = (1 - \Phi_T) \left[\left(F_{Ad} + Mg \sin \gamma_z \right) + \left(F_{Rf} + F_{Rr} \right) + Ma + \frac{J_{wf} a + J_{wr} a}{r_w^2} \right] - F_{Rf} - \frac{J_{wf} a}{r_w^2} \tag{5.46}$$

Following (3.32), it calculates the tractive force on the rear wheels

$$F_{xr} = \Phi_T \left[\left(F_{Ad} + Mg \sin \gamma_z \right) + \left(F_{Rf} + F_{Rr} \right) + Ma + \frac{J_{wf} a + J_{wr} a}{r_w^2} \right] - F_{Rr} - \frac{J_{wr} a}{r_w^2} \tag{5.47}$$

For the simplification, the lift force and the air drag neglected, and assume $\gamma_z = 0$. Based on (5.44) and using (5.46), (5.47), (3.48) and (3.50), the torque bias, which was introduced by (5.35), is calculated as

$$\Phi_T = \frac{M \dfrac{l_f}{l} + a_g \left(M \dfrac{h_s}{l} + \dfrac{J_{wf}}{r_w l} + \dfrac{J_{wr}}{r_w l} \right) + \dfrac{J_{wr}}{r_w^2}}{M + \dfrac{J_{wf} + J_{wr}}{r_w^2}} \tag{5.48}$$

Comparing (5.48) to (5.43), both are derived for the ideal tractive force distribution, where (5.48) is an extended result of (5.43) considering the rotational inertia of the wheels during acceleration. The key input of the calculation of the torque bias is acceleration a_g normalized to g, the the gravity acceleration.

In fact, torque bias Φ_T can be calculated from (5.48) for a target value of acceleration. The target acceleration of a vehicle under consideration can be computed using the vehicle's speed target that, in its turn, can be determined by the automated driving surrounding, for example autonomous cars. In today's vehicle standard technique, such as automatic cruise control and speed control, the speed target is determined automatically based on environmental data sensed by sensors. In Chapter 13, a vehicle speed controller will be described in detail.

Here, in this section, the torque bias calculation can be considered as an open-loop control as shown in Figure 5.18, where the second block represents the calculation by (5.48). The torque vectoring system in this figure splits the torque between the front and rear axles, and thus provides the optimal tractive forces at the front and rear wheels to guarantee the target acceleration.

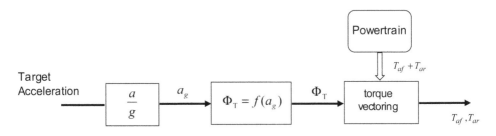

FIGURE 5.18 Open-loop control of torque distribution based on acceleration.

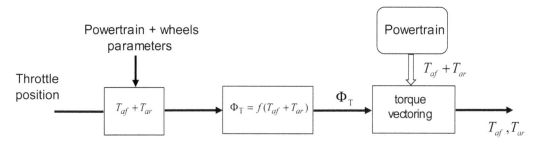

FIGURE 5.19 Open-loop control of torque distribution based on axle torque.

In the case that target acceleration is unknown, the total torque available at the engine output can be used. For this purpose, acceleration a_g in (5.48) is substituted by terms of $T_{af} + T_{ar}$ that is a component in (3.35):

$$\Phi_T = \frac{M\dfrac{l_f}{l} + \dfrac{J_{wr}}{r_w^2}}{M + \dfrac{J_{wf} + J_{wr}}{r_w^2}} + \frac{\left(T_{af} + T_{ar}\right)\left(M\dfrac{h_s}{l} + \dfrac{J_{wf}}{r_w l} + \dfrac{J_{wr}}{r_w l}\right)}{gr_w\left(M + \dfrac{J_{wf} + J_{wr}}{r_w^2}\right)^2} \tag{5.49}$$

where the same assumptions are made as for (5.48). Figure 5.19 describes the sequence of calculation steps, in which the second block represents (5.49). Normally, available axle torque $T_{af} + T_{ar}$ is calculated based on the throttle position signal, engine rotation speed and other powertrain and wheels parameters (gear ratios, rotational inertia, damping characteristics, effective rolling radius, etc.).

5.2.6 WHEEL TORQUE VECTORING

The previous section showed a method of a torque vectoring to manage the distribution of the driving torques between two drive axles. Torque vectoring can be used for various purposes, to maximize traction capability, to increase energy efficiency, to stabilize vehicles in critical driving situations, to support a driver's comfort, etc. Ultimately, it all comes down to the torque distribution between the drive wheels. This section focuses on the principle explanation of the torque vectoring functioning for improving traction capability. In the following, an approach of torque vectoring between two drive wheels is explained for a front-wheel-drive vehicle in the case of one of the drive wheels spinning while driving straight. The torque vectoring operates by applying a torque differential between the two wheels. When one of the wheels is on a low peak friction coefficient and the difference of the rotational velocities of the wheels is increasing, the torque vectoring automatically decreases the torque at this wheel and increases the torque at the other wheel with a higher peak friction coefficient. Hence, the wheel torque difference is created.

Let us consider the vehicle driven on a *split-μ* surface with a low peak friction coefficient on the left ($\mu_{Hl} < \mu_{Hr}$). The wheel dynamics is expressed in

$$\begin{aligned}
J_w\dot{\omega}_l + \mu_l F_N r_w &= T_{al} - T_{bl} \\
J_w\dot{\omega}_r + \mu_r F_N r_w &= T_{ar} - T_{br}
\end{aligned} \tag{5.50}$$

where F_N is the normal reaction, which is assumed the same at both wheels.

Building the torque difference, we have

$$J_{wf}\left(\dot{\omega}_r - \dot{\omega}_l\right) + \left(\mu_r - \mu_l\right)F_{Nw}r_w = \Delta T_a + \Delta T_b \tag{5.51}$$

with

$$\Delta T_a = T_{ar} - T_{al}$$

$$\Delta T_b = T_{bl} - T_{br}$$

When the left wheel starts to spin on the *low-μ* side of the road, the difference of the rotational speeds of the wheels becomes $\omega_l - \omega_r > 0$. While the left wheel spins, the torque provided by the driveline cannot be not fully utilized since it is used up to increase the rotational speed and inertia of the left wheel, hence, traction capability of both wheels is underutilized.

The question now is how to control the difference torque $\Delta T_a + \Delta T_b$ so that both wheels can improve their traction capability under the given μ-conditions? There are two options for such control. One is to engage the driveline torque vectoring system to increase the torque at the non-spinning wheel and decrease the torque at the spinning wheel. This is called *driving torque vectoring*, which distribute the driving torques between the wheels. If the wheels are individually controlled by electric motors, for example in electric vehicles, then the torque distribution is controlled through the motors directly. Chapter 7 will introduce the basic function description of an electric DC motor to understand the principle of the torque generation.

Another option to control the torque difference is the *braking torque vectoring*, which controls the brake of the spinning left wheel. Additional braking torque at this wheel will require a bigger input torque to the drive axle; this bigger torque will overcome the braking moment at the spinning wheel and equally increase the driving torque at the right wheel with better grip.

Therefore, the torque difference $\Delta T_a + \Delta T_b$ in (5.51) introduces the above-described two options of the torque vectoring. ΔT_a refers to the *driving torque vectoring* while ΔT_b relates to the *braking torque vectoring*. In case of driving torque vectoring, the braking torque on both wheels are set to zero. In case of the braking torque vectoring, there is no change on the driving torque distribution, i.e., $\Delta T_a = 0$, and only brakes are applied to the wheels. For vehicles without motor control of each individual wheel, a practical method here is to apply the brake on the *low-μ* side of the vehicle, i.e., $T_{bl} = \Delta T_b$ and $T_{br} = 0$. Note that the focus of this section here is to show the principle of the torque vectoring and control action needed to vector the torques rather than to explain a particular controller design. For this reason, a simple controller is chosen in the following to describe the torque needed.

In current vehicle's technology, the wheel rotational speed sensor has become a standard. Thus, the wheel speeds are measurable and known. The controller is defined to compute the torque difference to be proportional to the difference of the wheel angular velocities:

$$\Delta T_a + \Delta T_b = k\left(\omega_l - \omega_r\right) \tag{5.52}$$

where $k > 0$ is a controller parameter that needs to be chosen. By controlling the angular velocities, the controller's action results in the torque difference, which is actuated either as ΔT_a or ΔT_b depending on which system is in use: the driving torque vectoring or the braking torque vectoring.

Substituting the torque difference given by (5.52) in (5.51), we have the closed-loop dynamics for the angular velocity – torque control

$$J_{wf}\left(\dot{\omega}_r - \dot{\omega}_l\right) + k\left(\omega_r - \omega_l\right) = -\left(\mu_r - \mu_l\right)F_N r_w \tag{5.53}$$

Once *split-μ* occurs, the term on the right side of the above equation can be considered as a step input. Thus, the equation is linear in terms of the wheel rotational speed difference. Consider $\left(\mu_r - \mu_l\right)F_N r_w$ as the system input and the wheel speed difference $\omega_r - \omega_l$ as the output, (5.53) is a first–order differential equation, for which we can choose the controller parameter, k. With condition $k > 0$, the system pole is negative real, hence the control system (5.53) remains stable. On the other hand, the steady-state ratio between the output and input is determined by the final value theorem, which calculates the steady-state value of a function for $t \rightarrow \infty$:

$$\lim_{t \rightarrow \infty} \frac{\omega_l - \omega_r}{(\mu_r - \mu_l) F_{Nw} r_w} = \frac{1}{k} \tag{5.54}$$

The result above indicates that the steady control output $\omega_l - \omega_r$ depends on the parameter k. The larger k, the close the output $\omega_l - \omega_r$ is to zero. However, it always has a "deviation" depending on μ-difference and will never go to zero unless $k \rightarrow \infty$. If it could go to zero, then the controller would act exactly as a controller of a mechanism that locks the open differential and provides $\omega_l = \omega_r$. Chapter 12 will introduce an enhanced controller that meets the same purpose of controlling the angular velocities and is able to reach the optimal slips of both wheels to achieve maximum tractive forces.

Torque vectoring operates as an actuator to provide a torque differential between the front and rear axles or at the wheels of the same axle. It is a very practical tool to distribute driving and braking torques that are needed to control the vehicle dynamics. It is also beneficial for electric vehicles with individual electric motors in the wheels since the torque vectoring can be performed directly through the motors. In Chapters 12 to 14, various controls are designed utilizing torque vectoring as the actuation. For example, vehicle stability control and path-following control.

REFERENCES

1. J. E. Duffy and C. Johanson, Auto Drive Trains Technology, Illinois: The Goodheart-Willcox Company, 1995.
2. A. F. Andreev, V. I. Kabanau, V. V. Vantsevich, Driveline Systems of Ground Vehicles: Theory and Design, Boca Raton, London, New York: CRC Press, 2010.
3. A. Zomotor, Fahrwerktechnik: Fahrverhalten, Würzburg: Vogel Verlag, 1991.
4. U. Kiencke and L. Nielsen, Automotive Control Systems, Warrendale: SAE, 2000.
5. Vehicle Dynamics Terminology, SAE Standard J670, revised in 2008.
6. H. Fennel, R. Gutwein, A. Kohl, M. Latarnik and G. Roll, "Das modulare Regler- und Regelkonzept beim ESP von ITT Automotive," in *7th Achener Kolloquium Fahrzeug- und Motortechnik,* Achen, Okt. 1998.
7. A. van Zanten, R. Erhardt and G. Pfaff, "FDR-Die Fahrdynamikregelung von Bosch," *ATZ Automobiltechnische Zeitschrift,* vol. 96, no. 11, pp. 674–689, 1994.

6 Braking Mechanics

The brake systems play a vital role in making motor vehicles suitable for practical application. They are essential for ensuring safety, and are subject to strict official regulations. The brake is normally so designed that the vehicle may have the shortest possible braking distance while keeping the driving course. Braking or stopping distance refers to the distance a vehicle will travel from the point when its brakes are fully applied to when the vehicle comes to a complete stop. In the sense of active safety, however, the primary goal of braking dynamics is to maintain steerability and yaw stability. Active safety refers to design features that help avoid accidents such as good steering and braking. To understand the braking mechanics is important and necessary for designing active control systems that can contribute significantly in keeping the vehicle stable. Vehicle deceleration characteristics, analysis of the braking force distribution among the front and the rear wheels as a function of vehicle weight parameters, tire-road characteristics and braking system characteristics are presented in this chapter. Special attention is paid to lateral load transfer and semi-trailer dynamics. In addition, braking stability is also explained in a plausible way.

6.1 STRAIGHT-LINE BRAKING

Similar to acceleration performance described in the previous chapter, we apply the results obtained in Chapter 3 to investigate braking performance. We first consider braking in a straight-line motion. The thereby arising forces on the vehicle are illustrated in Figure 6.1. For the simplification, the road grade and the rotating mass inertia as well as the aerodynamic forces are neglected. The braking forces are summarized axle-wise since it is intended to analyze the brake behavior at each axle.

For the description, the following nomenclature is used:

M gross mass of the entire vehicle
a vehicle deceleration
g acceleration of gravity
F_{Bf} braking force on the front wheels
F_{Br} braking force on the rear wheels
F_{Nf} vertical force on the front wheels
F_{Nr} vertical force on the rear wheels
h_s height of CG (center of gravity)
l_r longitudinal distance between rear axle and CG
l_f longitudinal distance between front axle and CG

6.1.1 DECELERATION AND BRAKING EFFICIENCY

In a braking situation, the longitudinal tire force are presented by the braking forces, i.e.,

$$F_{xf} = -F_{Bf}$$
$$F_{xr} = -F_{Br}$$

(6.1)

From Figure 6.1, we have

$$F_{Bf} + F_{Br} = Ma$$

(6.2)

DOI: 10.1201/9781003134305-6

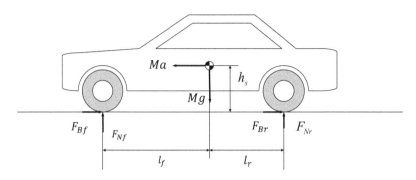

FIGURE 6.1   Braking forces acting on the vehicle.

Using the reference as (5.11)

$$a_g = \frac{a}{g} \text{ and } M_g = Mg \tag{6.3}$$

equation (6.2) becomes

$$F_{Bf} + F_{Br} = M_g a_g \tag{6.4}$$

Now a_g represents the normalized longitudinal deceleration, which is the longitudinal deceleration normalized to the gravity acceleration. The physical meaning of a_g is that this parameter characterizes a fraction of the gravity acceleration that is converted into the vehicle deceleration. For example, $a_g = 0.4$ means that 40% of the gravity acceleration is turned in vehicle deceleration.

As a measure of the vehicle braking performance, the braking efficiency is defined here as

$$\eta_B = \frac{a_g}{\mu_H} \tag{6.5}$$

The magnitude of the braking efficiency shows the percentage of the maximum possible braking force on a road with a given peak friction coefficient μ_H, which, i.e., the percentage, can be turned by the brake system of a vehicle in its actual braking force, and thus to develop deceleration a_g. A vehicle with two different brake systems will demonstrate different values of the braking efficiency. The higher the braking efficiency, the more effective the brake system is, hence, the shorter the vehicle braking distance will be. Theoretically, $\eta_B = 1$, implies

$$a_g = \mu_H \tag{6.6}$$

However, the above equation is not recommended for practical applications as this will be discussed later in next section. The braking efficiency given by (6.5) is a useful matrix to evaluate performance of a brake system and to assist the brake design.

The above-given analysis is suitable to contribute to an initial understanding of the braking dynamics. In engineering practice, however, more details should be included in the modeling of braking performance and then in control design. In particular, the peak friction coefficients at the front and rear wheels are not necessarily the same; there is a significant weight load shift between the vehicle axles during the braking; the braking torque distribution among the front and the rear wheels can be actively controlled to improve vehicle stability and safety. When varying the listed parameters and conditions, the contribution of the front and rear axles to the vehicle deceleration and braking can be assessed by examining the current friction coefficients developed by the tires of the front and rear wheels. This analysis begins in next section.

6.1.2 Braking Force Distribution

Vehicle braking dynamics is determined primarily by the braking force distribution among the wheels. The braking force distribution is also an important factor for designing the foundation brake components. Indeed, the braking force determines the braking torque that should be developed by the brake mechanism. The braking torque in its turn determines the physical size of the foundation brakes and the pressure of the air or fluid in the brake cylinders and in the brake actuation system in general. From the viewpoint of the hydraulic pressure in the brake system, the braking force distribution is also named as the braking force proportioning. Both terms are used in this book.

The braking force proportioning describes the relationship between the front and rear braking forces determined by the fluid pressure applied to each brake mechanism. The braking proportioning can be visualized graphically by plotting the braking forces at the rear wheels versus those at the front wheels. In the following, the description of this relationship is presented and analyzed.

First, we consider the braking forces at both axles of a vehicle and then relate the forces to the actual deceleration represented by a_g. The braking forces of the front and rear axles are mathematically represented by the equations

$$F_{Bf} = \mu_f F_{Nf} \tag{6.7}$$

$$F_{Br} = \mu_r F_{Nr} \tag{6.8}$$

where μ_f and μ_r are the current friction coefficients of the wheels at both axles, respectively. Substituting (6.7) and (6.8) into (6.4), we obtain

$$\mu_f F_{Nf} + \mu_r F_{Nr} = M_g a_g \tag{6.9}$$

from which we can then derive

$$\left(\mu_f - a_g\right) F_{Nf} + \left(\mu_r - a_g\right) F_{Nr} = 0 \tag{6.10}$$

since the sum of the normal reactions at the wheels is equal to vehicle weight M_g. As seen from (6.10), for a given deceleration a_g, a brake system should provide a braking force distribution that results in the same current friction coefficients at the front and rear tire, which are equal to the deceleration

$$\mu_f = \mu_r = a_g \tag{6.11}$$

Condition (6.11), which corresponds to vehicle dynamics described by (6.9), is called the ideal condition that indicates the optimum utilization of the potential braking capability of a vehicle.

Indeed, when applying braking torques that results in the current friction coefficients, which are smaller than the given deceleration, the brake potential is not utilized and the braking efficiency drops. If the applied braking torques are too high, the current friction coefficients can reach the peak friction coefficient, which limits the deceleration of the vehicle, a_g. As seen from Figure 4.4, reaching of the peak friction coefficient μ_H results in a high longitudinal skid of the wheel that drastically goes up to 100% (the wheel is not rotating any more, but just skidding along the road). Therefore, the point on the μ-curve with the coordinate of μ_H is unstable, i.e., when reaching the point, a wheel transits from rotation to skidding and becomes practically uncontrolled.

From the above-given consideration, two questions may be raised, namely, 1) how the braking force distribution can be expressed mathematically in terms of vehicle geometric parameters, the wheel normal forces and the current friction coefficients, and 2) what is the mathematical condition that should be imposed on the current friction coefficients of the front and rear tires to provide the ideal braking force distribution given by (6.11)? We start with the first question.

Figure 6.1 shows the forces acting on the vehicle in a straight-line braking situation. It was assumed that the front and the rear tires have the same road surface condition, the braking torque is defined for each axle (considering a vehicle bicycle model). For the force balance, we have already found (6.4), which presents the force relation in the longitudinal direction of the vehicle. As the second step, we obtain the moment balance equations with regard to the front and the rear axle, respectively.

$$F_{Nf}\left(l_f + l_r\right) = M_g l_r + a_g M_g h_s \tag{6.12}$$

$$F_{Nr}\left(l_f + l_r\right) = M_g l_f - a_g M_g h_s \tag{6.13}$$

Force $a_g M_g$, acting on the vehicle's center of gravity, causes a dynamic weight load transfer during the braking. The load is transferred from the rear axle to the front axle in proportion to the deceleration and the ratio of the CG height to the wheel base, $\left(l_f + l_r\right)$. By using the same definition as (5.18) and (5.19), then (6.12) and (6.13) can be rewritten in the form

$$F_{Nf} = M_g\left(1 - \zeta + a_g \chi\right) \tag{6.14}$$

$$F_{Nr} = M_g\left(\zeta - a_g \chi\right) \tag{6.15}$$

By applying (6.7), (6.8) to (6.14) and (6.15) respectively, and then using the expression of a_g from (6.4), we obtain

$$\frac{F_{Bf}}{M_g} = \mu_f\left(1 - \zeta + \frac{F_{Bf} + F_{Br}}{M_g}\chi\right) \tag{6.16}$$

$$\frac{F_{Br}}{M_g} = \mu_r\left(\zeta - \frac{F_{Bf} + F_{Br}}{M_g}\chi\right) \tag{6.17}$$

Dependencies (6.16) and (6.17) imply that the braking force proportioning depends only on the CG height and the wheel base for any desired combination of the current friction coefficients, which correspond to the braking forces summing up to $M_g a_g$ as seen from (6.4). Graphically, these equations are illustrated as two straight lines in Figure 6.2, in which the braking force distribution $\frac{F_{Br}}{M_g}$ vs. $\frac{F_{Bf}}{M_x}$ is projected for a particular set of the current friction coefficients, μ_f and μ_r, with constant values. The varying of values of μ_f in (6.16) can give a set of straight lines going out from the point with coordinates of $\left(0, -\frac{1-\zeta}{\chi}\right)$. Similarly, (6.17) can generate a set of straight lines going out from the point with coordinates of $\left(\frac{\zeta}{\chi}, 0\right)$ when parameter μ_r varies.

The intersection of the two lines in Figure 6.2, which is the joint solution of (6.16) and (6.17), determines the actual deceleration of the vehicle

$$a_g = \frac{\mu_f - \left(\mu_f - \mu_r\right)\zeta}{1 - \left(\mu_f - \mu_r\right)\chi} \tag{6.18}$$

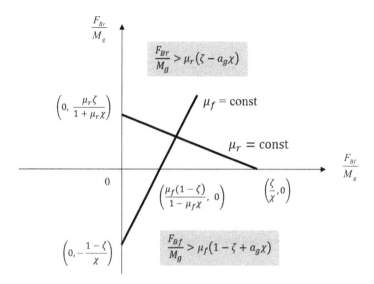

FIGURE 6.2 Braking force distribution dependencies.

The point with coordinates of $\left(0, \dfrac{\mu_r \zeta)}{1+\mu_r \chi}\right)$ displays the maximum achievable braking force of the

rear axle when the front axle brake fails, while the point of $\left(\dfrac{\mu_f\left(1-\zeta\right)}{1-\mu_f \chi}, 0\right)$ is the maximum front

braking force when the rear axle brake fails. The coordinates in the area above the μ_r-line match

condition $\dfrac{F_{Br}}{M_g} > \mu_r\left(\zeta - a_g \chi\right)$, which physically means that the rear braking force is higher than the

friction surface is capable to achieve when μ_f reaches the peak value μ_H, so the wheel is locked up.

The area below μ_f-line is covered by $\dfrac{F_{Bf}}{M_g} > \mu_f\left(1 - \zeta + a_g \chi\right)$ indicating the lockup of the front wheel

when μ_r equal to the peak friction coefficient. Note that both inequalities mentioned above in Figure
6.2 are derived using (6.16) and (6.17), in which a_g is a substitution coming from (6.4).

Ideal Braking Force Distribution
For the ideal braking force proportioning, the current friction coefficients of both axles are identical
and equal to a_g, such as given by (6.11). In this case (6.16) and (6.17) become

$$\frac{F_{Bf}}{M_g} = a_g\left(1 - \zeta + a_g \chi\right) \tag{6.19}$$

$$\frac{F_{Br}}{M_g} = a_g\left(\zeta - a_g \chi\right) \tag{6.20}$$

Substituting (6.4) into (6.20), we obtain the relationship between the ideal braking forces of the front
and the rear axles

$$\left(\frac{F_{Bf}}{M_g} + \frac{F_{Br}}{M_g}\right)^2 \chi - \left(\frac{F_{Bf}}{M_g} + \frac{F_{Br}}{M_g}\right)\zeta + \frac{F_{Br}}{M_g} = 0 \tag{6.21}$$

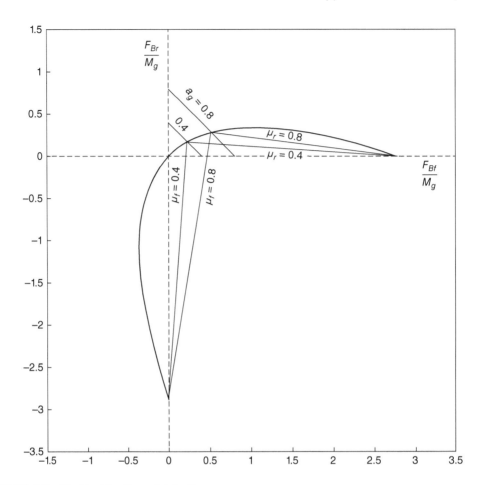

FIGURE 6.3   Ideal braking force distribution.

The relation between $\dfrac{F_{\mathrm{Br}}}{M_g}$ and $\dfrac{F_{\mathrm{Bf}}}{M_g}$ depends on the weight and the geometrical parameters of the vehicle, χ and ζ. Figure 6.3 illustrates (6.21) in a graphical form. The parabola is actually built by points that represent intersections of sets of three straight lines, which meet the condition given by (6.11). For example, there are two straight lines with $\mu_f = 0.4$ and $\mu_r = 0.4$ respectively that intersect in a point in Figure 6.3 (this intersection can be understandable if the reader looks at Figure 6.2). The third straight line in Figure 6.3 resulted from (6.11) at $a_g = 0.4$ goes through the same intersection point. Therefore, the intersection point is basically the concurrency point of the three straight lines that meet the condition (6.11). A similar intersection point is built for $a_g = 0.8$ in Figure 6.3. Any other points of the parabola in Figure 6.3 can be constructed as intersection points in the same way.

On the other hand, one can see that (6.21) differs from (5.30) only in the force directions, hence in the coordinate area shown in Figure 6.3 and Figure 5.13. This fact is reflected by the relation through (6.1). Combining (6.21) and (5.30), Figure 6.4 shows the common curve of the ideal longitudinal force distribution among the front and the rear wheels, where F_{xf} and F_{xr} represent the tractive and braking forces on the front and the rear axles.

As illustrated in Figure 6.3, the graphical implementation of (6.21) is a parabola. As long as the braking force distribution lies on this parabola, condition (6.11) is satisfied and the vehicle decelerates with a given deceleration a_g.

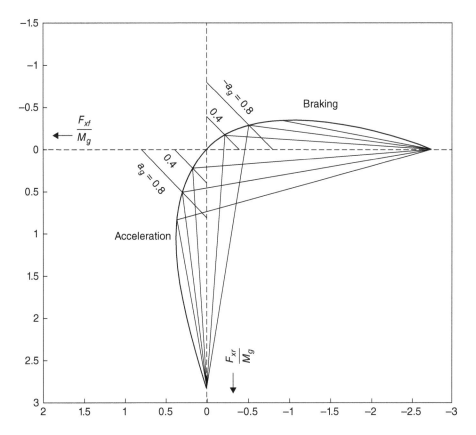

FIGURE 6.4 Ideal force distribution in acceleration and braking.

The ideal parabola can change depending on the presence or absence of the payload, which changes the position of the center of gravity. The road conditions may also change. First, we consider the influence of the payload on the braking force distribution. If the vehicle trunk is loaded, the normal force caused by the loading will be shifting the CG point back toward the rear axle, and hence parameter ζ increases. We view (6.19) and (6.20) again, the braking force ratio in the ideal distribution case becomes

$$\frac{F_{Br}}{F_{Bf}} = \frac{\zeta - a_g \chi}{1 - \zeta + a_g \chi} \tag{6.22}$$

In case of the payload presence, where ζ increases, the ratio in (6.22) will be also increasing. So, the parabola of the loaded vehicle moves up from the parabola computed for the curb weight, as illustrated in Figure 6.5.

The second effect on the ideal distribution is during braking in turn. The lateral acceleration plays a major role. In the presence of the centrifugal force, the longitudinal braking force cannot reach its maximum as in straight-line braking. The lateral transfer of the weight and elastokinematics have to be considered. Figure 6.5 shows the change of the ideal braking force distribution due to this cornering behavior. More details can be seen later in Section 6.2 of this chapter.

A further effect is rolling resistance, as introduced in Chapter 3. The rolling resistance, which is dependent on the speed and the tire slip angle that the vehicle is moving on a curve, can rise significantly. The vehicle will brake down in turn if it is not accelerated by the driver. During braking, therefore, there is an additional braking force acting on the vehicle, and the braking causes a

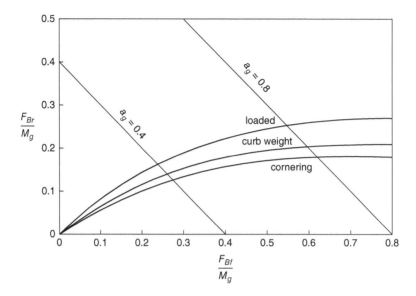

FIGURE 6.5 External influences on the ideal braking force distribution.

dynamic load transfer from the rear to the front. As indicated by (6.14) and (6.15), the dynamic load transfer results in an increase of the vertical force in the front and a reduction in the rear. In this case, the ratio in (6.22) is getting lower. So, the parabola of the cornering vehicle moves down from the parabola computed for the curb weight, as illustrated in Figure 6.5.

Fixed Braking Force Distribution

The ideal braking force distribution curve can be considered as a boundary that limits the optimum braking forces at the front and the rear axles for given values of the deceleration. When following the parabola, the optimum braking forces are reached on both axles. An attempt to brake a vehicle in the region above the parabola when the rear braking force goes higher than its boundary, the rear wheels will be locked up. Likewise, allowing the braking force in the region below the parabola, where the front braking force boundary is not satisfied, causes the front wheels to lockup. If the rear wheels are locked up before the front wheels, the vehicle is in an unstable condition. In other words, the brake system has to be designed that the wheels of the front axle always lock up first. The instability has huge impact on the vehicle dynamic behavior and must be carefully considered in brake design. More details about the braking stability will be described and discussed in a later section.

In the brake design practice, the ideal braking force distribution is not realizable by mechanical construction of the brake system. Thereby the primary challenge is the selecting of a proportioning ratio that will satisfy all design goals considering the variability in surface condition, weight distribution, CG height, and braking conditions such as in curve. A constant proportioning is a straight line on the graph starting at the origin and extending upward and to the right (see Figure 6.6). The slope is the ratio of the braking forces at the rear and the front axles.

To achieve a better performance, the straight line has to be close to and below the ideal parabolic curve. The standard brake design is to include a valve in the actuation hydraulic system, which adjusts the pressure portions between the rear and the front brake according to the slope. Such a valve is known as a pressure proportioning valve. Some pressure proportioning valves can split pressure to both the front and the rear brakes up to a certain pressure level, and then reduce the rate of pressure increase to one of the brakes thereafter [1, 2]. The braking forces are determined then by the brake dimensions. Figure 6.6 shows an example of the fixed braking force distribution. In this design, after accomplishing the proportional force distribution, the pressure is increased in the front brakes only.

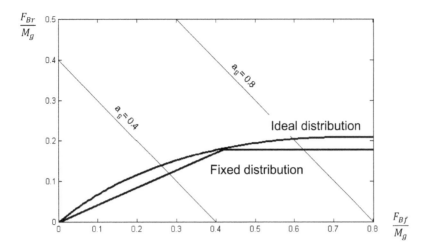

FIGURE 6.6   Fixed braking force distribution.

6.2   BRAKING IN TURN

In the previous section we considered straight-line braking. Lateral dynamics was assumed to be negligible so that the braking behavior of both the left and the right wheels at the same axle was considered identical. Based on the results obtained there, we will continue to discuss dynamic behaviors when vehicle braking occurs in a turn.

Mathematical modeling of braking dynamics in turn was investigated more than once [3]. On account of the complexity of the dynamics of vehicles in turns it is not possible to derive a comprehensive approach without highly complicated mathematical formulations. To understand it qualitatively, however, it is sufficient to discuss the essential influences on the turn dynamics. First, we make an assumption considering the vehicle as a rigid system. An additional consideration of the lateral load transfer including elastokinematics will give a better approximation of braking dynamics.

MORE ON BRAKING FORCE DISTRIBUTION

Under the assumption of a rigid system, the bicycle vehicle model can still be used as a starting point in this discussion. The forces during braking in a turn can be illustrated in Figure 6.7 showing the top view of a vehicle. Thereby the following notations in addition to Figure 6.1 are used:

a_y lateral acceleration
a_{yg} normalized lateral acceleration to the gravity acceleration
a_{xg} normalized longitudinal acceleration to the gravity acceleration
F_{yf} lateral tire force of the front axle
F_{yr} lateral tire force of the rear axle
d_t half wheel track
δ_f front steer angle

Applying the same principle used for (6.12) and (6.13), we have the moment balance equations in lateral direction,

$$F_{yf}\left(l_f + l_r\right)cos\delta_f = a_y M l_r \tag{6.23}$$

$$F_{yr}\left(l_f + l_r\right) = a_y M l_f \tag{6.24}$$

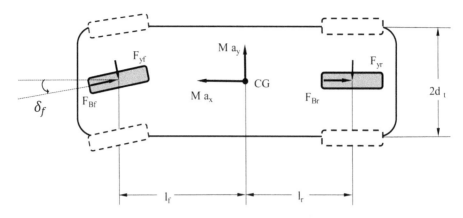

FIGURE 6.7   Vehicle braking force modeling in turn.

We adopt the notion given by (5.18) and (5.19), it follows

$$F_{yf} cos\delta_f = a_{yg} M_g (1-\zeta) \tag{6.25}$$

$$F_{yr} = a_{yg} M_g \zeta \tag{6.26}$$

The normal reaction determined in (6.14) and (6.15) are still applicable. From Kamm's circle presented by (3.20), the maximum braking force when vehicle is in turn is less than those in the straight-line motion since the lateral force arises,

$$F_{Bf} \le \sqrt{\mu_H^2 F_{Nf}^2 - F_{yf}^2} \tag{6.27}$$

$$F_{Br} \le \sqrt{\mu_H^2 F_{Nr}^2 - F_{yr}^2} \tag{6.28}$$

where μ_H is the peak friction coefficient for both the front and the rear tires. Unlike in straight-line braking, deceleration a_g cannot reach theoretically μ_H as its maximum. Apparently, the case where (6.27) and (6.28) are equal shows the ideal brake case, in which the maximum braking force is reached. Once again, we consider the ideal braking force distribution. Using (6.14), (6.15), (6.25) and (6.26), write (6.27) and (6.28) as

$$\frac{F_{Bf}}{M_g} = \sqrt{\mu_H^2 (1-\zeta+a_g\chi)^2 - a_{yg}^2 (1-\zeta)^2 / cos^2\delta_f} \tag{6.29}$$

$$\frac{F_{Br}}{M_g} = \sqrt{\mu_H^2 (\zeta-a_g\chi)^2 - a_{yg}^2 \zeta^2} \tag{6.30}$$

It is to point out, that the normalized acceleration, a_g, represents here the normalized resultant acceleration that consists of the longitudinal and lateral components. In order to have a qualitative assessment of the braking force distribution with a lateral acceleration comparing to the ideal distribution in the straight-line braking, it is assumed the front steer angle is small, $cos\delta_f \approx 1$. Substituting the above expressions into (6.4) gives a clear relation between deceleration a_{xg} and peak friction coefficient μ_H:

$$a_{xg} = \sqrt{\mu_H^2 (1-\zeta+a_{xg}\chi)^2 - a_{yg}^2 (1-\zeta)^2} + \sqrt{\mu_H^2 (\zeta-a_{xg}\chi)^2 - a_{yg}^2 \zeta^2} \tag{6.31}$$

For a given peak friction coefficient μ_H, the available maximum deceleration a_{xg} tends to decrease if lateral acceleration a_{yg} rises. The longitudinal deceleration cannot reach μ_H as long as the lateral acceleration exists, i.e., the lateral force lowers the maximum braking force that can be available for the vehicle braking. In the case $a_{yg} = 0$, it turns out that $a_{xg} = \mu_H$, which corresponds to peak friction coefficient in the straight-line braking. However, the reader should be referred here to comments on (6.6).

Looking at (6.29) and (6.30), it is better to see the relationship between $\dfrac{F_{Bf}}{M_g}$ and $\dfrac{F_{Br}}{M_g}$ clearly via a numerical solution of these equations. This will be done in next section.

Influence of Lateral Load Transfer

From Section 6.1 we know that during a straight-line braking a load transfer takes place from the rear axle to the front axle, and thus impacts on the braking force distribution. Similarly, the weight load transfer acts during a cornering. The normal reactions on the outer wheels increase at the expense of those on the inner wheels. That this weight load is transferred laterally is commonly called a lateral load transfer. Within this consideration, the effects of the movements of the sprung mass and the elastokinematics of the suspension need to be modeled to address the weight transfer more precisely. The elastokinematics describes suspension deflections and sprung mass movements, which cause variations of the tire forces as well as the moments between tires and road surface. Here, we directly take and present the results obtained from Chapter 10. The reader is advised to study further details from there.

The braking behavior of a vehicle with the lateral load transfer can be studied through either modeling or doing experimental tests. In reference [3], the lateral load transfer was demonstrated based on experimental measurements. Here, the vehicle braking performance in turn is approximated in a mathematical modeling. For this purpose, the influence of the braking force distribution on vehicle braking in turn is approximated with a model that represents the major lateral effects as considered below.

Vehicle roll is the gyration around the roll axis of the vehicle. The roll movement occurs in response to steering inputs or an external lateral force applied to the vehicle. Usually, it is mainly the result of the sprung mass roll caused by suspension and tire deflection, plus some axles roll from tire deflection. For simplification, we assume the roll angle of the sprung mass ϕ to be small. Further, it is assumed that the vehicle has identical suspensions at the front and the rear, and the roll moments act about so-called *roll center*. Figure 6.8 shows the rear-view diagram of a vehicle in left-hand cornering, where ΔF_{Nr} denotes the load transfer between the wheels of the rear axle.

As resulted from Chapter 10, the load transfers on both the front and the rear axles during steady-state cornering are

$$\Delta F_{Nf} = \frac{\left(c_{\phi f} + m_{vf}gh_r\right)m_v a_v \left(h_s - h_r\right)}{2d_t\left(\left(c_{\phi f} + c_{\phi r}\right) - m_v g\left(h_s - h_r\right)\right)} + \frac{m_{vf}a_v h_r}{2d_t} \tag{6.32}$$

$$\Delta F_{Nr} = \frac{\left(c_{\phi r} + m_{vr}gh_r\right)m_v a_v \left(h_s - h_r\right)}{2d_t\left(\left(c_{\phi f} + c_{\phi r}\right) - m_v g\left(h_s - h_r\right)\right)} + \frac{m_{vr}a_v h_r}{2d_t} \tag{6.33}$$

where $C_{\phi f}$, $C_{\phi r}$ denote front and rear roll stiffnesses of the sprung mass respectively. m_v and a_v are the sprung mass and its lateral acceleration. The lateral load transfer during a steady-state cornering is accomplished by two simultaneous compensation forces on the axle. The first term on the right-hand side of the above equations is the load transfer due to the suspension roll, which acts through the suspension springs imposing a moment about the roll center proportional to the suspension roll stiffness. This idealized elastokinematic effect determines the impact of the roll moment on the wheel normal reactions, and hence on the dynamics of the vehicle. The second term refers to the

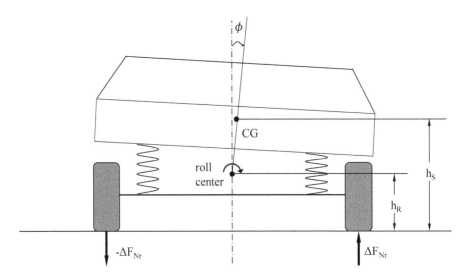

FIGURE 6.8 Rear view of a vehicle with a solid axle in left-hand cornering.

load transfer due to the centrifugal force acting on the vehicle in turn. This instantaneous load trans-fer mechanism arises from the lateral force imposed on the axle. As shown in the two equations above, the load transfers are explicitly expressed as dependent on vehicle parameters. Based on the known vehicle parameters, the lateral load transfer can be calculated straight forward.

The question now is how does the lateral load shift affects the braking force distribution? For the study of the braking force distribution in the previous section, we used the bicycle-model. The lateral load transfer cannot be studied by using that model. On the other hand, practical experi-ence shows that the braking stability during cornering is mostly dominated by the inner tires. This means, to get a qualitative verification it is sufficient to consider the load reduction on the inner wheels of the front and the rear axles. To this end, we rewrite (6.32) and (6.33) as the product of constant factors, $c_{\Delta f}$ and $c_{\Delta r}$, and the vehicle weight multiplied by the normalized lateral acceleration

$$\Delta F_{Nr} = c_{\Delta r}\ a_{yg}\ M_g \tag{6.34}$$

$$\Delta F_{Nf} = c_{\Delta f}\ a_{yg}\ M_g \tag{6.35}$$

The constant factors, $c_{\Delta f}$ and $c_{\Delta r}$, represent the proportionality of the change of the normal reaction to the lateral acceleration. These factors can be determined through vehicle parameters, both known from vehicle design and experimentally measured, using (6.32) and (6.33). By inserting (6.34) and (6.35) as to decrease normal reaction into the first component under the square root in (6.29) and (6.30), one can obtain

$$\frac{F_{Bf}}{M_g} = \sqrt{\mu_H^2 (1 - \zeta + a_g\chi - c_{\Delta f}\ a_{yg})^2 - a_{yg}^2 (1-\zeta)^2\ /\ cos^2\delta} \tag{6.36}$$

$$\frac{F_{Br}}{M_g} = \sqrt{\mu_H^2 (\zeta - a_g\chi - c_{\Delta r}\ a_{yg})^2 - a_{yg}^2\ \zeta^2} \tag{6.37}$$

This approach, which neglects the lateral load increase on the outer tires, showed a close approxi-mation with experimental results [3]. Figure 6.9 shows the ideal braking force distribution and the effect of the above-presented lateral load transfer, where the normalized resultant deceleration a_g varies from 0.2 to 0.9. The vehicle has the normalized lateral acceleration of $a_{yg} = 0$ and $a_{yg} = 0.2$, where the front steer angle $\delta \approx 0$. In order to keep the lateral acceleration constant, the vehicle needs to be steered on varying-radius turn. In case of $a_{yg} = 0$, the vehicle is driven straight ahead. One can see that the lateral acceleration causing the lateral load transfer will move the ideal curve down. Consequently, in case of $\mu \neq \mu_H$, the distribution curve will be slightly below both curves. The parameters from a passenger car are used in the computation: $c_{\Delta f} = 0.23$, $c_{\Delta r} = 0.16$, $l_r = 1.55$m, $l_f = 1.24$m, $h_s = 0.65$m, $\zeta = 0.444$ and $\chi = 0.233$.

Here an example is presented for the calculation of the braking forces. In most current vehicles, the braking force distribution between the front and the rear wheels is approximated to the ideal distribution with a possibility of a very limited adjustment through proportioning valves (see Section 6.1.2). However, we have seen that the braking forces of wheels must be controlled individually in order to reach the maximum braking performance. The basic idea is to calculate the required braking force based on the target deceleration of the vehicle. The example below illustrates the proposed method to compute individual braking forces that can provide a required deceleration.

Example 6.1

The same passenger car, which simulation results were previously discussed in Figure 6.9, is considered in this example. It is driven on a curved road made up of three sequential patches of different μ_H-values according to the table below. The vehicle runs continuously from time t_1 to t_3, and its normalized lateral deceleration is consistently of $a_{yg} = 0.2$. Assuming the braking force distribution follows the parabola shown in Figure 6.10, describe the braking force distributions when the vehicle deceleration is targeted at the normalized deceleration of $a_g = 0.6$.

Time	t_1	t_2	t_3
Road μ_H-value	$\mu_{H1} = 0.64$	$\mu_{H2} = 0.54$	$\mu_{H3} = 0.7$

From (5.18) and (5.19), we obtain $\chi = 0.233$ and $\zeta = 0.444$. To simplify the calculation, we assume that $c_{\Delta r} = c_{\Delta f} = 0$. At time t_1, the reverse calculation of (6.31) results in $\mu_H = 0.64$. This means, in order to get the deceleration $a_g = 0.6$ the peak friction coefficient should be at least 0.64, which is exactly equal to μ_{H1}. In Figure 6.10 point t_1 shows the braking force distribution, which correspond to $a_g = 0.6$ and $\mu_{H1} = 0.64$.

Then at time t_2, the road surface changes to the one with μ_{H2}, the wheels will lock up due to the high braking forces distributed as it was determined at time t_1. Due to the drop of the peak friction coefficient from $\mu_{H2} = 0.64$ to $\mu_{H2} = 0.54$ now, the achievable maximum deceleration decreases to 0.5, which is computed by (6.31), so the braking force distribution moves to point t_2 in Figure 6.10.

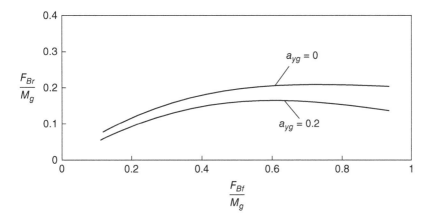

FIGURE 6.9 Ideal braking force distribution in straight line motion and during cornering.

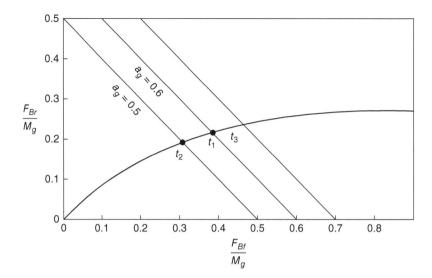

FIGURE 6.10   An example of calculating braking forces.

At time t_3, the braking forces are increased in order to achieve the target deceleration again. Since $\mu_{H3} > 0.64$, the braking operation point moves back to point t_1. With the braking forces distribution at t_1, the vehicle is able to achieve the target $a_g = 0.6$. Therefore, t_1 and t_3 time moments point to the same location in Figure 6.10.

❏

6.3   BRAKING STABILITY

In vehicle brake design, not only vehicle brake efficiency is a significant purpose but also the braking stability of the vehicle, which includes the yaw stability as well as the ability to keep a driving course. Stability plays also an important role for designing control systems to actively improve vehicle safety (known as active safety). Vehicle stability will be deeply investigated in relation with vehicle lateral dynamics later in Chapter 9. Here we explain the braking stability in a plausible way.

Braking stability is dependent on lockup of both the front and the rear wheels. Once one wheel locks up (i.e., the wheel does not rotate), it loses its ability to generate a lateral force in the tire-road patch needed to take an external lateral force that acts on the vehicle (i.e., a centrifugal force). In such case, the locked up wheel is not able to contribute other wheels, withholding the vehicle on the road in both the longitudinal and the lateral directions. There are three possible cases of the lookup: the front wheels lock up, the rear wheels lock up, or all wheels lock up. All three cases are analyzed below.

Lockup of the front wheels causes loss of ability to steer the vehicle, but it will stay in a stable condition as shown in Figure 6.11a. Once a yaw disturbance acts on the vehicle, which can be potentially caused by unwanted lateral forces (wind, road lateral grade, etc.), the resultant force acting on the vehicle is no longer along with the longitudinal direction of the vehicle. Consequently, the lateral forces (which are the reactions from the road) will be generated on the tires to keep the force balance on the vehicle. As the front wheels are locked up, they are unable to generate the lateral tire forces. However, the yaw moment resulted through the rear wheels is oriented against the yaw disturbance. In this case, the vehicle will normally continue its straight motion despite any steering inputs. Thus, the front lock-up affect can be mitigated in a straight-line deceleration. If the vehicle is in cornering, the vehicle may drift to the side out of the curve.

Further, Figure 6.11b shows the situation where the rear wheels lock up. Any yaw disturbances can initiate a rotation of the vehicle as it will be supported rather than compensated by the yaw moment developed through the front wheels. The vehicle may be placed in an unstable condition since it is no longer controllable. According to the experimental investigations [1], a typical driver

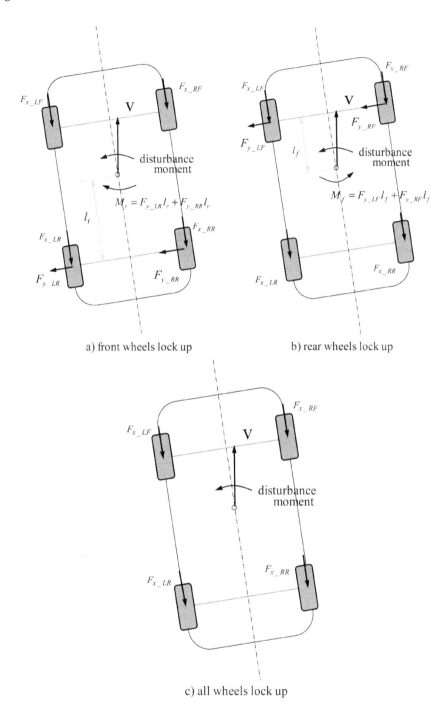

a) front wheels lock up   b) rear wheels lock up

c) all wheels lock up

FIGURE 6.11 Front and rear wheel lockup situations.

cannot readily control the vehicle in such driving situation. In Figure 6.11c the lockup of both the front and the rear wheels is illustrated. All wheels lose the lateral force, and the vehicle can drift to a side. Even the vehicle may have a rotation caused by yaw disturbances, this situation is less critical than an alone lockup of rear wheels since a correction is still possible. If the driver is able to keep the steering wheel straight and release the brake, the steering direction can be corrected.

From the braking stability point of view, the front brake bias constitutes the preferred brake design. However, the available braking torque would be wasted if the current friction coefficient is too far below the peak friction coefficient between the tire and road, so that the stopping distance will be longer. Therefore, a compromise is necessary between braking efficiency and stability, where stability is assured of more importance. Under all conditions, wheel lockup will result in a loss of either steerability or directional stability.

6.4 TRAILER INFLUENCE ON BRAKING

In the previous discussion so far, it has been recognized that the wheel lockup has a negative impact on braking stability of vehicles. Also, we learned that a vehicle weight load variation as well as braking-in-turn can change the wheel lockup behavior. Normally the braking force distribution is outlined for solo vehicle. Towing a trailer will definitely affect the braking forces on the vehicle front and the rear axles because of an additional force loads on the hitch. In this section, we look at the influences of a car semi-trailer on braking characteristics.

There are usually different kinds of trailers available for passenger cars and light pickup trucks. In the following, a single-axle semi-trailer is considered for the study. The results can be extended to a multiple-axle trailer with one hitch as long as the forces at the hitch are determinable. The common case is a single-axle semi-trailer having one hitch connection for towing, as illustrated in Figure 6.12. In addition to Figure 6.1, the following notations are used:

M_t gross mass of the trailer
F_{xt} longitudinal force on the hitch
F_{zt} normal force on the hitch
F_{Nt} normal force on the trailer wheels
F_{Bt} braking force on the trailer wheels
h_t height of the trailer CG (center of gravity)
h_h height of the hitch
l_s longitudinal distance between CG and trailer axle
l_t longitudinal distance between the hitch and trailer axle

For heavy duty trailers, it is required to be equipped with a trailer-brake. However, it is not a standard to have a brake system on light trailers. The trailer-brake system is usually so designed to generate a braking force on the trailer wheels proportional to longitudinal force F_{xt} on hitch [4], that is

$$F_{Bt} = i_{Bt}\, F_{xt} \tag{6.38}$$

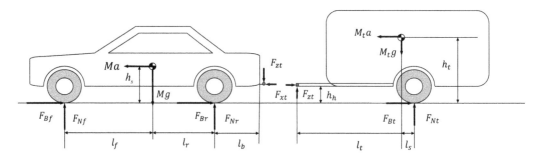

FIGURE 6.12 Braking forces on the wheels of a vehicle towing a trailer.

where i_{Bt} indicates the transfer ratio of the trailer braking force. If there is no brake system exists on the trailer, it becomes $i_{Bt} = 0$. As shown in the following, ratio i_{Bt} impacts the ideal braking force distribution. Higher ratio values will enlarge the normal force on the hitch, and thus move the ideal distribution parabola of a vehicle with a trailer away from the ideal distribution parabola of a single vehicle.

From the force balance on the trailer in the longitudinal direction, the force on the hitch can be calculated as

$$F_{xt} = \frac{M_{tg} a_g}{1 + i_{Bt}} \tag{6.39}$$

using a similar reference as (6.3)

$$a_g = \frac{a}{g} \quad \text{and} \quad M_{tg} = M_t g \tag{6.40}$$

The moment balance about the contact between the trailer tire and the ground yields the normal force on the hitch:

$$F_{zt} = M_{tg} a_g \left(h_t - \frac{h_h}{1 + i_{Bt}} \right) \frac{1}{l_t} + M_{tg} \frac{l_s}{l_t} \tag{6.41}$$

The last term in the above equation is the so-called tongue weight, which is the static load on the hitch. Then in the normal direction we have

$$F_{Nt} = M_{tg} \left[1 - \frac{a_g h_t}{l_t} + \frac{a_g h_h}{(1 + i_{Bt}) l_t} \right] - M_{tg} \frac{l_s}{l_t} \tag{6.42}$$

Now, in order to discuss the influence of the semi-trailer we consider a free-body diagram of the towing vehicle itself. There are two forces F_{xt} and F_{zt} added on the rear back of the vehicle, which act on the vehicle from the semi-trailer and need to be considered in the force and moment calculations. For this reason, (6.19) and (6.20) need to be extended to

$$\frac{F_{Bf}}{M_g} = a_g \left(1 - \zeta + a_g \chi \right) - \frac{F_{zt}}{M_g} \frac{l_b}{l_f + l_r} + \frac{F_{xt}}{M_g} \frac{h_h}{l_f + l_r} \tag{6.43}$$

$$\frac{F_{Br}}{M_g} = a_g \left(\zeta - a_g \chi \right) + \frac{F_{zt}}{M_g} \frac{l_b}{l_f + l_r} - \frac{F_{xt}}{M_g} \frac{h_h}{l_f + l_r} \tag{6.44}$$

Apparently, the ideal curve of the ideal braking force distribution can move up and down depending on F_{xt} and F_{zt} in relation with other vehicle parameters in the above equations. In case $l_b = h_h$, the ideal curve moves up relative to the ideal curve of the vehicle without the semi-trailer when $F_{zt} > F_{xt}$, and moves down when $F_{zt} < F_{xt}$.

Figure 6.13 shows an example of the ideal braking force distribution of a vehicle towing a semi-trailer. That the braking forces are not equal to zero at zero deceleration is a fact due to the tongue weight. When a semi-trailer brake exists and, for example, $i_{Bt} = 5$, the ideal curve moves up. The rear braking force gets strengthened by the hitch force for the same deceleration target. When the semi-trailer has no brakes, i.e., $i_{Bt} = 0$, the ideal curve moves down since the hitch force undermines the rear braking force. The front braking force needs to be increased to achieve the same target deceleration.

We look now at the effect of a fixed force distribution in the following example. Here, the same passenger car from Example 6.3 is considered as the towing vehicle. The semi-trailer parameters of

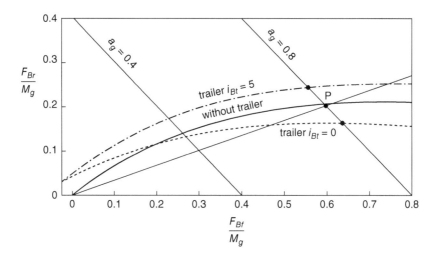

FIGURE 6.13 Braking force distribution for towing trailer.

$l_b = 0.8$, $h_h = 0.52$, $h_t = 0.8$, $l_t = 2.7$ and $\dfrac{M_{tg}}{M_g} = 0.5$ are used in the computation. In case the brakes of

a single vehicle are designed to reach the optimum braking condition at $a_g = 0.8$, Figure 6.13 shows this operation at position P (see the curve marked as without trailer). The towing of a semi-trailer with $i_{Bt} = 5$ at $a_g = 0.8$ will lock up the front wheels since position P stays below the ideal curve that marked as trailer $i_{Bt} = 5$. Changing to $i_{Bt} = 0$ (no brakes at the trailer) will lock up the rear wheels of the vehicle first so that the vehicle is getting unstable because position P becomes above the ideal curve that is the lowest in Figure 6.13.

REFERENCES

1. M. Mitschke, Dynamik der Kraftfahrzeuge, Band A: Antrieb und Bremsung, Berlin: Springer-Verlag, 1982.
2. Bosch, Automotive handbook, 8th ed., Plochingen: Robert Bosch GmbH, 2011.
3. E. C. Glasner von Ostenwall, Beitrag zur Auslegung von Kraffahrzeugbremsanlagen, Dissertation, Universität Stuttgart, Germany, 1973.
4. M. Mitschke and E. Sagan, Fahrdynamik von Pkw-Wohnnanhängerzügen, Köln: Verlag TÜV Rheinland, 1998.

7 Regenerative Braking

With advancements in the technology of *electric vehicles* (EV) and *hybrid electric vehicles* (HEV), *regenerative braking* has become an important aspect in the evaluation of vehicle braking performance as well as in the development of brake control technology. Electric vehicles use electric motors powered by batteries to drive axles and wheels while hybrid electric vehicles apply both electric motors and an internal combustion engine to deliver the propulsion power. Electric motors play here an essential role in energy conversion. One of the most important aspects of using an electric motor is the energy recovery – while braking, the vehicle kinetic energy can be recaptured back to the electric energy source. Such energy recovery is commonly called *regenerative braking*, which happens during the braking of a vehicle.

This chapter deals with the basic concepts of the electric motor operation and the principle of the energy regeneration, so the readers will be able to understand where and how the regeneration comes from and how the regenerative braking impacts braking performance of vehicles. The description is not intended to be thorough for the design and implementation, but should simply serve to highlight theoretical principles essential for control design.

We start with powertrain configurations of electric and hybrid electric vehicles. To describe the basic concept of the motor operation, a separately excited DC motor is taken as an example, in which the regeneration is a result of the generator mode of operation. The regeneration torque created during the braking as well as its impact on normal hydraulic braking (friction brake) will be characterized and explained. At the end, the consideration of the energy balance of a vehicle will close up the chapter.

7.1 EV AND HEV POWERTRAIN CONFIGURATION

Electric vehicles typically use electrochemical batteries as the source of energy. The batteries need to be charged to restore the energy once it is below the available range. Figure 7.1 shows a possible powertrain configuration. The electric motor converts electrical energy into mechanical energy of its shaft that is connected to a mechanical transmission, from which the power flow goes to the final drive, differential and, finally, to the drive wheels. A conventional mechanical transmission may be also used in the powertrain, where an adaptation is needed for the shaft coupling. However, since the electric motor has a much wider torque curve than the internal combustion engine (ICE), a complex transmission with automatic shifts is used much less, often only a simple constant-ratio gearbox is sufficient.

Hybrid electric vehicle normally has more than two energy sources to deliver the power to the drive wheels. The most common concept is to use an internal combustion engine as primary energy source while one or two electric motors are used as the secondary energy source. The ICE can ensure a long travel distance without a need of frequent refueling. However, the combustion is an irreversible process – the energy cannot be regained once the fuel is burned. Electric motors have a recuperative capacity that is demonstrated in the generator mode of the motor operation. There are mainly two different configurations of hybrid electric vehicle powertrains, *series* and *parallel configuration*, as shown in Figure 7.2.

In the *series configuration*, normally only one energy source provides the power for the propulsion. Here the internal combustion engine serves to charge the batteries. There is no any mechanical connection between the electric motor and the internal combustion engine. When the battery falls below a minimum required level, the internal combustion engine starts to charge the battery by driving an electric generator. When the battery is fully charged, the internal combustion engine

DOI: 10.1201/9781003134305-7

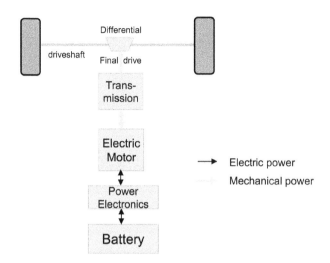

FIGURE 7.1   An EV powertrain configuration.

shuts off. In this configuration, the power required to propel the wheels is solely delivered by the electric motor, and the internal combustion engine runs at its most efficient region. Some series HEVs are designed that the motor draws electricity from both generator and battery when large amounts of power are required. The disadvantage of the series configuration is that the electric motor has to be sized for the maximum power requirement of the vehicle. Also, there are power losses during the conversion from fuel energy to electrical one.

In the *parallel configuration*, both the electric motor and the internal combustion engine are connected to the transmission and, thus, they drive the wheels according to preferred efficiency paths. The electric motor has lower power requirements since the ICE is able to complement for the total power demand. The drawback of this configuration is that the control complexity increases because the power flow needs to be regulated and blended from two parallel sources.

In both powertrain configurations, the electric motor has the same basic schematic for the energy conversion. For the vehicle propulsion, the motor converts electrical power from the battery to

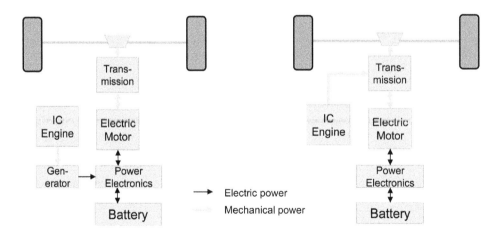

FIGURE 7.2   HEV powertrain configurations.

mechanical power that goes to the drive wheels. On the other hand, the motor is a reversible machine which can function as a *generator*, i.e., the power flows in reverse direction transforming mechanical energy to electrical one. This mode is referred to as *regeneration*. The regeneration takes place when the vehicle is braking, thereby electricity is generated and can be restored back to the battery. More electric vehicle powertrain architectures can be seen in reference [1].

7.2 ELECTRIC MOTOR

There is a variety of electric motor types used for electric vehicle and hybrid electric vehicle, they can be DC or AC type motors. AC motors have been increasingly used in electric vehicles, they are more rugged/robust and less expensive. For the sake of simplicity, a DC motor is chosen here to understand the basic principle of operation as well as the regeneration characteristics. Among the DC motors, there are different types such as shunt, series or separately excited. These motors all work on the same principles and are distinguished by the way in which the electromagnetic field and armature circuits are interconnected. In this section, a separately excited DC motor is taken as an example to view its characteristics.

EQUIVALENT CIRCUIT OF SEPARATELY EXCITED DC MOTOR

A DC motor consists of a stator with field windings and a rotor, to which armature winding is attached. Both parts operate through the interaction between the electric current through the armature and the magnetic flux yield by the stator to produce a rotational movement of the rotor and the torque. In the case of the separately excited motor, the field and armature windings are electrically separated from each other. The field winding is excited by a separate DC source. Figure 7.3 layouts the equivalent electrical circuit, where R_a and L_a represent the resistence and the inductance of the armature circuit. Denoting T_m as the total torque developed at the motor shaft, it is proportional to magnetic flux Φ and armature current i_a

$$T_m = K_a \Phi \, i_a \tag{7.1}$$

where K_a is constant for a given motor, it depends on the motor construction, winding and magnetic material properties. The field flux is a nonlinear function of field current i_f, that is

$$\Phi = f\left(i_f\right) \tag{7.2}$$

The nonlinearity is resulted from the DC motor magnetization characteristics exhibiting the relationship between the magnitude and intensity of the magnetic field. By varying the filed current i_f, the field flux can change accordingly.

The power developed at the mechanical output, i.e., the rotor, is given by the product of the torque and the rotational speed

$$P_m = T_m \, \omega_m \tag{7.3}$$

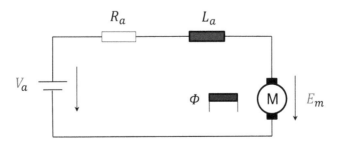

FIGURE 7.3   Equivalent electrical circuit of separately excited DC motor.

where ω_m is the rotor speed. Since the rotor turns, the armature conductors cut the magnetic flux and an *electromotive force* (EMF) is induced. The EMF, denoted as E_m, is proportional to the flux and motor speed ω_m

$$E_m = K_a \Phi \, \omega_m \tag{7.4}$$

In the motor operation, the EMF always opposes the applied voltage, and therefore it is also called *back EMF*. Applying the voltage equation (Kirchhoff's voltage law) to equivalent circuit Figure 7.3, we have

$$V_a = E_m + i_a R_a + L_a \frac{di_a}{dt} \tag{7.5}$$

A key part in the motor operation is the speed control. For this purpose, the transient dynamics and characteristics are essential for the control design, in order to have a quick response of the motor and a smooth transition of speed change. On the other hand, it is also important to know what torque is available and the variation in speed with the applied voltage, not the transient effects but rather in a steady operation. For this reason, the *steady-state characteristics* are considered. They are used to characterize motor performance as well as the capability of regeneration when the motor is in the generator mode.

In the steady state, the inductive effect of L_a is ignored. The voltage equation (7.5) becomes

$$V_a = i_a R_a + E_m \tag{7.6}$$

Using (7.4), the current is expressed as

$$i_a = \frac{V_a - K_a \Phi \, \omega_m}{R_a} \tag{7.7}$$

Hence, the total torque of the motor is given by

$$T_m = \frac{K_a \Phi}{R_a}(V_a - K_a \Phi \omega_m) \tag{7.8}$$

If the armature voltage and field flux are constant, the torque will be inversely proportional to the rotational speed. The torque–speed relationship is characterized by a straight line with a negative slope. Figure 7.4 graphically shows the torque–speed characteristic. The positive torque axis indicates the motoring while the negative axis represents the generating characteristic. At no-load speed, there is no torque at the rotor and the back EMF is equal to the applied voltage. The torque

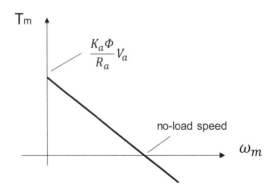

FIGURE 7.4 Torque-speed characteristic of the motor.

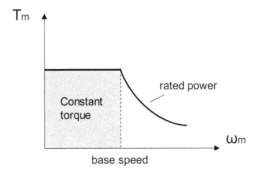

FIGURE 7.5 Torque-speed operation plane.

has the maximum at zero speed. With constant field flux Φ, the straight line can move up and down depending on an increase or a decrease of the armature voltage, V_a. So, the motor speed varies linearly with V_a.

Under steady driving conditions, the volt-drop on the resistors across the windings is very low, i.e., the most part of the applied voltage is across the circuit opposing to the EMF. At the starting condition, however, the EMF is at zero, hence the current could be extremely high if the full voltage is applied. For this reason, the current must be limited to a safe threshold known as a *rated current*. The motor can be overheated when operating above the *rated current*.

There are two control methods can be applied in order to vary the motor speed over a wide range: *armature voltage control* and *field flux control*. In armature voltage controls, the field flux is held constant, and the armature voltage is controlled to provide a desired motor speed. While the field flux is controlled, the voltage is held constant, and the motor speed changes proportionally in inverse to flux Φ. By applying these two controls, the torque-speed characteristic curve becomes an operation curve as shown in Figure 7.5. The motor can operate at the rated current up to a so-called *base speed*. The region between the zero and the base speed is referred as the "constant torque" area, where the torque has its maximum with the full flux. The armature voltage control is operated in this area until the base speed. According to (7.3), the mechanical power reaches the maximum at the based speed. Running faster than the base speed, the field flux needs to be reduced due to the inverse correlation as shown (7.8). The motor speed is controlled for a given current, the rated current, for example. In this case, the developed mechanical power remains the same (rated power) as at the base speed. This region is referred to as "field weakening". The maximum allowable speed in the field weakening area must be limited in order to avoid high back EMF, thus high current.

The torque-speed characteristic curve presented in Figure 7.5 is a desirable operation curve of an electric motor for electric vehicle applications. Indeed, while the start-and-stop driving, the motor is operated in the constant torque range. At higher speeds, the motor is operated in the constant power range.

REGENERATION BY GENERATOR OPERATION

From (7.5), we can see that the current flow can change depending on the difference the battery voltage and the back EMF

$$V_a - E_m = i_a R_a \tag{7.9}$$

When the battery voltage is higher than the EMF, the electric machine is motoring and running in the forward direction, i.e., the rotational speed of the rotor is in positive direction. When the battery voltage is lower than the back EMF, the machine is generating and running backwards. In this generator operation mode, the back EMF can be used to recharge the battery thereby recovering

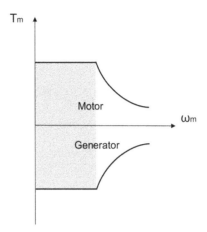

FIGURE 7.6   Motor-generator operation plane.

energy and adding to power efficiency. In order to make the machine to be capable of alternating the motor and the generator operations, the applied voltage is simply to be controlled to the extent, to which it is higher or lower than the back EMF. Figure 7.6 summarizes the motor and the generator operations into the torque-speed plane.

When the machine operates as a motor, the battery voltage is higher than the back EMF. The electric power drawn from the battery is positive and will be converted and supplied to the wheels. In generator operation, in which the battery voltage is lower than the back EMF, the current will reverse direction and hence the torque becomes negative. Now the power is supplied from the machine, so that the machine operates as a generator.

As can be concluded, the kinetic energy from the motor returns back to the power source during the generation. If we can do the process every time the vehicle is in deceleration, this will be an event of *regenerative braking* of the vehicle. It is important to note that the regeneration develops the counter torque, which acts similar to a brake mechanism because the current flows in reverse direction. In order to manage the different operations and to ensure the electric machine is in the desired mode when needed, an electronic control unit is required – the *power electronics unit.*

7.3   POWER ELECTRONICS UNIT

As we learned from the previous section, depending on a required operation mode, the electric machine can be a motor or a generator. The power electronics unit in conjunction with the electric machine is to provide forward and reverse motion with rapid regenerative braking in both directions. The power electronics unit primarily includes an electronic power converter and a drive controller, as shown in Figure 7.7. The power electronic converter consists of semiconductor devices and processes the power flow from the battery source to the motor input terminals. Those semiconductor devices can function as on-off electronic switches to transfer the applied voltage into variable voltage and frequency.

The drive controller includes a microcontroller and some signal electronics, which handle the sensor signal inputs as indicated by the dashed lines in Figure 7.7. The sensor signals can be voltage, current, motor rotor position, speed and torque as measured or calculated state variables. From the battery, the measured data can be the voltage or the current, which is used for the SoC (state of charge) calculation. The controller sets the charging profile based on the level of SoC. The power electronics provides the status of the switches and the current of the armature circuit. In order to control the motor torque, the motor rotor position or its speed is to feedback to the controller. The function of the microcontroller is to process the user commands and various sensor feedback

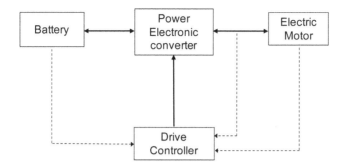

FIGURE 7.7   Power electronics conjunct with motor.

signals, and to follow a motor control algorithm. The user command can come from other electronic control units such as a brake control unit. In this way, the brake control unit is able to communicate the demand for its own control purpose during the regeneration process. Then the drive control algorithm determines the control input to the motor and generates the gate switching signals for the power semiconductor switches.

According to the operation plane in Figure 7.6, the motor operation implies the electric power drawn from battery. In regenerative braking, the electric power flow in the motor is reversed, and the torque is opposite to the direction of motion as well. The power electronic converter is designed to be able to handle this bi-directional power flow. A DC chopper is used as a typical DC to DC power electronic converter and is commonly used for DC electric motor drives. As an example, the four-quadrant chopper is chosen below to describe the regeneration operation process in the power electronics.

As shown in Figure 7.8, the four-quadrant chopper is often used to control motor in both forwards and backwards direction of rotation, corresponding to the motor-generator operation plane in Figure 7.6. In the basic circuit of the four-quadrant chopper in Figure 7.8, T_1, T_2, T_3 and T_4 symbolize the switches. For example, it can be thyristors, metal oxide semiconductor field effect transistors (MOSFET) or insulated gate bipolar transistors (IGBT). The unidirectional diode conducts current only in on direction and blocks voltage in the negative direction. The indices correspond to the quadrant respectively.

Figure 7.9 shows the function configuration of the chopper corresponding to the motor and the generator operation, respectively. In the motor operation, turning on T_1 and T_4 allows the current and the power to flow from the battery to the motor, which rotates in a forward direction, thus both voltage V_a and current i_a are positive. T_2 and T_3 remain continuously off, so the motor is in the acceleration mode. During the braking operation, the energy is supposed to be recovered through the

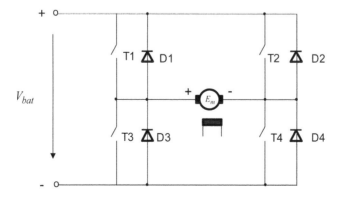

FIGURE 7.8   DC motor supplied by a four-quadrant chopper.

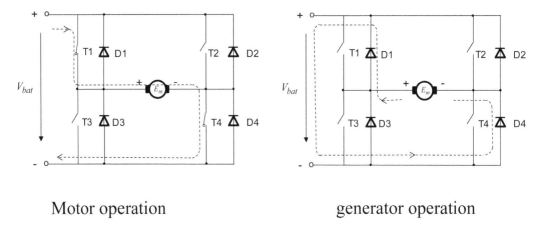

Motor operation   generator operation

FIGURE 7.9   Chopper function in motor operation and regeneration.

regeneration, corresponding to the operation as a generator. When the brake pedal is depressed, T_1 is kept off and T_3 turns on, which forces current i_a to become negative and to flow back through D_4. The motor acts as a generator producing an induced voltage across its terminals. At this moment, however, energy is only being dissipated in the switch and contact resistances. The induced voltage, E_m, will rise until it exceeds the cut-in voltage of diodes D_1 and D_4 above battery voltage V_{bat}, then D_1 and D_4 become forward biased, and start to conduct. At the same time, T_3 is switched off, the current is flowing into the battery. During this charging period, a counter torque, which brakes the motor, will be experienced as the charging current increases, that is so-called the *regeneration torque*.

In Figure 7.10, the diagram shows the steady status of the switches, diodes and the current curve during the regeneration phase. The battery is charging when D_1 conducts. Due to the inductance of the motor, the armature current is associated with the ripple so that the regeneration torque will

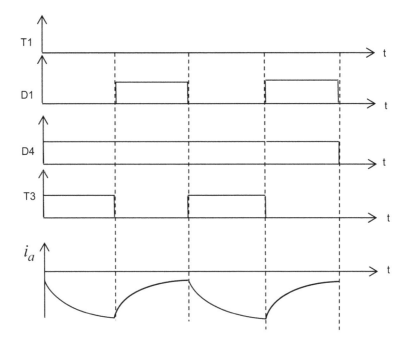

FIGURE 7.10   Chopper's signal diagram during regeneration.

have a ripple in the end. This effect certainly is undesirable in vehicles. In order to make the current smooth, one way is to pulse the switch to provide a low resistance path to the inductive current of the motor. High frequency switching can help smoothing the current into an almost DC current. For this reason, T_1 and T_3 are often controlled by PWM (pulse width modulation). T_4 can be left on if the current is high since the diode may lead to overheating the switch, such as MOSFET. In overall the diagram in Figure 7.10 basically concludes that the regeneration is achieved by changing the way the power switches are turned on and off.

7.4 REGENERATION TORQUE

The regeneration torque is developed while the electric machine acts as a generator. The created torque is opposite to the direction of rotor motion, and its amount relates to the rotational speed of the motor. In the generator operation mode, the regenerative torque is produced according to the profile as shown in the torque-speed plane in Figure 7.6. However, it is not desired to have the regeneration torque every time whenever possible. Indeed, to guarantee driver's comfort and to provide an acceptable vehicle drivability, a limit on the torque command in the regeneration is necessary. At very low vehicle speeds, the regenerative braking can also affect the driver's comfort, and thus is not desired. Therefore, the brake control unit sends a torque request command, that is a user command, to the power electronic unit controlling the regeneration. A study [2] showed the regeneration torque characteristics, which is representative for electric vehicle in braking. In general, the regeneration torque profile can be idealized as shown in Figure 7.11.

The horizontal axis maps the vehicle linear speed, which is proportional to the motor rotational speed. Around zero speed, there is no need in actuating the brakes, thus the regeneration is disabled. As the vehicle speed gets higher, the regeneration is gradually increased. A piecewise linear increase can smooth the vehicle jerk to improve the driver comfort while the torque command changes in the regeneration. The maximum allowed regeneration torque is normally limited by the capacity of the power electronics. However, the regeneration torque also has to be a function of the battery power state of charge (SoC). If the SoC already reaches a certain high level, the maximum allowed braking torque should be reduced. In the high vehicle speed area, where the "field weakening" of the generator takes place, the torque is limited by the power of the electric machine.

The purpose of the regeneration through the electric machine is to recover kinetic energy, which is normally dissipated as heat during vehicle braking. Consequently, the regeneration should be performed as much as possible. Based on the profile in Figure 7.11, the regeneration creates the torque, which slowing down the motor additionally. Now, an interesting problem arises: How should a vehicle deal with such additional braking torque? Without controlling the regeneration, the vehicle can be over braked by the regeneration. If the regeneration is technically managed and used for the vehicle braking, in many situations the amount of the torque may not be enough to cover driver

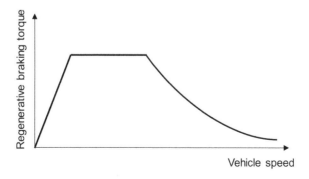

FIGURE 7.11 A typical torque-speed profile of regeneration.

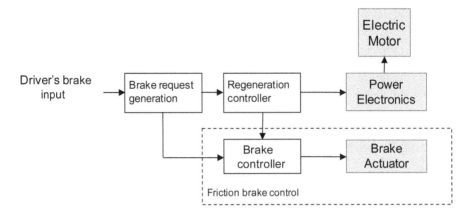

FIGURE 7.12   Function diagram of regeneration torque blending.

intended brake request. Therefore, it is beneficial to combine or *blend* both torques (i.e., the regeneration torque and the friction braking torque) to meet the braking torque request that comes from the driver. The braking torque *blending* technology has been developed, which involves electronic and mechanical components in co-action. The function diagram is displayed in Figure 7.12.

Electric or hybrid vehicles normally still use hydraulic pressure to apply a friction force to the brake rotors – the *friction brakes*. The driver's brake intent is sensed by a brake pedal travel sensor. Based on the sensor signal, the braking torque requested will be generated, which can be done by a pedal force simulator or a booster with processed data. The amount of deceleration is then proportional to how much the driver pushes the pedal. With the use of regeneration, the goal is to maintain the mentioned relationship between the vehicle deceleration and the brake pedal apply, which requires blending of the hydraulic friction braking and the regenerative braking.

The regeneration controller in Figure 7.12 calculates the amount of the braking torque provided by the regeneration based on the voltage feedback from the generator and the motor speed. The friction brake control system subtracts the regeneration torque from the total braking torque request. The leftover amount of the torque is then produced by the friction brakes. Of course, if the driver's request is lower than the regeneration can create, then the regeneration controller will request to cut the torque demand so that the regeneration torque meets the driver's request. Since the amount of pressure on the friction brakes is no longer dependent of the driver's demand on the pedal, the brake controller in Figure 7.12 must have a pressure modulation to ensure an accurate pressure and a torque for the friction brakes. Figure 7.13 shows two scenarios of the torque blending. In both scenarios, the driver applies the brake at a high speed of the vehicle until it is stopped and then the driver holds the brake at standstill. In Figure 7.13a, the braking torque request is higher than the regeneration capacity, hence the friction brakes will have to fill the gap. Figure 7.13b shows a case,

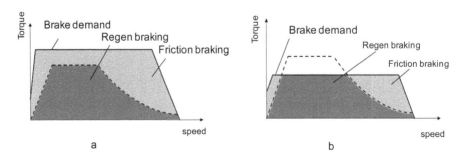

FIGURE 7.13   Regeneration capacity (a) lower (b) higher than the brake demand.

in which at times the torque demand is lower than the maximum the regeneration can deliver, and the regeneration alone is controlled to meet the brake demand.

7.5 VEHICLE ENERGY BALANCE IN BRAKING

Now we take a look at how kinetic energy in a vehicle is balanced during the braking. As seen from (3.35), when decelerating and the brakes at the front and rear wheels are applied, the brake torques are not the only source that contributes to the deceleration. The aerodynamic drag, the road grade and the rolling resistance make some considerable contribution. Equation (3.35) is written down here for the braking as follows:

$$T_{bf} + T_{br} + \left(F_{Ad} + Mg\sin\gamma_z \right) r_w + \left(F_{Rf} + F_{Rr} \right) r_w = -Mar_w - \frac{J_{wf}a + J_{wr}a}{r_w} \tag{7.10}$$

For the simplification in the description, we combine the front and the rear (index f and r). The braking power is obtained by the braking force multiplying the vehicle actual speed, i.e.,

$$P_b = \frac{T_b}{r_w} v$$

$$= -Mav - \frac{J_w a v}{r_w^2} - F_{Ad}v - Mg\sin\gamma_z v - F_R v \tag{7.11}$$

The braking energy, i.e., the energy that should be developed by the brake mechanisms to decelerate the vehicle, will be the integration over the time period of the braking process. On level ground, if the vehicle with a speed of v_0 starts to brake over time t_b to a stop, the brake energy results in

$$E_b = \int_0^{t_b} P_b dt = \frac{1}{2} M v_0^2 + \frac{1}{2} J_w \frac{v_0^2}{r_w^2} - \int_0^{t_b} \left(F_{Ad}v + F_R v \right) dt \tag{7.12}$$

where (7.11) is applied here. The vehicle energy in translational motion (the first term in the above equation) is the greatest component at the beginning of the deceleration because of the mass and the speed. The second term of the above equation is the rotational energy, which is normally small and ignorable because moment of inertia J_w is much smaller comparing to the amount of the vehicle mass for passenger cars. The term $\frac{v_0}{r_w}$ represents actually the rotational speed of the tire. These two energy components will be absorbed by the brakes and dissipated in heat through convection. The last negative term in (7.12) is the energy component that is consumed in aerodynamic drag and rolling resistance, and cannot be recaptured. If the vehicle is braking on an uphill slope, kinetic energy will be transferred to the potential energy. E_b is basically the energy, which can be recovered by the regeneration.

Example 7.1

A vehicle with a gross mass of 2350 kg is driven down a $4°$ incline at a speed of $v_0 = 60$ km/h when the brakes are applied, causing a deceleration of $a = \frac{4m}{s^2}$. Assume that the aerodynamic drag can be neglected, $f_R = 0.015$, $J_w = \frac{3kg}{m^2}$ and $r_w = 0.37$ m. How much energy will be absorbed by the brakes before the vehicle comes to a complete stop? Calculate the travel of the vehicle to the full stop. Solve the problem under the assumption made in the above-discussed material.

Using (7.11), the brake power is calculated as

$$P_b = 2350 * 4v + 21.9 * 4v - 1606.5v - 345.5v \qquad (7.13)$$

The brake power varies with the vehicle speed and keeps declining while the speed decreases. Since the deceleration is considered constant, the braking time period $t_b = \dfrac{v_0}{a} = 4.16$ s, and the distance s_b traveled is

$$s_b = a \int_0^{t_b} t \, \mathrm{d}t = 34.7 \text{ m}$$

Thus, the energy absorbed by brakes will be

$$
\begin{aligned}
E_b \quad &= \int_0^{t_b} \left(2350 * 4v + 21.9 * 4v - 1606.5v - 345.5v \right) \mathrm{d}t \\
&= 261 \text{ kJ}
\end{aligned}
\qquad (7.14)
$$

One can see that the rotation energy in the second term on the right side of the above equation is considerably smaller than kinetic energy given by the first term.

❑

In this chapter, the theoretical principles of regenerative braking were highlighted. It was understood that regenerative braking contributes to the vehicle braking. Indeed, the regenerative braking is a braking source additional to the friction brake as well as an instrument for energy saving. Since the brake control is an essential part in control applications of vehicle dynamics discussed later in this book, the regenerative braking can be introduced as a component of the brake control design.

REFERENCES

1. Iqbal Husain, Eletric and HYbrid Vehicles: Design Fundamentals, 2nd ed., Boca, London, New York: CRS Press, Taylor & Francis Group, 2011.
2. Y. Gao, L. Chen and M. Ehsani, Investigation of the Effectiveness of Regenerative Braking for EV and HEV, SAE International Conference, Costa Mesa, California, 1999.

8 Vehicle Lateral Dynamics

So far the study of vehicle dynamics has focused on the longitudinal direction of vehicle motion. In this chapter we consider equations in the lateral direction. Lateral dynamics of a vehicle refers to its response to turning and directional change inputs. The lateral motion is influenced by a driver commands and environmental inputs, such as wind and road disturbances, including other cars and pedestrians. The major purposes of modeling lateral dynamics are to describe

- Vehicle lateral and yaw motions.
- Vehicle transient behavior following a steering input or a notional small disturbance.
- Formulation of vehicle models suitable for control design.

There are two technical issues related to lateral dynamics that we address in this chapter. The first one is the cornering geometry of a vehicle with steered front wheels. As long as the wheels operate under normal conditions, i.e., the wheels are not locked up, the turning of the steered wheels will produce tire side forces that affect the lateral behavior of the vehicle and its rotational movement about the vertical axis. The thereby created steering geometry associated with vehicle kinematic parameters determines the lateral acceleration and the yaw rate of the vehicle. In the second problem under consideration, the state variables of the lateral, the yaw and longitudinal velocity are given, and we wish to find required steering inputs as well as the tire forces. This formulation, which is in fact an inverse dynamics formulation, is useful for solving problems of controlling vehicle lateral dynamics. In the following we will layout the steering geometry and the determination of the kinematic parameters first and then examine vehicle models with regard to the operation conditions, in which the model assumptions and disturbances are made. At the end, the lateral dynamics model will be extended to dynamics of the vehicle towing a trailer.

8.1 STEERING GEOMETRY

Steering is usually affected by changing the heading of the front wheels through the steering system. To improve steering performance in relation to the vehicle stability, the rear wheels can also be steered. However, a rear-wheel steering alone is not applicable since it may cause instability of the vehicle. Also, in some situations the rear-wheel steering alone is unconventional and impractical. Basically, there are two kinds of steering systems: front-wheel steering (FWS) and four-wheel steering (4WS), where the rear-wheel steering of limited angular range enhances steering performance of vehicles. Here, the steering geometry is introduced first while the impacts on the vehicle dynamic behavior are described in Chapter 9.

FRONT-WHEEL STEERING

The steering geometry from a FWS is illustrated in Figure 8.1. The vehicle is in a left turn, where the front wheels have inside turn angle δ_{LF} (left front wheel) and outside turn angle δ_{RF} (right front wheel). In order to understand the relation of the steering angle parameters, we consider the steady-state response to steering input of the vehicle at low speeds, thereby negating the centrifugal force. At very low speeds with negligible tire slip angles (i.e., considering the wheels rigid), the vehicle can only corner precisely when the wheels follow the circular paths with different turn radii originating from a common instantaneous point, called the *turn center*. Since in the front-wheel steering

DOI: 10.1201/9781003134305-8

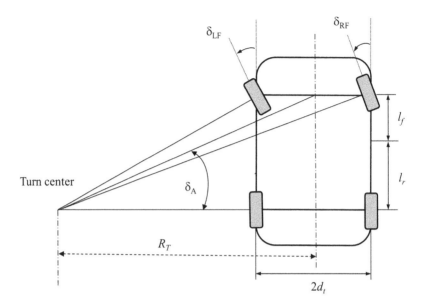

FIGURE 8.1 Front-wheel steering geometry.

configuration the rear wheels are not steered, the turn center is the intersection of the rear axle axis and the front wheel axes. R_T is the theoretical turn radius of the vehicle (as it will be discussed later, the vehicle does not turn on this radius, but on an actual turn radius). This steering configuration requires to steer the inside wheel to a greater angle than the outside wheel, and establish a proper relationship between both angles. From the analysis of the triangles, it can be readily shown that the inside angle and outside angle satisfy the relationship:

$$cot\ \delta_{RF} = cot\delta_{LF} + \frac{2d_t}{l_f + l_r} \tag{8.1}$$

The steering geometry satisfies (8.1) for front-wheel-steering is referred to as *Ackermann steering geometry*. It basically expresses that for turning the vehicle around a center the front wheels angles need to be turned so, that their axes of rotation intersect on the axis of rotation of the rear wheels. The steered wheel angle satisfies this condition has come to be known as Ackermann angle.

Using the geometry in Figure 8.1, the front steer angles have the following relations [1]:

$$\delta_{LF} = arctan\left\{ \frac{l_f + l_r}{\dfrac{l_f + l_r}{tan\delta_A} - d_t} \right\}$$

$$\delta_{RF} = arctan\left\{ \frac{l_f + l_r}{\dfrac{l_f + l_r}{tan\delta_A} + d_t} \right\} \tag{8.2}$$

For small angles, as are typical of turning, the turn radius is far larger than the wheel track $2d_t$. So, we can justify the approximation:

$$\delta_A = \frac{l_f + l_r}{R_T} = \delta_{LF} = \delta_{RF} \tag{8.3}$$

δ_A is the so-called *Ackerman angle*.

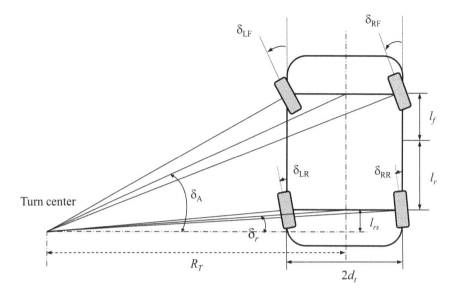

FIGURE 8.2 In-phase rear-wheel steering.

At high speeds, the vehicle is in the driving operational mode, and the effect of the centrifugal force acting on the center of gravity cannot longer be neglected. The determination of the vehicle kinematic parameters will be discussed as a general case considering lateral forces in the next section.

FOUR-WHEEL STEERING

In a 4WS system, the rear wheels will be turned additionally to the turning of the front wheels. The basic idea is to turn the rear wheels in the same direction shown in Figure 8.2. In this case, the vehicle is able to make a wide turn, thereby increasing the turn radius. As we will discuss later, this steering can improve stability of the vehicle in high-speed turning.

With respect to this four-wheel-steering configuration, we consider the Ackermann angle once again. In the sense of Ackermann geometry, the axes of all four wheels intersect in the turn center. Since the rear steer angles are usually very small, we assume that they are the same, denoted as δ_r Similar to (8.1), the front steer angles are given by

$$cot\ \delta_{LF} = cot\delta_{RF} + \frac{2d_t}{l_f + l_r + l_{rs}} \tag{8.4}$$

Similar to (8.2), the rear steer angles are determined by

$$\delta_{LR} = arctan\left\{ \frac{l_{rs}}{\dfrac{l_{rs}}{tan\delta_r} - d_t} \right\}$$

$$\delta_{RR} = arctan\left\{ \frac{l_{rs}}{\dfrac{l_{rs}}{tan\delta_r} + d_t} \right\} \tag{8.5}$$

Taking the same approximation as in (8.3), we obtain

$$\delta_A = \frac{l_f + l_r + l_{rs}}{R_T}$$

which is

$$\delta_A = \frac{l_f + l_r}{R_T} + \delta_r \tag{8.6}$$

Another option for the rear-wheel steering is to steer the rear wheels out-of-phase with the front wheels to reduce the turn radius. Thus, as shown in Figure 8.3 the steering configuration improves maneuverability by providing tighter turns, especially at low speed. The turn radius with the rear-wheel steer is determined using the turn radius of the front wheels because the rear steer is limited in a small range.

Applying the same approach as for (8.4), it leads to the same result as (8.6) in which the rear steer angle is negative, $\delta_r < 0$.

To conclude, the rear-wheel steering may change the Ackermann angle presented by (8.6). The Ackermann angle increases when the rear wheels are steered in-phase and decreases for the out-of-phase rear-wheel-steering, which are characterized by positive or negative sign of the rear steer angle respectively. For the safety reasons, the rear steer angle cannot be arbitrary and should be operated in a small range. More details on the four-wheel steering including its influence on the vehicle dynamic behavior will be discussed in Chapter 9.

8.2 KINEMATIC PARAMETERS

The steering geometry has been introduced; the kinematic parameters associated with steering response of the vehicle will be determined in this section. For a general case, the 4WS in high-speed turning is taken into account.

Here we still consider the steady-state response. The steering geometry at high-speed is more complex than at low speed. When the vehicle is negotiating a turn at high-speed, the effect of the centrifugal force will be considered. To balance the centrifugal force, the tires must develop

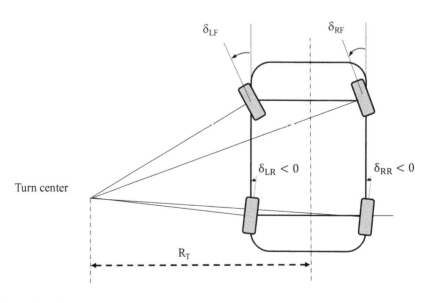

FIGURE 8.3 Out-of-phase rear steering.

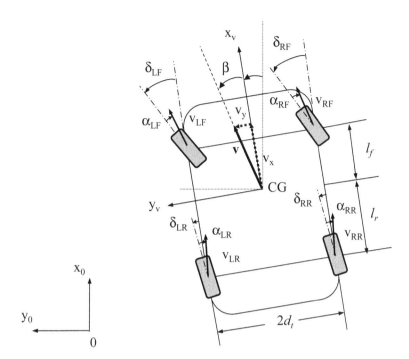

FIGURE 8.4   Vehicle kinematic parameters in turn.

the appropriate lateral forces, thus the wheel slip angles. The kinematic features are presented in Figure 8.4 which shows the vehicle in a left turn. The following notation is used:

a vehicle acceleration vector
v vehicle velocity vector
β vehicle sideslip angle
ψ vehicle yaw angle
α_{LF} slip angle of the left front wheel
α_{RF} slip angle of the right front wheel
α_{LR} slip angle of the left rear wheel
α_{RR} slip angle of the right rear wheel
δ_{LF} steer angle of the left front wheel
δ_{RF} steer angle of the right front wheel
δ_{LR} steer angle of the left rear wheel
δ_{RR} steer angle of the right rear wheel
d_t half wheel track
l_f distance from CG to the front axle
l_r distance from CG to the rear axle.

Referring to the above graph, The vehicle is moving with velocity $\mathbf{v} = \left(v_x, v_y \right)^T$ in the earth-fixed inertial coordinate system 0_{xyz} while simultaneously rotating about z-axis with an angular velocity $\dot{\psi}$, which is also called vehicle *yaw rate*. The yaw angle ψ is angle between the vehicle x_v-axis of the vehicle-fixed coordinate system and the x_0-axis of the inertial coordinate system 0_{xyz}. The x_v-axis is aligned with the vehicle heading direction. The *sideslip* angle, β, is the angle between the vehicle x_v-axis of the vehicle-fixed coordinate system and the vehicle CG velocity vector. It is defined as positive in this case because the velocity vector is oriented counterclockwise from the x_v-axis following the right-hand rule of the coordinate system.

The wheel slip angle was introduced in Chapter 3 and is known as the angle between the direction of the wheel heading and the direction of travel. On the other hand, the vehicle velocity can be directed into x and y components through sideslip angle β:

$$v_x = v\cos\beta \tag{8.7}$$
$$v_y = v\sin\beta$$

From (2.46) and using the vehicle diagram in Figure 8.4, we derive the following relationship between the steer and slip angles of the left front wheel and the vehicle sideslip angle, the vehicle velocity and the yaw rate:

$$tan(\delta_{LF} - \alpha_{LF}) = \frac{v\sin\beta + l_f\dot{\psi}}{v\cos\beta - d_t\dot{\psi}} \tag{8.8}$$

Then the wheel slip angle can be computed as

$$\alpha_{LF} = \delta_{LF} - arctan\left(\frac{v\sin\beta + l_f\dot{\psi}}{v\cos\beta - d_t\dot{\psi}}\right) \tag{8.9}$$

Using (2.48) through (2.50), we compute each of other three wheel slip angles in the same way:

$$\alpha_{RF} = \delta_{RF} - arctan\left(\frac{v\sin\beta + l_f\dot{\psi}}{v\cos\beta + d_t\dot{\psi}}\right) \tag{8.10}$$

$$\alpha_{LR} = \delta_{LR} - arctan\left(\frac{v\sin\beta - l_r\dot{\psi}}{v\cos\beta - d_t\dot{\psi}}\right) \tag{8.11}$$

$$\alpha_{RR} = \delta_{RR} - arctan\left(\frac{v\sin\beta - l_r\dot{\psi}}{v\cos\beta + d_t\dot{\psi}}\right) \tag{8.12}$$

For simplifying the steering effect in the range of small angles according to (8.3), the left wheel turn angle can be seen the same as the right one so that we can set

$$\delta_f = \delta_A = \delta_{LF} = \delta_{RF} \tag{8.13}$$
$$\delta_r = \delta_{LR} = \delta_{RR}$$

The conditions above are widely valid in the practice. The difference between the inside and outside turn angles is usually adjusted smaller than the Ackermann-steering to allow more tire side forces. The vehicle measurements show that the difference remains below 10% [2].

If the angles are small, we can approximately use Figure 8.1 to compute some vehicle kinematic parameters. The turn center can be interpreted as an instantaneous zero-speed point. The velocity of the turn center, referred to the vehicle velocity, is then equal to zero. So, using (2.44), we get

$$\begin{bmatrix} v\cos\beta - R_T\dot{\psi} \\ v\sin\beta - l_r\dot{\psi} \end{bmatrix} = \begin{bmatrix} 0 \\ 0 \end{bmatrix} \tag{8.14}$$

and then solve the equation for v:

$$v = R_T \, \dot{\psi} \tag{8.15}$$

as $\cos \beta \approx 1$.

8.3   NONLINEAR TWO-TRACK MODEL

The nonlinear two-track model considered in this section describes vehicle dynamics in a horizontal two-dimensional space, i.e., a planer motion in the road plane. The model possesses three degrees of freedom, which can be expressed in three state variables: the vehicle speed, the vehicle sideslip angle and the yaw rate. These state variables will be specified in order to derive equations of dynamics. For the modeling, we make the following assumptions:

- The roll and pitch movements are neglected. The vehicle's center of gravity is on the ground level.
- The lateral tire force and the slip angle are linear over the cornering stiffness.
- No compliance effect in the steering mechanism and the vehicle kinematics.
- No aerodynamics effects.

Figure 8.5 shows significant forces and kinematic parameters while the vehicle makes a left turn. In addition to parameters shown in Figure 8.4, the forces on the tires are defined:

F_{x_LF} longitudinal tire force of the left front wheel
F_{y_LF} lateral tire force of the left front wheel
F_{x_RF} longitudinal tire force of the right front wheel
F_{y_RF} lateral tire force of the right front wheel
F_{x_LR} longitudinal tire force of the left rear wheel
F_{y_LR} lateral tire force of the left rear wheel
F_{x_RR} longitudinal tire force of the right rear wheel
F_{y_RR} lateral tire force of the right rear wheel

Referring to Figure 8.5, we take the vehicle acceleration calculated in (2.52) and apply Newton's equation (2.64) to the forces balance on the vehicle:

$$M \begin{bmatrix} \dot{v} \cos \beta - v \sin \beta \left(\dot{\beta} + \dot{\psi} \right) \\ \dot{v} \sin \beta + v \cos \beta \left(\dot{\beta} + \dot{\psi} \right) \\ 0 \end{bmatrix} = \begin{bmatrix} F_{x\_veh} \\ F_{y\_veh} \\ 0 \end{bmatrix} \tag{8.16}$$

where F_{x_veh}, F_{y_veh} denote the total forces in x and y directions respectively, and they are determined by

$$
\begin{aligned}
F_{x\_veh} &= F_{x\_LF} \cos \delta_{LF} - F_{y\_LF} \sin \delta_{LF} + F_{x\_RF} \cos \delta_{RF} - F_{y\_RF} \sin \delta_{RF} + \\
&\quad F_{x\_LR} \cos \delta_{LR} - F_{y\_LR} \sin \delta_{LR} + F_{x\_RR} \cos \delta_{RR} - F_{y\_RR} \sin \delta_{RR} \\
F_{y\_veh} &= F_{y\_LF} \cos \delta_{LF} - F_{x\_LF} \sin \delta_{LF} + F_{y\_RF} \cos \delta_{RF} + F_{x\_RF} \sin \delta_{RF} + \\
&\quad F_{y\_LR} \cos \delta_{LR} + F_{x\_LR} \sin \delta_{LR} + F_{y\_RR} \cos \delta_{RR} + F_{x\_RR} \sin \delta_{RR}
\end{aligned} \tag{8.17}
$$

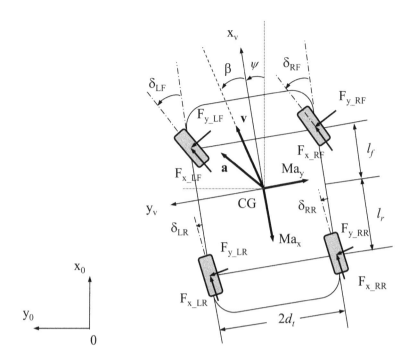

FIGURE 8.5 Free-body vehicle diagram.

To express $\dot{v}$ and $\dot{\beta}$ in the regular differential equation form, we solve (8.16) to obtain

$$\dot{v} = \frac{F_{x\_veh}}{M} \cos \beta + \frac{F_{y\_veh}}{M} \sin \beta \qquad (8.18)$$

$$\dot{\beta} = \frac{F_{y\_veh}}{Mv} \cos \beta - \frac{F_{x\_veh}}{Mv} \sin \beta - \dot{\psi} \qquad (8.19)$$

In addition, the rotational yaw movement about the z-axis is determined through Euler's equation (2.70), that is

$$
\begin{aligned}
I_z \ddot{\psi} =\ & \left(F_{y\_LF} \cos \delta_{LF} + F_{x\_LF} \sin \delta_{LF} + F_{y\_RF} \cos \delta_{RF} + F_{x\_RF} \sin \delta_{RF} \right) l_f - \\
& \left(F_{y\_LR} \cos \delta_{LR} + F_{x\_LR} \sin \delta_{LR} + F_{y\_RR} \cos \delta_{RR} + F_{x\_RR} \sin \delta_{RR} \right) l_r + \\
& \left(F_{x\_RF} \cos \delta_{RF} - F_{y\_RF} \sin \delta_{RF} + F_{x\_RR} \cos \delta_{RR} - F_{y\_RR} \sin \delta_{RR} \right) d_t - \\
& \left(F_{x\_LR} \cos \delta_{LR} - F_{y\_LR} \sin \delta_{LR} + F_{x\_LF} \cos \delta_{LF} - F_{y\_LF} \sin \delta_{LF} \right) d_t
\end{aligned}
\qquad (8.20)
$$

I_z is the rotational inertia of the vehicle about the z-axis. Here we assumed that the lateral tire force acts in the center of the contact patch. The pneumatic trail (see Figure 3.3), usually in a range from 4 to 7 cm, is neglected since it is far smaller than the dimension parameters l_f and l_r. Furthermore, from Chapter 3 we have learned that the lateral force acting on the tire is approximately proportional to the corresponding slip angle over the stiffness as given by (3.17). Applying this relationship to all the four wheels, we have

$$
\begin{aligned}
F_{y\_LF} &= C_{\alpha\_LF}\ \alpha_{LF} \\
F_{y\_RF} &= C_{\alpha\_RF}\ \alpha_{RF} \\
F_{y\_LR} &= C_{\alpha\_LR}\ \alpha_{LR} \\
F_{y\_RR} &= C_{\alpha\_RR}\ \alpha_{RR}
\end{aligned}
\qquad (8.21)
$$

Now the equations derived above will be rewritten in a technically convenient form, in which the dynamic as well as kinematic parameters are readily expressed. Later on, they will be considered from a point of view of control design. We substitute (8.13) and (8.21) into (8.18) through (8.20) and use the trigonometric relations to obtain

$$\dot{v} = \frac{1}{M}(C_{\alpha\_LF}\,\alpha_{LF} + C_{\alpha\_RF}\alpha_{RF})\,sin\,(\beta - \delta_f) + \frac{1}{M}(F_{x\_LF} + F_{x\_RF})\,cos\,(\delta_f - \beta) +$$

$$\frac{1}{M}(C_{\alpha\_LR}\alpha_{LR} + C_{\alpha\_RR}\alpha_{RR})\,sin\,(\beta - \delta_r) + \frac{1}{M}(F_{x\_LR} + F_{x\_RR})\,cos\,(\delta_r - \beta) \qquad (8.22)$$

$$\dot{\beta} = \frac{1}{Mv}(C_{\alpha\_LF}\alpha_{LF} + C_{\alpha\_RF}\alpha_{RF})\,cos\,(\delta_f - \beta) + \frac{1}{Mv}(F_{x\_LF} + F_{x\_RF})\,sin\,(\delta_f - \beta) +$$

$$\frac{1}{Mv}(C_{\alpha\_LR}\alpha_{LR} + C_{\alpha\_RR}\alpha_{RR})\,cos\,(\delta_r - \beta) + \frac{1}{Mv}(F_{x\_LR} + F_{x\_RR})\,sin\,(\delta_r - \beta) - \dot{\psi} \qquad (8.23)$$

$$I_z\ddot{\psi} = C_{\alpha\_LF}\,\alpha_{LF}(l_f\,cos\,\delta_f + d_t\,sin\,\delta_f) + C_{\alpha\_RF}\,\alpha_{RF}(l_f\,cos\,\delta_f - d_t\,sin\,\delta_f)$$

$$- C_{\alpha\_LR}\,\alpha_{LR}(l_r\,cos\,\delta_r - d_t\,sin\,\delta_r) - C_{\alpha\_RR}\,\alpha_{RR}(l_r\,cos\,\delta_r + d_t\,sin\,\delta_r)$$

$$+ F_{x\_LF}\,(l_f\,sin\,\delta_f - d_t\,cos\,\delta_f) + F_{x\_RF}\,(l_f\,sin\,\delta_f + d_t\,cos\,\delta_f) \qquad (8.24)$$

$$- F_{x\_LR}\,(l_r\,sin\,\delta_r + d_t\,cos\,\delta_r) + F_{x\_RR}\,(l_r\,sin\,\delta_r - d_t\,cos\,\delta_r)$$

Equations (8.22), (8.23) and (8.24) describe the nonlinear two-track model. In the model, the longitudinal tire forces F_{x_LF}, F_{x_RF}, F_{x_LR}, F_{x_RR} and steer angles δ_f, δ_r are interpreted as system inputs, and v, β, $\dot{\psi}$ correspond to the state variables. Equations (8.22) or (8.18) shows the dynamics along the vehicle longitudinal direction. In the case that the vehicle is moving straight and sideslip angle is really small, that equation presents then the standard vehicle longitudinal dynamics, which has already been detailed in Chapter 3. In the following, we focus on (8.23) and (8.24), which is the basis to describe vehicle lateral dynamics.

8.4 SINGLE-TRACK MODEL

For the control design purpose, it is convenient to use a simple model to reduce the complexity in the system analysis and the control implementation. The aim is now, to simplify the above-discussed model without sacrificing too much significant dynamic effects. From the preceding section we know that the two-track model is a nonlinear time-varying system. In nonlinear theory there is no general approach to reduce nonlinearities. The solution is usually dependent on what motion variables are those to be essentially considered. For the investigation of lateral dynamics, the left and right wheels will be combined to one, and then a linearization approach is used to derive a linear model which is an approximation of the nonlinear equations in the neighborhood of an operating point. Exactly speaking, based on the nonlinear two-track model we need to assume:

- Each axle can be represented by one wheel.
- The wheel steer angles, the tire slip angles and the vehicle sideslip angle are sufficiently small.

Combing the two wheels of the axle to one wheel, as shown in Figure 8.6, we define an average slip angle for each axle

$$\alpha_f = \alpha_{LF} = \alpha_{RF}$$
$$\alpha_r = \alpha_{LR} = \alpha_{RR} \qquad (8.25)$$

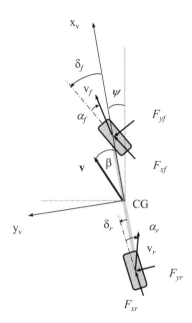

FIGURE 8.6   Single-track model.

and a doubled cornering stiffness for each wheel

$$C_{\alpha f} = \left(C_{\alpha\_LF} + C_{\alpha\_RF} \right)$$
$$C_{\alpha r} = \left(C_{\alpha\_LR} + C_{\alpha\_RR} \right)$$

$$(8.26)$$

Also, in this case the wheel track vanishes, i.e., $d_t = 0$. Substituting (8.9) to (8.12) into (8.25), we obtain

$$\alpha_f = \delta_f - arctan\left(\frac{v sin\beta + l_f \dot\psi}{v cos\beta} \right) \tag{8.27}$$

$$\alpha_r = \delta_r - arctan\left(\frac{v sin\beta - l_f \dot\psi}{v cos\beta} \right) \tag{8.28}$$

Since all angles are small, thus the nonlinear functions can be linearized so that $sin\beta = \beta$ and $cos\beta = 1$. Then the above equations result in

$$\alpha_f = \delta_f - \left(\beta + \frac{l_f \dot\psi}{v} \right) \tag{8.29}$$

$$\alpha_r = \delta_r - \left(\beta - \frac{l_r \dot\psi}{v} \right) \tag{8.30}$$

By eliminating β, the two equations above lead to

$$\delta_f - \delta_r = \alpha_f - \alpha_r + \frac{\left(l_f + l_r \right) \dot\psi}{v} \tag{8.31}$$

Now, rewrite (8.31) by substituting (8.15) and obtain

$$\delta_f = \frac{\left(l_f + l_r\right)}{R_T} + \alpha_f - \alpha_r + \delta_r \tag{8.32}$$

NONLINEAR MODEL

Further, the longitudinal tire force along the wheel heading direction and the lateral forces are the sums from both wheels so that for each axle we have

$$
\begin{aligned}
F_{xf} &= \left(F_{x\_LF} + F_{x\_RF}\right) \\
F_{xr} &= \left(F_{x\_LR} + F_{x\_RR}\right) \\
F_{yf} &= \left(F_{y\_LF} + F_{y\_RF}\right) \\
F_{yr} &= \left(F_{y\_LR} + F_{y\_RR}\right)
\end{aligned}
\tag{8.33}
$$

Assuming the angles are small, and taking (8.25), (8.26) and (8.33) into (8.23) and (8.24), they become

$$\dot{\beta} = \frac{1}{Mv}C_{\alpha f}\alpha_f + \frac{1}{Mv}C_{\alpha r}\alpha_r + \frac{1}{Mv}F_{xf}\left(\delta_f - \beta\right) + \frac{1}{Mv}F_{xr}\left(\delta_r - \beta\right) - \dot{\psi} \tag{8.34}$$

$$I_z\ddot{\psi} = C_{\alpha f}\alpha_f l_f - C_{\alpha r}\alpha_r l_r + F_{xf}l_f\delta_f - F_{xr}l_r\delta_r \tag{8.35}$$

Finally, substituting (8.29) and (8.30), we get

$$\dot{\beta} + \left(1 + \frac{C_{\alpha f}l_f - C_{\alpha r}l_r}{Mv^2}\right)\dot{\psi} + \frac{C_{\alpha f} + C_{\alpha r}}{Mv}\beta - \frac{C_{\alpha f}}{Mv}\delta_f - \frac{C_{\alpha r}}{Mv}\delta_r - \frac{\delta_f - \beta}{Mv}F_{xf} - \frac{\delta_r - \beta}{Mv}F_{xr} = 0 \tag{8.36}$$

$$\ddot{\psi} + \frac{C_{\alpha f}l_f^2 + C_{\alpha r}l_r^2}{I_z v}\dot{\psi} + \frac{C_{\alpha f}l_f - C_{\alpha r}l_r}{I_z}\beta - \frac{C_{\alpha f}l_f}{I_z}\delta_f + \frac{C_{\alpha r}l_r}{I_z}\delta_r - \frac{l_f\delta_f}{I_z}F_{xf} + \frac{l_r\delta_r}{I_z}F_{xr} = 0 \tag{8.37}$$

which also can be formulated in the state space form

$$
\begin{bmatrix} \dot{\beta} \\ \ddot{\psi} \end{bmatrix} =
\begin{bmatrix}
-\dfrac{C_{\alpha f} + C_{\alpha r}}{Mv} & -1 - \dfrac{C_{\alpha f}l_f - C_{\alpha r}l_r}{Mv^2} \\
-\dfrac{C_{\alpha f}l_f - C_{\alpha r}l_r}{I_z} & -\dfrac{C_{\alpha f}l_f^2 + C_{\alpha r}l_r^2}{I_z v}
\end{bmatrix}
\begin{bmatrix} \beta \\ \dot{\psi} \end{bmatrix}
$$
$$
+ \begin{bmatrix}
\dfrac{C_{\alpha f}}{Mv} & \dfrac{C_{\alpha r}}{Mv} \\
\dfrac{C_{\alpha f}l_f}{I_z} & -\dfrac{C_{\alpha r}l_r}{I_z}
\end{bmatrix}
\begin{bmatrix} \delta_f \\ \delta_r \end{bmatrix}
+ \begin{bmatrix}
\dfrac{\delta_f - \beta}{Mv} & \dfrac{\delta_r - \beta}{Mv} \\
\dfrac{l_f\delta_f}{I_z} & -\dfrac{l_r\delta_r}{I_z}
\end{bmatrix}
\begin{bmatrix} F_{xf} \\ F_{xr} \end{bmatrix}
\tag{8.38}
$$

In the last term of the above equation, F_{xf} and F_{xr} associate with the steer angles and slip angle in a nonlinear manner. Such nonlinearity will impact the dynamic behavior significantly when the vehicle is accelerated or braked during cornering. From control system design point of the view, steer angles δ_f and δ_r are obviously the input variables. In addition, longitudinal tire forces F_{xf} and F_{xr} can also be viewed system inputs when they are ones transferred from the drive or brake torque. The response of the vehicle model, expressed in terms of yaw rate $\dot{\psi}$ and sideslip angle β as function of time, will be determined by the steer angles and the longitudinal tire forces.

Due to the condition expressed by (8.33), a disadvantage of the model given by (8.38) is that the braking forces at each left and right wheel are summed up to one single wheel. Later in this book, when designing a control, we will see that the controller needs to decide on which wheel should provide the braking force required. For this reason, it is beneficial to keep the braking force of each individual wheel (the same for the traction mode of operation). Considering the longitudinal forces in (8.23) and (8.24), model (8.38) is changed to

$$
\begin{pmatrix} \dot{\beta} \\ \ddot{\psi} \end{pmatrix} = \begin{bmatrix} -\dfrac{c_{\alpha f}+c_{\alpha r}}{Mv} & -1-\dfrac{c_{\alpha f}l_f-c_{\alpha r}l_r}{Mv^2} \\ -\dfrac{c_{\alpha f}l_f-c_{\alpha r}l_r}{I_z} & -\dfrac{c_{\alpha f}l_f^2+c_{\alpha r}l_r^2}{I_z v} \end{bmatrix} \begin{bmatrix} \beta \\ \dot{\psi} \end{bmatrix} + \begin{bmatrix} \dfrac{c_{\alpha f}}{Mv} & \dfrac{c_{\alpha r}}{Mv} \\ \dfrac{c_{\alpha f}l_f}{I_z} & -\dfrac{c_{\alpha r}l_r}{I_z} \end{bmatrix} \begin{bmatrix} \delta_f \\ \delta_r \end{bmatrix}
$$

$$
+ \begin{bmatrix} \dfrac{\delta_f-\beta}{Mv} & \dfrac{\delta_r-\beta}{Mv} \\ \dfrac{l_f\delta_f-d_t}{I_z} & -\dfrac{l_r\delta_r+d_t}{I_z} \end{bmatrix} \begin{bmatrix} F_{x\_LF} \\ F_{x\_LR} \end{bmatrix} + \begin{bmatrix} \dfrac{\delta_f-\beta}{Mv} & \dfrac{\delta_r-\beta}{Mv} \\ \dfrac{l_f\delta_f+d_t}{I_z} & -\dfrac{l_r\delta_r-d_t}{I_z} \end{bmatrix} \begin{bmatrix} F_{x\_RF} \\ F_{x\_RR} \end{bmatrix}
$$

$$(8.39)$$

where the individual longitudinal force at each wheel is an input to the system model. The longitudinal forces play a part of the nonlinear correlation with steering angles as well the sideslip angle. The equation is used in control design where those longitudinal forces are controllable, i.e., in vehicles with torque vectoring systems, electric vehicles with individual electric drives for the drive wheels, in vehicles equipped with controllable braking force systems. At the same time, in many control applications, those angles in (8.39) can be assumed small, and hence the impact of the longitudinal forces is less significant than the impact of the lateral forces.

LINEAR MODEL

As mentioned above, the influence of the longitudinal tire forces on the lateral dynamics can be much smaller than the impact of the lateral tire forces when the steer and slip angles are small so that terms with F_{xf} and F_{xr} in (8.39) can be excluded from consideration. This condition is especially true when there is no drive or braking torque applied. Following (8.38) we have then a linear model

$$
\begin{bmatrix} \dot{\beta} \\ \ddot{\psi} \end{bmatrix} = \begin{bmatrix} -\dfrac{C_{\alpha f}+C_{\alpha r}}{Mv} & -1-\dfrac{C_{\alpha f}l_f-C_{\alpha r}l_r}{Mv^2} \\ -\dfrac{C_{\alpha f}l_f-C_{\alpha r}l_r}{I_z} & -\dfrac{C_{\alpha f}l_f^2+C_{\alpha r}l_r^2}{I_z v} \end{bmatrix} \begin{bmatrix} \beta \\ \dot{\psi} \end{bmatrix} + \begin{bmatrix} \dfrac{C_{\alpha f}}{Mv} & \dfrac{C_{\alpha r}}{Mv} \\ \dfrac{C_{\alpha f}l_f}{I_z} & -\dfrac{C_{\alpha r}l_r}{I_z} \end{bmatrix} \begin{bmatrix} \delta_f \\ \delta_r \end{bmatrix}
$$

$$(8.40)$$

with time-varying parameters. The model will describe a linear time-invariant system if the vehicle is moving at a constant speed.

8.5   BICYCLE MODEL

The bicycle model is a special case of the linear single-track model, in which no rear steering exists, i.e., $\delta_r = 0$ as in most common automotive practice. It follows directly from (8.40)

$$\begin{bmatrix} \dot{\beta} \\ \ddot{\psi} \end{bmatrix} = \begin{bmatrix} -\dfrac{C_{\alpha f} + C_{\alpha r}}{Mv} & -1 - \dfrac{C_{\alpha f} l_f - C_{\alpha r} l_r}{Mv^2} \\ -\dfrac{C_{\alpha f} l_f - C_{\alpha r} l_r}{I_z} & -\dfrac{C_{\alpha f} l_f^2 + C_{\alpha r} l_r^2}{I_z v} \end{bmatrix} \begin{bmatrix} \beta \\ \dot{\psi} \end{bmatrix} + \begin{bmatrix} \dfrac{C_{\alpha f}}{Mv} \\ \dfrac{C_{\alpha f} l_f}{I_z} \end{bmatrix} \delta_f \qquad (8.41)$$

which can be easily identified with Figure 8.7.

The bicycle model was introduced by Riekert and Schunck in 1940. Since then it has been used widely in literature and has become a standard of linear theory in vehicle dynamics. For the investigation of vehicle cornering behavior, as a response to the steering inputs, the linear bicycle model is a useful platform for control design and analysis. Example 9.2 given in Chapter 9 will show vehicle transient behaviors with regard to system parameter variations.

The validity of the above-derived model can be confirmed by comparing with experimental vehicle data. The validation shows that the linear single-track model yields good performance in representation of vehicle lateral dynamics [3]. The measured signals, such as the yaw rate, the lateral acceleration and the sideslip angle can be sufficiently approached by the developed model.

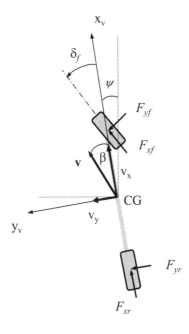

FIGURE 8.7   Bicycle model.

Another representation form of the bicycle model is to use the velocity, denoted as v_y, along the y_v-axis instead of sideslip angle β. Considering small β, thus the approximation $\dot{v}_y \approx v\dot{\beta}$, (8.41) can be transferred to

$$
\begin{bmatrix} \dot{v}_y \\ \ddot{\psi} \end{bmatrix} = \begin{bmatrix} -\dfrac{C_{\alpha f}+C_{\alpha r}}{Mv} & -v-\dfrac{C_{\alpha f}l_f-C_{\alpha r}l_r}{Mv} \\ -\dfrac{C_{\alpha f}l_f-C_{\alpha r}l_r}{I_z v} & -\dfrac{C_{\alpha f}l_f^2+C_{\alpha r}l_r^2}{I_z v} \end{bmatrix} \begin{bmatrix} v_y \\ \dot{\psi} \end{bmatrix} + \begin{bmatrix} \dfrac{C_{\alpha f}}{M} \\ \dfrac{C_{\alpha f}l_f}{I_z} \end{bmatrix} \delta_f \qquad (8.42)
$$

In some control applications this form may be directly used.

8.6   INFLUENCE OF CROSSWIND

One typical disturbance on vehicle lateral dynamics is a crosswind. In straight running the vehicle may be subjected to a side wind, giving a lateral force and a yaw moment. These additional lateral force and yaw moment will be considered in vehicle dynamics models. In a straight way, we use the bicycle model to present the effect of the crosswind.

As shown in Figure 8.8, the wind imposes a force on the left side of the vehicle acting on the center of pressure (CP). This results in a yaw moment due to the distance between CP and CG. From Chapter 3 we have (3.38) and (3.41) of the aerodynamic lateral force and yaw moment:

$$
F_{Al} = \frac{\rho}{2}c_{Al}Av_{ad}^2
$$

$$
M_{Al} = \frac{\rho}{2}c_{Al}l_{Al}Av_{ad}^2
$$

Since coefficient c_{Al} already encompasses the dependence of the force magnitude on wind angle τ_w, so the lateral force F_{Al} is set perpendicular to the centreline of the vehicle (see Figure 3.9). Considering the bicycle model Figure 8.8, we insert (3.38) and (3.41) into (8.34) and (8.35) respectively, and this finally results in the following

$$
\begin{bmatrix} \dot{\beta} \\ \ddot{\psi} \end{bmatrix} = \begin{bmatrix} -\dfrac{C_{\alpha f}+C_{\alpha r}}{Mv} & -1-\dfrac{C_{\alpha f}l_f-C_{\alpha r}l_r}{Mv^2} \\ -\dfrac{C_{\alpha f}l_f-C_{\alpha r}l_r}{I_z} & -\dfrac{C_{\alpha f}l_f^2+C_{\alpha r}l_r^2}{I_z v} \end{bmatrix} \begin{bmatrix} \beta \\ \dot{\psi} \end{bmatrix}
$$
$$
+ \begin{bmatrix} \dfrac{C_{\alpha f}}{Mv} \\ \dfrac{C_{\alpha f}l_f}{I_z} \end{bmatrix} \delta_f - \begin{bmatrix} \dfrac{\rho}{2Mv}c_{Al}A \\ \dfrac{\rho}{2I_z}c_{Al}l_{Al}A \end{bmatrix} v_{ad}^2 \qquad (8.43)
$$

In the above equation the wind effect appears either as an input to stimulate the system, or a disturbed term to the linearity of the system. Given the model configuration in Figure 8.8, the wind opposes the vehicle lateral acceleration and yaw motion. The stronger the wind magnitude is, the more steering input the vehicle requires to compensate for the wind. In order to quantify the steering compensation during a steady crosswind, we look at the straight-line driving. In this case, the signal conditions are:

$$
\ddot{\psi} = 0, \quad \dot{\psi} = 0, \quad \dot{\beta} = 0 \qquad (8.44)
$$

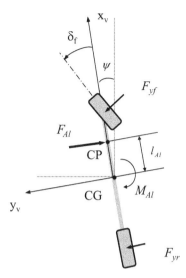

FIGURE 8.8 Bicycle model with crosswind.

These conditions lead (8.43) to

$$\left(C_{\alpha f}+C_{\alpha r}\right)\beta-C_{\alpha f}\delta_f+\frac{\rho}{2}c_{Al}Av_{ad}^2=0$$

$$\left(C_{\alpha f}l_f-C_{\alpha r}l_r\right)\beta-C_{\alpha f}l_f\delta_f+\frac{\rho}{2}c_{Al}l_{Al}Av_{ad}^2=0$$

(8.45)

so that the steering angle can be solved as follows:

$$\delta_f=\frac{\rho}{2}c_{Al}Av_{ad}^2\frac{C_{\alpha r}\left(l_r+l_{Al}\right)-C_{\alpha f}\left(l_f-l_{Al}\right)}{\left(C_{\alpha f}+C_{\alpha r}\right)\left(l_r+l_f\right)}$$

(8.46)

The required steering compensation is proportional to the wind speed and the vehicle cross section area, A. On the other hand, a zero compensation can be reached if

$$C_{\alpha r}\left(l_r+l_{Al}\right)-C_{\alpha f}\left(l_f-l_{Al}\right)=0$$

(8.47)

The CP distance, which corresponds to zero steering compensation, results then in

$$l_{Al}=\frac{C_{\alpha f}l_f-C_{\alpha r}l_r}{C_{\alpha r}+C_{\alpha f}}$$

(8.48)

We will introduce some vehicle model characteristics in the next chapter. From there, we will know that all the vehicles are designed to understeer, i.e. $C_{\alpha r}l_r>C_{\alpha f}l_f$, satisfying the stable motion condition. To satisfy the above-given inequality, CP distance l_{Al} in (8.48) is supposed to be negative. This kind of vehicle design requires a CP shift to the rear, for example by designing a huge tailfin. Another way to reduce the steering compensation is to shift CG to the front as front-wheel-drive cars usually have. At the same time, experiments show that drivers are easily able to correct the steering change through the crosswind, where the vehicle yaw rate is a more significant factor [4].

8.7 VEHICLE-TRAILER MODEL

The purpose of the investigation on vehicle-trailer dynamics is to understand vehicle-trailer combinations and how the coupling or interaction between the two units affects the vehicle stability. Although various types of trailers have been manufactured in automotive industry, the dynamics modeling can be accomplished using the same principle to be described below. Based on the single-track model, we study a combination of the vehicle with a single-axle semi-trailer. The approach will give a basic idea in modeling of a vehicle-trailer system.

LINEAR VEHICLE-TRAILER MODEL

We still consider a plane motion on the horizontal level. Figure 8.9 shows a vehicle-trailer system, where the single-axle semi-trailer is abstracted in a single-track model similar to the towing vehicle. In the vehicle diagram, hitch force F_h acts at the hitch point of the towing vehicle and semi-trailer in the opposed directions (friction at hitch is not considered). In addition to those in Figure 8.6, we define:

F_h hitch force with x-component F_{xh} and y-component F_{yh}
ψ_h hitch articulation angle
v_t trailer's wheel velocity
F_{xt} tire force in x-direction of the trailer wheel
F_{yt} lateral tire force in y-direction of the trailer wheel
α_t slip angle of the trailer wheel
M_t gross mass of the trailer
I_t rotational inertia of the trailer about z-axis
l_t distance from the hitch to the trailer CG
l_k distance from the trailer's CG to the wheel.

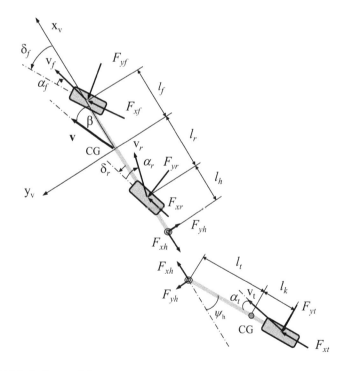

FIGURE 8.9 Vehicle-trailer model.

First, we look at the wheel velocity of the trailer. Following Example 2.4 in Chapter 2, the velocity in the plane of motion is given by

$$\mathbf{v}_t = \begin{bmatrix} v\cos\beta + (l_t + l_k)(\dot{\psi} + \dot{\psi}_h)\sin\psi_h \\ v\sin\beta - (l_r + l_h + (l_t + l_k)\cos\psi_h)\dot{\psi} - (l_t + l_k)\dot{\psi}_h\cos\psi_h \end{bmatrix}$$

Since β and ψ_h are assumed to be small, from the trigonometrical relationship we obtain

$$\tan(\psi_h - \alpha_t) = \frac{v\sin\beta - (l_r + l_h + l_t + l_k)\dot{\psi} - (l_t + l_k)\dot{\psi}_h}{v\cos\beta} \tag{8.49}$$

and the trailer wheel slip angle α_t is computed as

$$\alpha_t = \psi_h - \beta + \frac{(l_r + l_h + l_t + l_k)\dot{\psi} + (l_t + l_k)\dot{\psi}_h}{v} \tag{8.50}$$

Second, we consider the force and moment balances of the lumped trailer model, where the lumped mass is at the trailer's center of gravity. For the calculation of the trailer lateral acceleration, we adopt the result of Example 2.4 in Chapter 2. From Newton's and Euler's equations, it yields

$$M_t\left[v\dot{\beta} - (l_r + l_h + l_t)\ddot{\psi} - l_t\ddot{\psi}_h + v\dot{\psi}\right] = F_{yt}\cos\psi_h - F_{yh} \tag{8.51}$$

$$I_t\left(\ddot{\psi} + \ddot{\psi}_h\right) = -F_{yt}l_k - F_{yh}\cos\psi_h l_t \tag{8.52}$$

Note that the dynamic change in longitudinal direction is neglected since we focus on the contribution of the lateral forces. For further model analysis, the hitch angle is assumed to be small, i.e., $\cos\psi_h = 1$. The lateral tire force of the trailer is simplified as a linear function of the slip angle, that is

$$F_{yt} = C_{\alpha t}\,\alpha_t \tag{8.53}$$

In this manner, we wish to find a linear expression of the vehicle model in terms of the state variables. Referring to Figure 8.9, we recall (8.34) and (8.35) to derive equations for the towing vehicle. By neglecting the longitudinal forces F_{xf}, F_{xr}, and adding the hitch force yields

$$\dot{\beta} = \frac{1}{Mv}C_{\alpha f}\alpha_f + \frac{1}{Mv}C_{\alpha r}\alpha_r + \frac{1}{Mv}F_{yh} - \dot{\psi} \tag{8.54}$$

$$I_z\ddot{\psi} = C_{\alpha f}\alpha_f l_f - C_{\alpha r}\alpha_r l_r - F_{yh}(l_r + l_h)I_z\ddot{\psi} = C_{\alpha f}\alpha_f l_f - C_{\alpha r}\alpha_r l_r - F_{yh}(l_r + l_h) \tag{8.55}$$

Now, (8.51) through (8.55) have been derived to describe the complete vehicle-trailer dynamics. In order to obtain the linear equations in terms of state variables β, $\dot{\psi}$, $\dot{\psi}_h$ and ψ_h, we eliminate first F_{yh} and F_{yt} among (8.51) through (8.55). Then we substitute α_f, α_r and α_t by (8.29), (8.30) and (8.50) respectively. As the result, three independent equations are created. We obtain the state space form

$$\mathbf{M}_T\dot{\mathbf{x}} = \mathbf{A}_T\mathbf{x} + \mathbf{B}_T\mathbf{u} \tag{8.56}$$

where $\mathbf{x} = \begin{bmatrix} \beta & \dot{\psi} & \dot{\psi}_h & \psi_h \end{bmatrix}^T$ is the state vector and $\mathbf{u} = \begin{bmatrix} \delta_f & \delta_r \end{bmatrix}^T$ is the input vector.

The associated system matrices are defined as

$$
\mathbf{M}_T = \begin{bmatrix}
(M+M_t)\mathrm{v} & -M_t(l_r+l_h+l_t) & -M_t l_t & 0 \\
M\mathrm{v}(l_r+l_h) & I_z & 0 & 0 \\
M_t\mathrm{v}\,l_t & -M_t l_t(l_r+l_h+l_t)-I_t & -\left(M_t l_t^2+I_t\right) & 0 \\
0 & 0 & 0 & 1
\end{bmatrix},
$$

$$
\mathbf{A}_T = \begin{bmatrix}
-(C_{\alpha f}+C_{\alpha r}+C_{\alpha t}) & -(M+M_t)\mathrm{v}-\dfrac{C_{\alpha f}l_f-C_{\alpha r}l_r-(l_k+l_r+l_h+l_t)C_{\alpha t}}{\mathrm{v}} & \dfrac{(l_k+l_t)C_{\alpha t}}{\mathrm{v}} & C_{\alpha t} \\
-C_{\alpha f}(l_r+l_h+l_f)-C_{\alpha r}l_h & -\dfrac{C_{\alpha f}l_f(l_f+l_r+l_h)-C_{\alpha r}l_r l_h}{\mathrm{v}}-M\mathrm{v}(l_r+l_h) & 0 & 0 \\
-C_{\alpha t}(l_k+l_t) & \dfrac{C_{\alpha t}(l_t+l_r+l_h+l_k)(l_t+l_k)}{\mathrm{v}}-M_t\mathrm{v}l_t & \dfrac{(l_k+l_t)^2 C_{\alpha t}}{\mathrm{v}} & (l_k+l_t)C_{\alpha t} \\
0 & 0 & 1 & 0
\end{bmatrix}
$$

$$
\mathbf{B}_T = \begin{bmatrix}
C_{\alpha f} & C_{\alpha f}(l_f+l_r+l_h) & 0 & 0 \\
C_{\alpha r} & C_{\alpha r}l_h & 0 & 0
\end{bmatrix}^T.
$$

Equation (8.56) presents a linear system, i.e., as long as the vehicle system operates under small angle conditions, such as during a turn with large radius or with small perturbations during a straight-line motion, the vehicle-trailer combination behaves as a linear system.

TOWING VEHICLE MODEL WITH DISTURBANCE

From the system component point of view, the trailer can also be considered an additional disturbance component, which is not specified in its dynamic nature. Exactly speaking, for some control applications it is technically convenient to model the towing vehicle with an unknown trailer disturbance.

By substituting (8.29) and (8.30), we have from (8.54) and (8.55) that

$$
\begin{bmatrix} \dot{\beta} \\ \ddot{\psi} \end{bmatrix} =
\begin{bmatrix}
-\dfrac{C_{\alpha f}+C_{\alpha r}}{M\mathrm{v}} & -1-\dfrac{C_{\alpha f}l_f-C_{\alpha r}l_r}{M\mathrm{v}^2} \\
-\dfrac{C_{\alpha f}l_f-C_{\alpha r}l_r}{I_z} & -\dfrac{C_{\alpha f}l_f^2+C_{\alpha r}l_r^2}{I_z \mathrm{v}}
\end{bmatrix}
\begin{bmatrix} \beta \\ \dot{\psi} \end{bmatrix}
$$

$$
+\begin{bmatrix}
\dfrac{C_{\alpha f}}{M\mathrm{v}} & \dfrac{C_{\alpha r}}{M\mathrm{v}} \\
\dfrac{C_{\alpha f}l_f}{I_z} & -\dfrac{C_{\alpha r}l_r}{I_z}
\end{bmatrix}
\begin{bmatrix} \delta_f \\ \delta_r \end{bmatrix}
+\begin{bmatrix}
\dfrac{1}{M\mathrm{v}} \\
-\dfrac{l_r+l_h}{I_z}
\end{bmatrix} F_{yh}
\tag{8.57}
$$

In this equation, β and $\dot{\psi}$ are the state variables of the system while δ_f, δ_r are the control input signals. Hitch force F_{yh} represents an input in the system. It is considered as an additional dynamic load to the towing vehicle system, or simply a system disturbance.

Oscillation Characteristics

Vehicle-trailer dynamics is significantly influenced by trailer oscillations. The oscillations can be initiated by a lateral movement that displaces the trailer from its equilibrium position, and trailer keeps moving back and forth across its position of equilibrium. Depending on the vehicle speed, the system will result in oscillations with different amplitudes and frequencies. To characterize the oscillations, we consider the trailer itself while assuming the towing vehicle is in a steady-state without lateral movement, i.e., $\ddot{\psi} = 0$, $\dot{\psi} = 0$, $\dot{\beta} = 0$ and $\beta = 0$. From (8.56), we get

$$\left(M_t l_t^2 + I_t\right) v \ddot{\psi}_h + C_{\alpha t}(l_t + l_k)^2 \dot{\psi}_h + C_{\alpha t}(l_t + l_k) v \psi_h = 0 \tag{8.58}$$

Hence, the open loop dynamics of the trailer system is described by a second-order linear differential equation with time-varying coefficients. Equation (8.58) features a typical mechanical damped oscillating system. The damping is characterized by the fact that the part of the lateral tire force is proportional to the angular velocity of the hitch articulation [see the second component in (8.58)].

The trailer oscillating system might exhibit several different characteristic motions depending upon the trailer parameters. From the study of differential equations, for example *Laplace transforms*, we know that the solution of an equation of form (8.58) depends on the roots of its *characteristic equation*, which is

$$\left(M_t l_t^2 + I_t\right) v \ddot{s} + C_{\alpha t}(l_t + l_k)^2 \dot{s} + C_{\alpha t}(l_t + l_k) v = 0 \tag{8.59}$$

A common way of describing such oscillatory second-order systems is in terms of the *damping ratio* and the *natural frequency*. For this reason, we define the damping ratio as

$$\zeta = \frac{1}{2v} \sqrt{\frac{C_{\alpha t}}{M_t \dfrac{l_t^2}{(l_t + l_k)^3} + I_t \dfrac{1}{(l_t + l_k)^3}}} \tag{8.60}$$

and the natural frequency as

$$\omega_o = \sqrt{\frac{C_{\alpha t}(l_t + l_k)}{M_t l_t^2 + I_t}} \tag{8.61}$$

Hence, by parameterizing the characteristic equation (8.59), it is translated to

$$\ddot{s} + 2\zeta\omega_o \dot{s} + \omega_o^2 = 0 \tag{8.62}$$

Now, the roots (sometimes called poles) of the characteristic equation are

$$s_1 = -\zeta\omega_o + \omega_o\sqrt{\zeta^2 - 1}$$
$$s_2 = -\zeta\omega_o - \omega_o\sqrt{\zeta^2 - 1} \tag{8.63}$$

If $\zeta > 1$, the motion of the system, i.e., the trailer, is sluggish and nonoscillatory. The system is over-damped and actually regains its equilibrium position after a finite time. If $\zeta < 1$, then the motion of the system is oscillatory and underdamped. In the case $\zeta = 1$, the motion obtained is again nonoscillatory. The system is critically damped and regains its equilibrium position in the shortest possible time without oscillation.

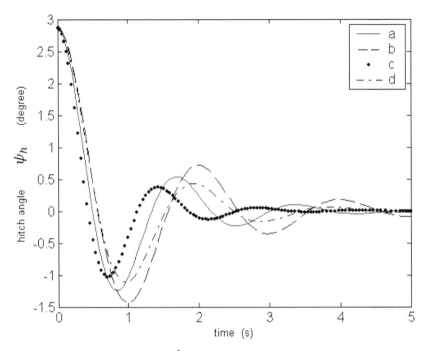

a. $M_t = 2000$ kg, $I_t = 4234$ kgm^2, $C_{\alpha t} = 90000$ N/rad, $l_t = 3.0$ m, $l_k = 0.7$ m

b. $M_t = 3000$ kg, $I_t = 4834$ kgm^2, $C_{\alpha t} = 90000$ N/rad, $l_t = 3.0$ m, $l_k = 0.7$ m

c. $M_t = 2000$ kg, $I_t = 4234$ kgm^2, $C_{\alpha t} = 130000$ N/rad, $l_t = 3.0$ m, $l_k = 0.7$ m

d. $M_t = 2000$ kg, $I_t = 4234$ kgm^2, $C_{\alpha t} = 90000$ N/rad, $l_t = 4.0$ m, $l_k = 0.7$ m

FIGURE 8.10   Hitch angle's time response to different trailer parameters.

Now it is easy to see how the trailer oscillation may behave when parameters $M_t, I_t, C_{\alpha t}, l_t, l_k$ change. In Figure 8.10 some simulation results of different trailer parameters are illustrated to deliver a qualitative assessment. Each simulation shows the amplitude of the hitch angle and the corresponding time response to the same nonzero initial condition, which is the vehicle speed assumed to be at $v = 100$ km/h. Figure 8.10a presents a stable case, the trailer exhibits an over-damped motion. Both the increased trailer mass and the rotational inertia will decrease the damping ratio and the natural frequency, and thus move the poles to the right in the s-plane, as shown in (8.63). In this case the trailer will exhibit more oscillatory tendency, as shown in Figure 8.10b. If the trailer tire has a better lateral characteristic and a higher tire inflation pressure, i.e., stronger stiffness $C_{\alpha t}$, it will increase the damping ratio and make the trailer more stable as shown in Figure 8.10c. Also Figure 8.10d shows that longer l_t or l_k will damp the oscillation amplitude.

REFERENCES

1. A. F. Andreev, V. I. Kabanau, V. V. Vantsevich, Driveline Systems of Ground Vehicles: Theory and Design, Boca Raton, London, New York: CRC Press, 2010.
2. M. Mitschke, Dynamik der Kraftfahrzeuge, Band C: Fahrverhalten, Berlin: Springer-Verlag, 1990.
3. U. Kiencke and L. Nielsen, Automotive Control Systems, Warrendale: SAE, 2000.
4. A. Zomotor, Fahrwerktechnik: Fahrverhalten, Würzburg: Vogel Verlag, 1991.

9 System Characteristics of Lateral Dynamics

In the previous chapter, vehicle lateral dynamics was studied for the purpose of having a mathematical description needed for modeling and designing control systems. The analysis was dealing with the following technical question: what is the *performance* of a given vehicle system in response to changes of inputs or disturbances? Here, the system refers to models of vehicle lateral dynamics. Control design concerns how to improve the performance without changing the system to be controlled. The term performance is used to summarize several aspects of the dynamic behavior, such as transient period, oscillatory and overshoot behavior during the transient state, steady-state errors as well as the robustness versus parameter variations and disturbance inputs. For those reasons, the attention in this chapter is turned to characteristics of the system models, including *transient, frequency* and *steady-state responses* so that one can visualize how the system characteristics might be changed to modify the responses in some desirable direction.

Transient response is the system response in time domain while a transient state occurs. The transient state takes place when the motion characteristics change in response to system inputs or external disturbances, such as external forces acting on the vehicle (e.g., a wind, a bump on the road), steering angle change. In the case of a sudden change of the inputs and disturbances, a certain period of time will be required for transient response terms to decay and for the system outputs to level-off at new values. Steady-state *frequency response* is an equivalent approach to conduct the model analysis. The frequency response data is obtained from the steady-state response to sinusoidal input with various frequencies.

According to SAE definition, the *steady state* exists when a periodic (or a constant) vehicle response to periodic (or constant) control and/or disturbance inputs does not change over an arbitrarily long time. The vehicle motion responses in steady-state are referred to as *steady-state responses*. The consideration of steady-state behavior due to control inputs and disturbances has been an important design component in control design methods.

Control design for acceptable steady-state errors can be thought of as placing a lower bound on the very-low-frequency portion of the system. We will deal with the responses to steady-state control inputs, such as constant path radius of curvature, constant steering angle and lateral acceleration. In this manner, the preferred vehicle characteristics will be determined to indicate how to achieve the optimum through the dynamics control. Readers not familiar with linear control systems are recommended to review Chapter 11 first, which provides some basics of control theory,

9.1 STEERING CHARACTERISTICS

Following the introduction of the steering geometry in the previous chapter, we present here main steering characteristics, which are directly determined from the steering wheel angles. They play significant roles in understanding the steady-state as well as the transient state of vehicle lateral dynamics.

Steering Ratio

Up to this point, the *steer angle*, which is the angle between the wheel plane and the vehicle heading direction, has been used as a term throughout all the equations. In the physical build there exist the steering gearbox and some mechanical linkages from the vehicle wheel to the driver's steering

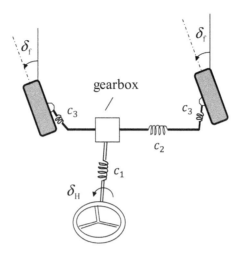

FIGURE 9.1   Steering linkages model.

wheel. Consequently, the steer angle is related to the *steering wheel angle* via a so-called *steering ratio.*

The steering ratio refers to the change of steering wheel angle at a given steady-state steering wheel position, with respect to the change of steer angle. For a front-wheel-steering car, the *steering wheel angle*, δ_H, is then given by

$$\delta_H = i_s \, \delta_f \tag{9.1}$$

where i_s denotes the steering ratio that is provided by the gearing and the dimension of the mechanical linkages of the steering system. It is assumed that the backlash between the gear teeth is zero. Thus, a constant i_s in (9.1) is the ideal condition, meaning no other physical effects on the steering ratio come from the steering linkage, the gearing, and the steering column. In actual designs, however, the linkage, gears and the steering column exhibit compliance effects influenced by steering and moment that load the steering system from the tire-road patches to the driver's steering wheel. These significant physical elasticity properties can be taken into account as the composite values between the steering gearbox and the wheels, as shown in Figure 9.1. The compliance reduces the effective steer angles and the driver has to increase the steering wheel angle to reach the intended steer angles. Given a steering moment, M_s, caused by the stiffness between the steering wheel and the road wheels, the definition (9.1) is extended to

$$\delta_H = i_s \, \delta_f + i_s \, \frac{M_s}{c_s} \tag{9.2}$$

where c_s is the composite stiffness between steering wheel and road wheel that is determined based on the torsional stiffness c_1 of the steering column, stiffness value c_2 of the tire rod and c_3 for two steering arms (depending on the steering system design, there are some other links that can be included in the computation of the total compliance). Among the motor vehicles steering ratio i_s normally ranges from 16 to 24 on passenger cars, and from 23 to 36 on trucks.

Steering Characteristics

There have been different steering characteristic terminologies introduced in literature to characterize the relation of the steering wheel angle versus the lateral acceleration [1]. These definitions basically characterize the same steady-state behavior. One of them, *steering wheel angle gradient*, is chosen here as a basic term. The steering wheel angle gradient refers to a change in the steering wheel angle with respect to a change in the steady-state lateral acceleration on a level road at a given *trim* and test conditions. *Trim* is the steady-state condition of the vehicle with a constant input which can be the turn radius or vehicle speed. Hence, the steering wheel angle gradient would have the form $d\delta_H/da_y$ with $a_y = v^2/R_T$.

In order to study the relation between the steer angle and the lateral acceleration, we recall (8.32):

$$\delta_f = \frac{l_f + l_r}{R_T} + \alpha_f - \alpha_r + \delta_r \tag{9.3}$$

In Section 8.1 we described two cornering conditions, namely a low-speed turning and a high-speed turning. The above equation covers the second condition, in which the lateral acceleration and the wheel slip angles are not neglected. Setting the wheel slip angle to zero, the steer angle will turn into the Ackermann angle given by (9.4):

$$\delta_A = \frac{l_f + l_r}{R_T} + \delta_r \tag{9.4}$$

which corresponds to the low-speed turning condition.

Following (9.3), the steering angle cannot be kept constant when the vehicle is turning on a constant radius. Due to its dependence on the wheel slip angles the steering angle must be adjusted by the driver to keep the vehicle on the turning radius. Otherwise, the turning radius will enlarge or decrease depending on the difference of the steer angles. This fact leads to a significant question: how does the steering characteristics influence vehicle lateral dynamics? The following section will introduce the understeer gradient as a measure of the steering performance at a given trim state.

9.2   UNDERSTEER/OVERSTEER GRADIENT

Among the steady-state characteristics of vehicle lateral dynamics, the *understeer gradient* is a frequently used and important index for the vehicle stability consideration. The common definition for the understeer gradient is [2]:

$$U_g = \frac{1}{i_s} \frac{d\delta_H}{da_y} - \frac{d\delta_A}{da_y} \tag{9.5}$$

with unit $\text{rad}/\text{ms}^{-2}$. Steering ratio i_s is involved in the definition since it provides a perceptible view to the driver to use the steering wheel angle. The definition basically expresses the difference of steering wheel angle gradient, reduced to the steered wheels, and the Ackermann angle gradient, with respect to the lateral acceleration.

For the steady-state motion, the differentials in (9.5) can be changed to the parameters themselves, and thus the required steering wheel angle in a linear range leads to the form

$$\delta_H = i_s \, \delta_A + i_s \, U_g a_y \tag{9.6}$$

Using the steer angle definition from (9.1), (9.6) can be rewritten in

$$\frac{\delta_H}{i_s} = \delta_A + U_g \, a_y \tag{9.7}$$

As seen from the above equation, the understeer gradient determines the quantity of the steering angle needed additionally to the Ackermann angle. Condition (9.7) expresses the characteristics of vehicle understeer and oversteer behaviour. Depending on the sign of the understeer gradient, the vehicle will demonstrate different behaviours.

In case of

$$U_g > 0 : \textit{understeer} \tag{9.8}$$

This condition means equivalently that ratio of steering wheel angle gradient $d\delta_H / da_y$ to the steering ratio in (9.5) is greater than the Ackerman steer angle gradient. Should the vehicle speed increase, then a_y, followed by steering wheel angle δ_H, must increase as well to keep the vehicle driving on the same turning radius. If not, the vehicle will pursue a larger radius, i.e., is the vehicle is in an understeer maneuver.

In case of

$$U_g < 0 : \textit{oversteer} \tag{9.9}$$

This condition means equivalently that the ratio of the steering wheel angle gradient $d\delta_H / da_y$ to the steering ratio is less than the Ackerman steer angle gradient. Should the vehicle speed increase, then a_y, followed by steering wheel angle δ_H, must decrease to maintain the same turning radius. If not, the vehicle will pursue a smaller radius demonstrating its oversteer.

$$U_g = 0 : \textit{neutral steer} \tag{9.10}$$

Finally, the *neutral steer* means equivalently that the ratio of the steering wheel angle gradient $d\delta_H / da_y$ to the steering ratio equals the Ackerman steer angle gradient. Should the vehicle speed increase, the turning radius remains the same without additional steering. It is an intermediate condition between under- and oversteer.

Back to (9.3), we see that the above definitions conclude the conditions over the wheel slip angles, which are used for the definition of the understeer gradient in some technical references:

$$\alpha_f - \alpha_r \begin{cases} > 0 \ \textit{understeer} \\ < 0 \ \textit{oversteer} \\ = 0 \ \textit{neutral steer} \end{cases} \tag{9.11}$$

The above equation explains clearly the physical content of the definition. The reader can apply (9.11) to Figure 8.4 in order to see the changes of the turning radius.

In practice, we are interested to measure and investigate the understeer, oversteer and neutral steer behaviours for different vehicles. There are two basic ways, in which steering wheel angle changes can be measured. One way is to vary the vehicle speed while keeping the vehicle on a constant turn-radius by changing the steering wheel angle, and the other method is to maintain a constant speed while varying the turn radius by changing the steering wheel angle. For describing these two methods, we substitute (9.4) into (9.7) and obtain

$$\frac{\delta_H}{i_s} = \frac{l_f + l_r}{R_T} + U_g a_y + \delta_r \tag{9.12}$$

or

$$\frac{\delta_H}{i_s} = \frac{l_f + l_r}{v^2} a_y + U_g a_y + \delta_r \tag{9.13}$$

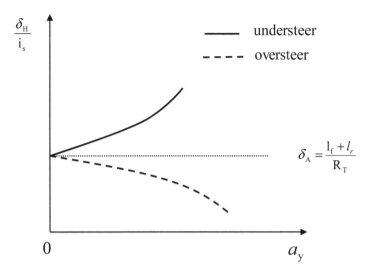

FIGURE 9.2 Understeer gradient for a constant turning radius.

Both methods can be readily illustrated in a graph plotting the steering wheel angle vs. the lateral acceleration. For simplification in the graphical illustration, the rear steer angle is set to zero. Figure 9.2 illustrates (9.12) while the turning radius is kept constant. The vehicle understeer behaviour is illustrated by the solid curve. When the vehicle speed is increasing and so do the lateral acceleration. To keep the vehicle on the same turning radius, the steering wheel angle should keep increasing above the Ackermann angle that is specified for the neutral steer maneuverer. In the same way, the vehicle oversteer is illustrated by the dash line. The steering wheel angle keeps decreasing below the specified Ackermann angle.

Figure 9.3 presents a graphical interpretation of (9.13) for positive and negative values of the understeer gradient while the vehicle speed is maintained constant. The Ackermann angle propagates linear with the lateral acceleration corresponding to the vehicle neutral steer at a specified radius of turn.

Another fact to be noted is the linearity in the operating range of the lateral acceleration. In the range of $a_y < 4\,\mathrm{m/s^2}$, understeer gradient U_g is constant and the steering wheel angle in (9.12) and

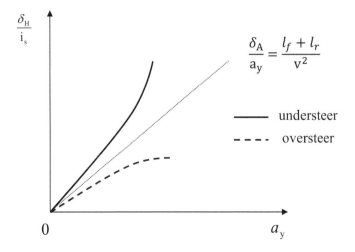

FIGURE 9.3 Understeer gradient for a constant vehicle speed.

(9.13) behaves linear to the lateral acceleration when the speed and the turning radius vary. This result, generally confirmed by the experimental measurements [1], corresponds to the approximation we made in Chapter 3, that the lateral tire force is proportional to the wheel slip angle over the cornering stiffness. The linear range is also the most considered operating range for the vehicle dynamics analysis and control.

As described above, the understeer gradient was defined through the steering geometry analysis. On the other hand, it is also possible to view the relation of the understeer gradient to vehicle parameters. In the steady-state turn of a vehicle being with constant radius at a constant forward speed, the body slip angle and yaw rate of the vehicle remain steady, i.e. $\dot{\beta} = 0$ and $\ddot{\psi} = 0$. Taking these conditions into account, (8.40) can be solved for front steer angle by eliminating the sideslip angle

$$\delta_f = \frac{(l_f + l_r)\dot{\psi}}{v} + \frac{Mv\left(C_{\alpha r}l_r - C_{\alpha f}l_f\right)\dot{\psi}}{(l_f + l_r)C_{\alpha f}C_{\alpha r}} + \delta_r \tag{9.14}$$

Since $v^2 / R_T = a_y$, it follows

$$\delta_f = \frac{(l_f + l_r)}{R_T} + \frac{M\left(C_{\alpha r}l_r - C_{\alpha f}l_f\right)}{(l_f + l_r)C_{\alpha f}C_{\alpha r}} a_y + \delta_r \tag{9.15}$$

Comparing (9.3) and (9.15) yields

$$\alpha_f - \alpha_r = \frac{C_{\alpha r}l_r - C_{\alpha f}l_f}{(l_f + l_r)C_{\alpha f}C_{\alpha r}} Ma_y \tag{9.16}$$

Comparing (9.15) with (9.12), the understeer gradient can be identified as

$$U_g = \frac{M\left(C_{\alpha r}l_r - C_{\alpha f}l_f\right)}{(l_f + l_r)C_{\alpha f}C_{\alpha r}} \tag{9.17}$$

Analogous to (9.11), the understeer conditions can be equivalently expressed by using (9.17):

$$C_{\alpha r}l_r - C_{\alpha f}l_f \begin{cases} > 0 \ understeer \\ < 0 \ oversteer \\ = 0 \ neutral \ steer \end{cases} \tag{9.18}$$

As seen here, the understeer gradient is physically affected by the vehicle parameters such as the gross mass, the wheelbase and the CG location as well as the tire cornering stiffness. The understeer gradient is constant in the linear range ($a_y < 4 \ ms^2$) since the cornering stiffness is approximately a constant value. Changing the vehicle parameters may vary the understeer gradient significantly. From (9.17) we see, the more mass the vehicle has, the higher the understeer gradient magnitude is. The longer the wheelbase is, the smaller the magnitude of the understeer gradient will be. At the same time, the influence of the mass and the wheelbase on the vehicle maneuverer is considerably tied to the magnitudes of the cornering stiffness of the front and rear tires and to the location of the gravity center. Indeed, the vehicle CG location can shift due to the weight distribution of the vehicle. The shifting of the CG to the front will facilitate an increase of the understeer gradient, and thus contribute to vehicle understeer if $C_{\alpha r}l_r \geq C_{\alpha f}l_f$. Vice versa, the moving of the CG to the back will decrease the understeer gradient. A large shift of the CG from the front to the back could turn the understeer vehicle into oversteer. This phenomenon can be, for example observed at FWD vehicles, which usually have the CG far in the front. By loading the rear of the car, the CG moves to the back so that it tends to oversteer. In the case that vehicle is subject to a high lateral acceleration, i.e., the vehicle dynamics is in the nonlinear range, we still can consider a linearized operation point of the cornering stiffness. In this situation, it can happen that the rear wheel slip angle is smaller than in the front, so the vehicle understeers.

The application of a driving or braking torque to the tire during a turn will also affect the cornering stiffness. For a FWD vehicle, the driving torque during a turn increases the side slip angle and reduces the cornering stiffness of the front tires, so that it produces the understeering effect. For a RWD vehicle, the driving torque during a turn reduces the stiffness of the rear tires, thus introducing oversteering effect. With a 4WD vehicle, the driving torque during a turn can also produce understeer or oversteer behaviour depending on the torque distribution among the front and rear wheels. For example, if the center differential between the front and the rear axles is locked, the vehicle can have a severe understeering effect compared to the unlocked central differential. When the center differential is locked and the steer angle is big, the front wheels can develop a negative longitudinal force, the front lateral tire force changes its direction and the front slip angle can drastically go up [3]. The unlocked differential allows for positive longitudinal forces at the front and rear tires and thus is more favourable for vehicle turn.

Example 9.1

A vehicle is accelerating while turning into a curve on a dry asphalt road. How may the vehicle behave if it has a) a front-wheel drive, b) a rear-wheel drive or c) an all-wheel drive system?

During the acceleration, the drive wheels have an increased side slip angle and reduce the cornering stiffness in order to maintain the given lateral forces. If the drive wheels begin to slip, the vehicle will develop a situation similar to one described in Example 5.1.

a. Figure 9.4 shows a FWD car turning left into the curve. Since the front wheels are driven, the longitudinal tire forces are increasing so that the front tire cornering stiffness is reducing and the side slip angles of both front wheels are becoming larger. Per definition given by (9.11), the vehicle is more *understeer*. At this moment the front end tends to drift out of the curve, and the driver has to increase the steering wheel angle. Example 5.1a showed that the FWD car in the same driving situation may become to understeer

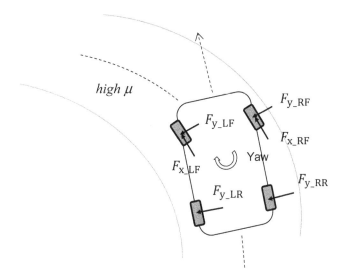

FIGURE 9.4 FWD vehicle turning into a curve on high-μ surface.

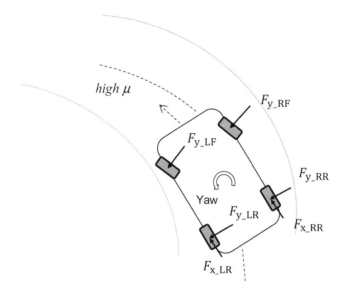

FIGURE 9.5   RWD vehicle turning into a curve on high-μ surface.

  when the front wheels slip. As a result, a front-wheel-drive vehicle tends to understeer when driving a curve.

b. Figure 9.5 shows a RWD car turning left into a curve. Since the rear wheels are driven, the longitudinal tire forces are increasing so that the slip angles of both rear wheels are becoming larger. So, the vehicle tends to *oversteer*. The rear end of the car tends to swing away. Example 5.1b showed that the RWD car in the same driving situation may become to oversteer when the rear wheels slip. As a result, a rear-wheel-drive vehicle tends to oversteer when driving along a curve.

c. For AWD vehicle, the maneuver depends on how the driving torque is distributed among two drive axles. When the total driving torque is distributed to both axles in a way that the longitudinal forces at the front and rear wheel are positive and more or less equal, the car tends to neutral steer. At the same time, the neutral steer is not necessarily a desired maneuver since the actual turning radius is still greater than the theoretical one even the side slip angles are equal at the front and rear tires. However, the vehicle is more stable on the road. A more exact answer to question c) can come from a detailed modeling of the car.

❏

9.3 VEHICLE DYNAMIC RESPONSE TO STEERING INPUT

Up to now, we have introduced the understeer gradient and learned what the driver has to adjust the steering wheel angle to maintain the vehicle on a desired turning radius. In this section, it is to investigate the response of a vehicle to a given steering angle input. For this purpose, the vehicle's yaw rate and sideslip angle are considered as the system outputs. Depending on the input function, there are two different ways to analyze this response. The one is called *transient response*, which is considered in the time domain while the other one, *frequency response*, is measured in the frequency domain.

  The transient response is one of the important aspects of control system performance. In the case of a sudden change of the steering input, a certain period of time is required for transient response terms to decay and for the output to level off at the new value. For control systems, the key feature of this transient period is that this period should be sufficiently short and should not be excessively oscillatory.

Theory of vehicle transient behaviour is complex and can be highly mathematical; therefore, a linearized dynamics model will be taken here to serve as an introduction and overview. From the theory of differential equations, we know that *Laplace transformation* is able to turn a linear differential equation into an input and output related form with the so-called *transfer function*. The transfer function provides a direct view of dynamic characteristics of a system and their relationship in the system. Mathematically it is easier to obtain the transient response through the transfer function. The correlation between the transfer function and the response characteristics is developed in terms of positions of the system poles and zeros in the *s*-plane. We start with the transfer function in the following and then view the transient response behavior. Afterwards the *frequency response* method will be discussed.

YAW RATE AND SIDESLIP ANGLE TRANSFER FUNCTIONS

Recall linearized model (8.40) derived in the preceding chapter,

$$
\begin{bmatrix} \dot{\beta} \\ \ddot{\psi} \end{bmatrix} = \begin{bmatrix} -\dfrac{C_{\alpha f}+C_{\alpha r}}{Mv} & -1-\dfrac{C_{\alpha f}l_f - C_{\alpha r}l_r}{Mv^2} \\ -\dfrac{C_{\alpha f}l_f - C_{\alpha r}l_r}{I_z} & -\dfrac{C_{\alpha f}l_f^2 + C_{\alpha r}l_r^2}{I_z v} \end{bmatrix} \begin{bmatrix} \beta \\ \dot{\psi} \end{bmatrix} + \begin{bmatrix} \dfrac{C_{\alpha f}}{Mv} & \dfrac{C_{\alpha r}}{Mv} \\ \dfrac{C_{\alpha f}l_f}{I_z} & -\dfrac{C_{\alpha r}l_r}{I_z} \end{bmatrix} \begin{bmatrix} \delta_f \\ \delta_r \end{bmatrix}
$$

(9.19)

Let the above equation refer to the state space form

$$
\dot{\mathbf{x}} = \mathbf{A}\mathbf{x} + \mathbf{B}\mathbf{u}
$$

where $\mathbf{x} = \begin{bmatrix} \beta & \dot{\psi} \end{bmatrix}^T$ is the state vector for the sideslip angle and the yaw rate, and $\mathbf{u} = \begin{bmatrix} \delta_f & \delta_r \end{bmatrix}^T$ is the steering input vector. $\mathbf{A}$, $\mathbf{B}$ refer to the matrices in (9.19) correspondingly. According to (11.18) from Chapter 11, the Laplace transform is applied to state space model (9.19):

$$
\begin{bmatrix} s\beta(s) \\ s\dot{\psi}(s) \end{bmatrix} = \begin{bmatrix} -\dfrac{C_{\alpha f}+C_{\alpha r}}{Mv} & -1-\dfrac{C_{\alpha f}l_f - C_{\alpha r}l_r}{Mv^2} \\ -\dfrac{C_{\alpha f}l_f - C_{\alpha r}l_r}{I_z} & -\dfrac{C_{\alpha f}l_f^2 + C_{\alpha r}l_r^2}{I_z v} \end{bmatrix} \begin{bmatrix} \beta(s) \\ \dot{\psi}(s) \end{bmatrix}
$$

(9.20)

$$
+ \begin{bmatrix} \dfrac{C_{\alpha f}}{Mv} & \dfrac{C_{\alpha r}}{Mv} \\ \dfrac{C_{\alpha f}l_f}{I_z} & -\dfrac{C_{\alpha r}l_r}{I_z} \end{bmatrix} \begin{bmatrix} \delta_f(s) \\ \delta_r(s) \end{bmatrix}
$$

where the vehicle speed, v, is considered constant and the initial conditions of the state variables are set to zero. The system outputs, $\dot{\psi}(s)$ and $\beta(s)$, are now more transparent to the steering inputs $\delta_f(s)$ as well as $\delta_r(s)$.

Denoting

$$
|s\mathbf{I} - \mathbf{A}| = s^2 + \frac{I_z\left(C_{\alpha f}+C_{\alpha r}\right)+M\left(C_{\alpha f}l_f^2 + C_{\alpha r}l_r^2\right)}{I_z Mv}s
$$

$$
+ \frac{Mv^2\left(C_{\alpha r}l_r - C_{\alpha f}l_f\right)+C_{\alpha f}C_{\alpha r}\left(l_f+l_r\right)^2}{I_z Mv^2}
$$

(9.21)

as the characteristic equation of (9.20) in which $\mathbf{I}$ denotes the unit matrix, and using (11.21), the transformed system output is given by

$$\begin{bmatrix} \beta(s) \\ \dot\psi(s) \end{bmatrix} = \frac{1}{I_z M v\,|s\mathbf{I} - \mathbf{A}|} \cdot$$

$$\begin{bmatrix} I_z C_{\alpha f} s - C_{\alpha f} l_f M v + \dfrac{C_{\alpha f} C_{\alpha r}\left(l_f + l_r\right) l_r}{v} & I_z C_{\alpha r} s + C_{\alpha r} l_r M v + \dfrac{C_{\alpha f} C_{\alpha r}\left(l_f + l_r\right) l_f}{v} \\ C_{\alpha f} l_f M v\ s + C_{\alpha f} C_{\alpha r}\left(l_f + l_r\right) & -C_{\alpha r} l_r M v\ s - C_{\alpha f} C_{\alpha r}\left(l_f + l_r\right) \end{bmatrix} \begin{bmatrix} \delta_f(s) \\ \delta_r(s) \end{bmatrix}$$

$$(9.22)$$

It is readily to see that there are two ordinary transfer functions relating the transforms of the inputs and outputs. Each of them delivers a part of the output due to the corresponding input. All transfer functions have the same denominator – the characteristic equation.

NATURAL FREQUENCY, DAMPING AND STABILITY

The characteristic equation essentially influences and characterizes the transient response and the steady-state behavior of a system. In Chapter 11, a standard characteristic equation form of a second-order linear systems is given by

$$\ddot{s} + 2\zeta\omega_o \dot{s} + \omega_o^2 = 0 \tag{9.23}$$

where ζ is the damping ratio, and ω_o is the natural frequency. Referring to the above form, the characteristic equation (9.21) yields the following parameterization:

$$2\zeta\omega_o = \frac{I_z\left(C_{\alpha f} + C_{\alpha r}\right) + M\left(C_{\alpha f} l_f^2 + C_{\alpha r} l_r^2\right)}{I_z M v}$$

$$\omega_o^2 = \frac{M v^2\left(C_{\alpha r} l_r - C_{\alpha f} l_f\right) + C_{\alpha f} C_{\alpha r}\left(l_f + l_r\right)^2}{I_z M v^2} \tag{9.24}$$

As described in Chapter 11, which presents some basics of control theory, the roots of the characteristic equation are computed by the quadratic formula to be

$$s_1 = -\zeta\omega_o + \sqrt{\zeta^2\omega_0^2 - \omega_0^2}$$

$$s_2 = -\zeta\omega_o - \sqrt{\zeta^2\omega_0^2 - \omega_0^2} \tag{9.25}$$

From the stability theory, we know that system (9.19) is stable if and only if the *eigenvalues* of the matrix $\mathbf{A}$, i.e., the roots of $|s\mathbf{I} - \mathbf{A}| = 0$, all lie in the left-half s-plane. In fact, that the term of $\zeta\omega_o$ presented by (9.24) is always positive, the location of the roots is decided through ω_0^2. The imaginary part of the poles, $\omega_o\sqrt{1 - \zeta^2}$, is the so-called *damped natural frequency*.

Considering the understeer gradient given by (9.17), ω_0^2 can be rewritten

$$\omega_0^2 = \frac{C_{\alpha f} C_{\alpha r} \left(l_f + l_r \right) \left(v^2 U_g + \left(l_f + l_r \right) \right)}{I_z M v^2} \tag{9.26}$$

If the vehicle is in understeer, i.e., $U_g > 0$, ω_0^2 is always positive. The poles lie in the left-half s-plane. So, the system is stable and nonoscillatory. For an oversteer vehicle, i.e., $U_g < 0$, ω_0^2 could result in zero at a particular speed. This speed is called *critical speed* and defined as

$$v_{crit}^2 = -\frac{l_f + l_r}{U_g} \tag{9.27}$$

At the critical speed, the roots of the characteristic equation $\left| s\mathbf{I} - \mathbf{A} \right|$ are $s_1 = 0$ and $s_2 = -2\zeta\omega_o$, which present the dynamic characteristics of an integrator and lag second-order system. Due to the unstable pole, the yaw rate as the system output will tend to the infinite when $t \to \infty$. Chapter 11 shows the details of this kind of unstable system. As a conclusion, the oversteer vehicle is stable up to the critical speed. We will consider the critical speed regarding the steady-state conditions later again in this section.

Given (9.24), ω_0^2 can be positive or negative. There are three possibilities for the location of the poles. Figure 9.6 will show the significance of these three pole locations. In this figure, the vehicle initially is traveling in a straight line at a constant speed, and a step input is made to the steering wheel angle after 5 seconds of the motion. The positions of the poles are analyzed here:

a. $\zeta^2\omega_0^2 > \omega_0^2 > 0$

In this case, both system poles are negative real. For a constant steering input, the yaw rate as well as vehicle sideslip angle are approaching a steady-state without excessive overshoot. The transient response corresponds to (11.9). In addition, the third plot in Figure 9.6 displays an aerial view of the trajectory traversed by the vehicle CG when it is in the steady-state. As equation (9.7) explains, the turning radius is determined by the steering wheel angle and the vehicle speed. If the driver holds the steering wheel at a constant angle and keeps the speed unchanged, the trajectory is a circle with a constant radius.

b. $\omega_0^2 > \zeta^2\omega_0^2 > 0$

For this condition, (11.6) yields a complex conjugate pair of poles. The transient response corresponds to (11.8) exhibiting oscillatory behavior. However, the vehicle keeps moving along a circular course as well. Of course, the turning radius varies with the vehicle speed. Increasing the speed will turn the sideslip to negative values and thus enlarge the circle.

c. $\omega_0^2 < 0$

To meet this condition, the vehicle must be oversteered and the vehicle speed must go beyond the critical speed. One of the system poles is positive real while the other one is negative real, which corresponds to (11.12) showing unstable behavior. Both yaw rate and the sideslip angle are out of control. In this unstable situation, the vehicle speed and the turning radius are not able to match the steering wheel angle controlled by the driver, so that the turning radius becomes smaller and smaller, and the vehicle will go off the course in the end.

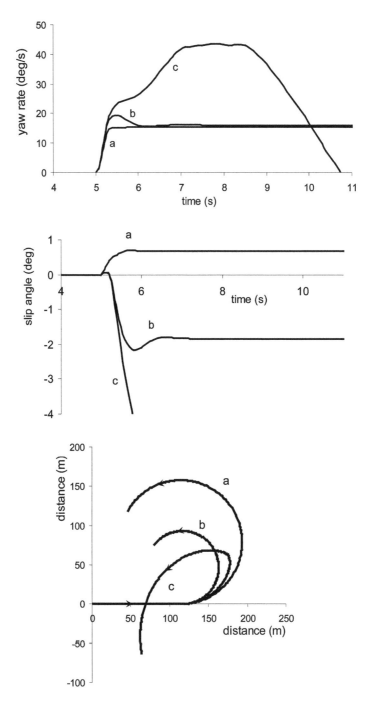

FIGURE 9.6 Time responses and vehicle trajectories corresponding to pole locations.

Apparently any variation of the system parameters, such as the vehicle mass, the cornering stiffness, the speed, the wheelbase and the moment of inertia, will move the location of the system poles and thus affect the dynamics in the transient response. The following example in transient response will demonstrate the impact of these parameters.

STEERING RESPONSE CHARACTERISTICS

We have seen above that the system characteristic equation determines the transient response, which impacts vehicle reactions to the steering inputs resulting in understeer and oversteer maneuvers. The next question for consideration will be, is it possible to derive certain conclusions on vehicle performance from the transient response? In Example 9.2, step inputs were applied to the system. A Step function is a common test input to evaluate the performance of the control systems. Referring to (9.19), the step input means a sudden change of the steering wheel angle. Clearly, the vehicle's response behavior to this kind of input provides the status about the speed of the response, vehicle stability under existing conditions as well as for the stiffness characteristics of the steering system.

For the purpose of theoretical analysis, using a mathematical step function is common and sufficient as described in Chapter 11. However, the step function cannot be used in vehicle testing because the infinitely steep slope cannot be implemented in actual tests. For this reason, the ramp function is normally used for the vehicle testing. Figure 9.7 shows the steering ramp input and the transient response of the yaw rate, respectively.

In ISO7401 [4], the steering input and the test procedure are standardized. For example, from straight-line driving at a constant speed of approximately 80 km/h the steering wheel is moved as fast as possible to the angle position that will result in a lateral acceleration of 4 m/s² as the vehicle now begins to corner. The speed of the steering wheel angle is set at 500°/s for tests.

The transient response can be used for viewing stability of the vehicle reaction to the steering input. On the other hand, the transient response exhibits the vehicle dynamic characteristics, which are essential factors for driver's subjective perception on how good or bad the vehicle performs. Hence, the driver's perception and then an action becomes a part of vehicle control systems resulting

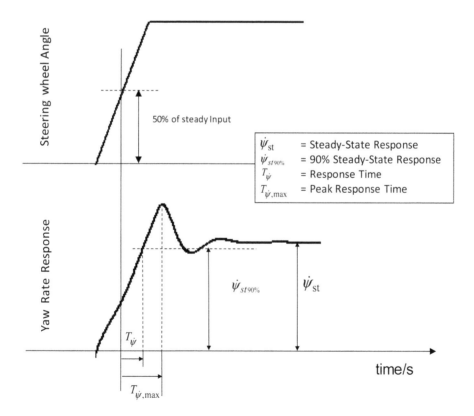

FIGURE 9.7 Steering input and its transient response.

in a certain vehicle behavior, i.e., in the performance of the vehicle. In practice, the following characteristics can be used to evaluate the vehicle performance in a subjective way [1]

- *Yaw rate gain factor*
- The *peak response time*
- The *maximum percent overshoot*

The *yaw rate gain factor*, the quotient of the yaw rate and the steering wheel angle, is a measure of how much of the steering angle the driver needs in order to generate a certain yaw response. It is denoted as K_ψ and calculated by (9.29) in the next section of the book. A precise steering system is characterized by a large gain factor, but if it is too large, a small unintentional steering could cause vehicle reaction.

Referring to Figure 9.7, the *peak response time* is the time at the maximum peak of the overshoot. A large $T_{\dot\psi max}$ means a major phase delay between the steering wheel input and the yaw speed response, so the vehicle can be perceived as inert by possessing poor cornering ability. The *maximum percent overshoot* is the difference between the maximum yaw rate and the steady yaw rate normalized on the steady yaw rate, and it can be used as an indicator of vehicle stability. Higher values of the maximum overshoot with a long transient period can alleviate vehicle stability.

Note that terms "large, short and small" used in the above descriptions are relative and subjective. There are no defined values to be referenced. Engineers usually make the evaluation based on their experience. As an example, reference [1] showed some measurement data on passenger cars: $K_\psi = 0.2 \cdots 0.3$; $T_{\dot\psi max} = 0.3 \ldots 0.5 \mathrm{s}$; $(\dot\psi_{max} - \dot\psi_{st}) / \dot\psi_{st} = 13\% \cdots 27\%$, where the steering rate is at $200° / \mathrm{s}$ and the vehicle speed is of 100 km/h.

Example 9.2

Use LabVIEW® to set up a simulation of model (8.41) based on the following vehicle parameters: $M = 2280$ kg, $I_z = 5050$ kgm^2, $l_f = 1.39$ m, $l_r = 1.81$ m, $C_{\alpha f} = 100000$ N/rad, $C_{ar} = 160000$ N/rad. The vehicle speed is at $v = 50$ km/h.

Analyze the transient responses of the yaw rate and sideslip angle respectively to a step input of 90° of the steering wheel angle. Note that the step function in LabVIEW is simulated as a sudden change step. What change of the transient response is to observe when a) the vehicle speed increases, b) the mass and the moment of inertia increase, c) CG shifts to the rear?

The simulation is programmed in LabVIEW, and the program setup is presented in Appendix A.3. Here the linear bicycle model (8.41) is implemented by MathScript in the block diagram. The While-loop and Event-Structure are used to manage the parameter change and to run the simulation several times when needed. The model parameters are to insert in the front panel.

Figure 9.8 shows the simulation results of the vehicle sideslip angle and yaw rate responses to a step input of 90° at the steering wheel, and the vehicle data is listed below the figure. To view the transient responses, the vehicle parameters are changed in an order. Figure 9.8a shows the transient responses of the yaw rate and the sideslip angle with the initial parameters. As the simulations shown, the maneuvers contribute a sideslip angle around 2°. With increasing vehicle speed v, both $\zeta\omega_o$ and natural frequency ω_o decrease, which degrades the damping of the system oscillation. Therefore, the vehicle tends to oscillate easily at high speed, as shown by Figure 9.8b.

From the study of the tire characteristics in Chapter 3, the cornering stiffness changes when the normal tire force varies as shown in Figure 3.4. Therefore, an increase of the vehicle mass shall result in a slight increase of the cornering stiffness. Ultimately the mass change affects the coefficients of the entire transfer function (9.22), hence impacts the poles position as well. The calculation of the poles proves that $\zeta\omega_o$ and the damped natural frequency are becoming higher. The oscillation effect due to the position change of the poles cannot be clearly seen in Figure 9.8c. Figure 9.8d shows the effect of the understeer gradient. Normally a change of the weight

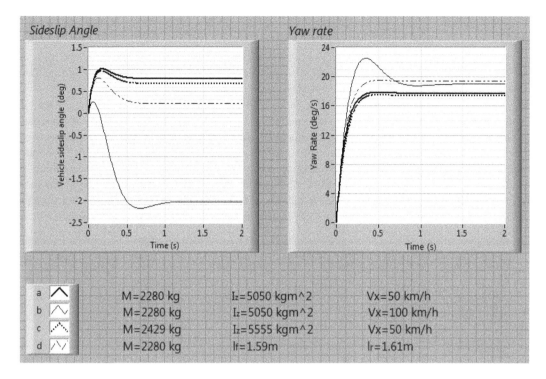

FIGURE 9.8 Transient responses of the sideslip angle and yaw rate.

distribution will cause a shift of the CG position, which in turn influences the understeer gradient of the vehicle, hence the transient response. If the weight transfer is biased to the rear wheels, the CG position will follow it rearward, and the understeer gradient will decrease. If the vehicle is less understeered, applying the same steering wheel angle leads to a higher lateral acceleration and receives a strong yaw response.

The example illustrates different system responses when vehicle parameters are changing. This fact needs to be considered in the control design, i.e., the controller should be sufficiently robust to the variation of vehicle parameters.

❑

FREQUENCY RESPONSE

Another popular practice for analysis and control design is the frequency-response method. The system behavior is evaluated from the steady-forced response to a sinusoidal input, such as $sin\omega t$. The frequency response testing, over a range of frequencies ω, is a convenient method for assessing system dynamics. Applying a sinusoidal input with amplitude A_δ to the steering angle gives

$$\delta(t) = A_\delta sin\omega t,$$

the vehicle dynamic system responds with yaw rate

$$\dot{\psi}(t) = A_\psi sin(\omega t - \phi_\psi),$$

where A_ψ is the amplitude and ϕ_ψ is the *phase angle* between the steering angle input and the yaw rate output. As the standard test procedure defined in ISO7401, the steering wheel angular position or its amplitude should be previously measured during a steady-state circular motion so that the vehicle develops a lateral acceleration of 4 m/s².

In general, the primary reason of using the frequency-response method is controller design in the face of uncertainty in plants, such as systems with poorly known or varying high-frequency resonances. A feedback compensation can alleviate the effects of those uncertainties. Specifically, applying the frequency response method to vehicle dynamics allows using experimental data for control design purposes, for example, for designing filters that pass only input signal components in a selected range of frequencies. This relates to the problem of designing systems for a specified bandwidth. It should be also pointed out that for some vehicle testing at high speeds, which generate high lateral accelerations, and thus require large area for maneuver. The transient response method cannot be used since it requires high effort to control the vehicle. This method can be used to analyze vehicle dynamics with uncertainties in control systems, when for instance considering s driver in the control loop. Also, for analysis of vehicle vertical dynamics, the frequency-response method is a useful method to examine vibrations produced on the sprung mass as a result of inputs from road roughness, arising from road stochastic nonuniformities, or forces in suspension.

With respect to the steering input, only the frequency range of 0.5–2.0Hz is relevant for simulations and tests with a human driver because the driver can merely perform the steering in this range. The following example shows a typical frequency response of vehicle lateral dynamics, in which the linear bicycle model is assumed.

Example 9.3

Use the same simulation setup of Example 9.2 to measure the frequency response. According to the standard test procedure defined in ISO7401, the steering wheel angular position or its amplitude should be previously measured during a steady-state circular motion so that the vehicle develops a lateral acceleration of 4 m/s². Now, consider the vehicle is driven straight-line at speed of $v = 80$ km / h. Analyze the frequency responses of the yaw rate and the sideslip angle respectively. What a change of the frequency response is to be observed when the understeer gradient changes? Such change of the understeer gradient can be created by moving the CG location.

The simulation is programmed in LabVIEW, and the program setup is presented in Appendix A.4. The amplitude of the sinus function is measured in the simulation at a lateral acceleration of 4 m/s². From a straight-line driving at a constant speed of approximately 80 km/h, the steering wheel begins to perform sinusoidal move with a pre-defined frequency. Figure 9.9 shows the results of a vehicle with two different understeer gradients, respectively. The understeer gradients of the vehicle in this example are (a) $U_g = 0.0067$ (b) $U_g = 0.0044$. The difference of the gradients is caused by different positions of the gravity center. The stronger the understeer, the larger gain the frequency response receives in order to achieve the same lateral acceleration. The *break frequency* (also called *corner frequency*) is around 1.0 Hz, where the yaw rate phase angle falls at −45°. The phase angle becomes larger while the frequency is increasing, and is independent of the understeer gradient.

❏

In frequency response, a common measure for system performance is the so-called *bandwidth*, which is the range of frequencies over which the gain of response makes at least 0.707 of its value at the zero-frequency, and starts to fall with a higher frequency. Within the bandwidth the response is considered satisfactory. From the example above, we can see that the bandwidth is close to the break frequency 1Hz.

Similar to the response time in the transient response, the bandwidth is a measure of speed of the response. A larger bandwidth implies that a system is able to respond quicker. Therefore, the response time of the transient response can be estimated from the bandwidth. The overshoot in frequency response indicates similar sense as in the transient case. Thus, the same information is essentially contained in both the transient and the frequency responses.

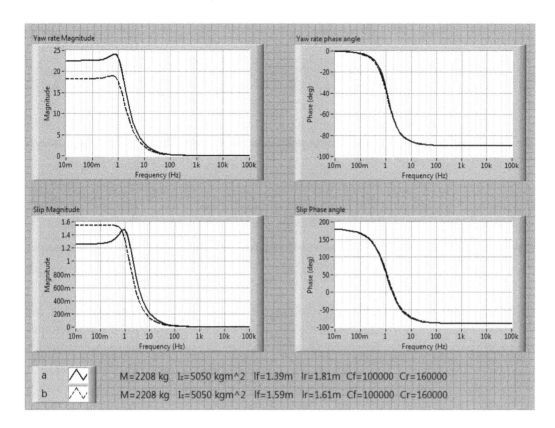

FIGURE 9.9 Frequency response of vehicle with different understeer gradients.

9.4 STEADY-STATE GAINS

Substantial analysis of lateral dynamics of a motor vehicle moving on a curve can be made by considering steady-state motion.

The steady-state gain, defined for $t \to \infty$, is the ratio of a motion variable, i.e., an output, with respect to a constant input. It can be determined using the *final value theorem*. Following (9.22), we have the yaw rate transfer function to the front and rear steer angles:

$$\dot{\psi}(s) = \frac{C_{\alpha f} l_f M v \ s + C_{\alpha f} C_{\alpha r}\left(l_f + l_r\right)}{I_z M v |s\mathbf{I} - \mathbf{A}|} \delta_f(s) + \frac{-C_{\alpha r} l_r M v \ s - C_{\alpha f} C_{\alpha r}\left(l_f + l_r\right)}{I_z M v |s\mathbf{I} - \mathbf{A}|} \delta_r(s) \quad (9.28)$$

Using the final value theorem, the steady-state gain with respect to the front steer angle and the rear steer angle can be calculated as:

$$K_{\psi f} = \lim_{t \to \infty} \frac{\dot{\psi}(t)}{\delta_f(t)} = \lim_{s \to 0} \frac{\dot{\psi}(s)}{\delta_f(s)} = \frac{v}{v^2 U_g + l_f + l_r} \quad (9.29)$$

$$K_{\psi r} = \lim_{t \to \infty} \frac{\dot{\psi}(t)}{\delta_r(t)} = \lim_{s \to 0} \frac{\dot{\psi}(s)}{\delta_r(s)} = -\frac{v}{v^2 U_g + l_f + l_r} \quad (9.30)$$

The steady-state gain of the yaw rate resulted from the rear steering is just the negative one from the front steering if the front and rear wheels are steered in the same direction.

In the same way, we obtain the sideslip angle transfer function:

$$\beta(s) = \frac{I_z C_{\alpha f} s - C_{\alpha f} l_f M v + \dfrac{C_{\alpha f} C_{\alpha r}\left(l_f + l_r\right) l_r}{v}}{I_z M v \left| s\mathbf{I} - \mathbf{A}\right|}\delta_f(s) + \frac{I_z C_{\alpha r} s + C_{\alpha r} l_r M v + \dfrac{C_{\alpha f} C_{\alpha r}\left(l_f + l_r\right) l_f}{v}}{I_z M v \left| s\mathbf{I} - \mathbf{A}\right|}\delta_r(s)$$

(9.31)

The steady-state gains of the sideslip angle with respect to the front steer angle and the rear steer angle are determined, respectively:

$$K_{\beta f} = \lim_{t\to\infty}\frac{\beta(t)}{\delta_f(t)} = \lim_{s\to 0}\frac{\beta(s)}{\delta_f(s)} = \frac{-l_f M v^2 + C_{\alpha r}\left(l_f + l_r\right) l_r}{C_{\alpha r}\left(l_f + l_r\right)\left(v^2 U_g + l_f + l_r\right)}$$

(9.32)

$$K_{\beta r} = \lim_{t\to\infty}\frac{\beta(t)}{\delta_r(t)} = \lim_{s\to 0}\frac{\beta(s)}{\delta_r(s)} = \frac{l_r M v^2 + C_{\alpha f}\left(l_f + l_r\right) l_f}{C_{\alpha f}\left(l_f + l_r\right)\left(v^2 U_g + l_f + l_r\right)}$$

(9.33)

On the other hand, the lateral acceleration is a very useful signal for vehicle dynamics control, and the lateral acceleration sensor is widely used in today's vehicle technology. The lateral acceleration gain, defined as the ratio of the steady-state lateral acceleration to the steer angle, is another commonly used parameter for evaluating the steering response of a vehicle. From (2.53), we know that the lateral acceleration is related to the sideslip angle and the yaw rate through

$$a_y = v\left(\dot{\beta} + \dot{\psi}\right)$$

(9.34)

Taking the Laplace transform of (9.34) and using (9.28) and (9.31), we obtain the lateral acceleration transfer function:

$$a_y(s) = \frac{C_{\alpha f} I_z v\, s^2 + C_{\alpha f} C_{\alpha r}\left(l_f + l_r\right)\left(l_r s + v\right)}{I_z M v \left| s\mathbf{I} - \mathbf{A}\right|}\delta_f(s)$$

$$+ \frac{C_{\alpha f} I_z v\, s^2 + C_{\alpha f} C_{\alpha r}\left(l_f + l_r\right)\left(l_f s - v\right)}{I_z M v \left| s\mathbf{I} - \mathbf{A}\right|}\delta_r(s)$$

(9.35)

Likewise, the acceleration steady-state gain with respect to the front steer angle and the rear steer angle are given, respectively:

$$K_{af} = \lim_{t\to\infty}\frac{a_y(t)}{\delta_f(t)} = \lim_{s\to 0}\frac{a_y(s)}{\delta_f(s)} = \frac{v^2}{v^2 U_g + l_f + l_r}$$

(9.36)

$$K_{ar} = \lim_{t\to\infty}\frac{a_y(t)}{\delta_r(t)} = \lim_{s\to 0}\frac{a_y(s)}{\delta_r(s)} = -\frac{v^2}{v^2 U_g + l_f + l_r}$$

(9.37)

A comparison of (9.13) to above equation (9.36) shows the identical form, in which (9.36) is the Laplace transform of (9.13).

From the above results, the steady-state gains of the yaw rate, the sideslip angle and the lateral acceleration with respect to the steer angle show a fundamental characteristic, which is a function determined by the understeer gradient and the vehicle speed. Basically, in the steady-state the system outputs are determined by the understeer gradient relative to the speed for a given steering input. Once the driver steers the front or rear wheels, the vehicle can maintain the driver's intended turning radius only when the yaw rate and lateral acceleration correspond to the steady-state form. In this context, the dependence of the steady-state gain upon the understeer gradient will be further elaborated in Section 9.6.

STEADY-STATE YAW RATE

From the steady-state gain of the yaw rate, the steady-state yaw rate can be directly derived. On the other hand, the explanation from geometric point of view would provide a better physical understanding. For simplification, the rear steering angle is set to zero in the following.

For a constant steering, the vehicle in steady-state moves at a constant speed (the throttle is in a constant position, too) and turns on a circular path with a constant radius, as previously shown in Figure 9.2 The turning radius can be calculated from (9.12):

$$\frac{1}{R_T} = \frac{\delta_f}{v^2 U_g + l_f + l_r} \tag{9.38}$$

The steady-state yaw rate follows by using (8.15):

$$\dot{\psi}_{st} = \frac{v\,\delta_H}{i_s(v^2 U_g + l_f + l_r)} \tag{9.39}$$

which is the same as (9.29) that was derived from the yaw rate transfer function.

Figure 9.10 displays the vehicle in a turn. The yaw angle, ψ, describes the orientation of the vehicle referring to the initial coordinate system 0_{xyz}. In order to keep the driving course, the velocity at the CG has to make angle β relative to axis x_v, i.e., to stay tangent to the trajectory path. Therefore, the *course angle*, denoted as φ,

$$\varphi = \psi + \beta \tag{9.40}$$

keeps the vehicle on the course. Note that the angles follow the right-hand rule, and hence the sideslip angle is negative. In the steady state, the derivative of the course angle is equal to the yaw rate since $\dot{\beta} = 0$:

$$\dot{\varphi}_{st} = \dot{\psi}_{st} \tag{9.41}$$

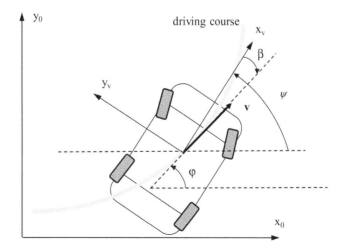

FIGURE 9.10 Vehicle orientation at turning.

From (9.32), we can determine the steady sideslip angle as well, that is

$$\beta_{st} = \frac{-l_f M v^2 + C_{\alpha r}\left(l_f + l_r\right)l_r}{C_{\alpha r}\left(l_f + l_r\right)\left(v^2 U_g + l_f + l_r\right)}\delta_f \tag{9.42}$$

For vehicle dynamics control, the steady value of the yaw rate is often considered as a desirable control target to improve the vehicle stability and handling during a turn. We will study it again in the control design of Chapter 14.

9.5 CHARACTERISTIC AND CRITICAL SPEEDS

In order to derive the relation of the understeer gradient along with the vehicle speed in steady state, we review the yaw rate steady-state gain one more time. The approach provides the same understanding for the sideslip angle and the acceleration steady-state gains.

The yaw rate steady-state gain with respect to the front steer angle for understeered vehicle has a maximum. The *characteristic speed*, v_{ch}, is the speed at that maximum. It can be determined by setting

$$\frac{dK_\psi}{dv} = 0 \tag{9.43}$$

Using (9.29), it follows

$$v_{ch}^2 = \frac{l_f + l_r}{U_g} \tag{9.44}$$

The yaw rate steady-state gain presented by (9.29) becomes

$$K_\psi = \frac{v_{ch}}{2(l_f + l_r)} \tag{9.45}$$

Figure 9.11 shows the graph of the steady-state gain of the yaw rate with respect to the front steer angle along with the speed, where the characteristic speed v_{ch} will be detailed in the next section.

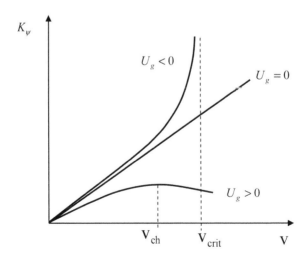

FIGURE 9.11 Yaw rate steady-state gain with different understeer gradients.

From (9.44), we see that the characteristic speed directly related to the understeer gradient. Evaluating (9.12) at v_{ch} by using $a_y = v_{ch}^2 / R_T$, the front steer angle at the characteristic speed results in

$$\delta_f = \frac{2(l_f + l_r)}{R_T} + \delta_r \tag{9.46}$$

Without the rear steering ($\delta_r = 0$), the front steer angle is the double Ackermann angle. Since the above equation was obtained for the characteristic speed, the equation can be utilized also for experimental determining v_{ch}. To measure the characteristic speed under steady-state conditions, various types of test can be conducted on a large, flat, paved area. From (9.12) and (9.13), we know that two tests can be distinguished: one test is conducted at a constant turning radius and the other one is with a constant vehicle speed. While in the first test the vehicle is driven along a curve with a constant radius at various speeds, in the second test the vehicle is driven at a constant forward speed and various turning radii. In both cases, the steering wheel angle and the lateral acceleration are measured. The constant forward speed test has a disadvantage that it requires a large test area, which is in practice not always available. Therefore, the constant radius test is usually preferred.

Due to the nonlinear behavior of tires, vehicle dynamics and the effect of the tire forces, the value of characteristic speed varies with the operating conditions. Figure 9.12 shows a typical measurement, in which the test vehicle is driven along a curve with a constant radius at various speeds. A curve rather than a straight line represents the steer angle as a function of the lateral acceleration.

Starting at the Ackermann angle, a straight line can be approximated through the curve up to $a_y = 4$ m/s^2 because of the linear behaviour. In its extension until to an acceleration a_{y_ch}, the approximated straight line reaches the value of the twice Ackermann angle, which is associated with the characteristic speed for a given radius. Since a_{y_ch} is determinable in the graphic and the turning radius is a constant to be known, the characteristic speed is calculated:

$$v_{ch} = \sqrt{a_{y\_ch} R_T} \tag{9.47}$$

Hence, the understeer gradient is obtained from (9.44)

$$U_g = \frac{l_f + l_r}{v_{ch}^2} \tag{9.48}$$

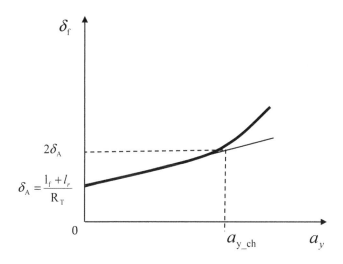

FIGURE 9.12 Measurement of the characteristic speed.

This calculation can be verified with the slop of the approximated straight line in Figure 9.12. The lower the characteristic speed, the more the vehicle tends to understeer.

In the previous section, the critical speed, as shown by (9.27), was introduced when the transient response of the system was studied. Now, under the steady-state conditions, the critical speed leads to the same physical phenomenon for the yaw rate, the sideslip angle and the lateral acceleration. Considering steady-state gains of the above-listed parameters, it can be stated that they all tend to the infinite if

$$v^2 U_g + l_f + l_r = 0 \tag{9.49}$$

Hence, the vehicle becomes unstable. As seen from Figure 9.11, the understeer gradient is negative and drastically increases its magnitude. This happens only at the critical speed when the vehicle goes in severe oversteer

$$v_{crit}^2 = -\frac{l_f + l_r}{U_g} \tag{9.50}$$

It can be noticed in conclusion that the critical speed only has a real meaning when vehicle is oversteered while the characteristic speed is a parameter for understeered vehicles.

9.6 STABILITY CONSIDERATION

Consider the characteristic speed again and place its expression from (9.50) into (9.26), we have

$$\omega_0^2 = \frac{C_{\alpha f} C_{\alpha r} (l_f + l_r)^2}{I_z M v^2} \left(\frac{v^2}{v_{ch}^2} + 1 \right) \tag{9.51}$$

From the discussion about term ω_0^2 of natural frequency, we know that it can be positive or negative. Since v_{ch}^2 is positive, ω_0^2 will be always positive and hence the vehicle is stable. In this case, the understeer gradient is positive as seen from (9.48). Therefore, understeered vehicle is stable. If a vehicle is oversteered, i.e., $U_g < 0$, ω_0^2 is positive until the critical speed v_{crit} as defined in (9.50). Beyond the critical speed, ω_0^2 becomes negative. As a result, an oversteered vehicle is only stable until to a certain speed, which is the critical speed, and becomes unstable beyond this speed.

Basically, the critical speed has only a theoretical meaning in the linear range since all the vehicles are designed to understeer satisfying the stability condition. However, the critical speed does have a meaning in the nonlinear range of vehicle dynamics, especially when a vehicle is in its unsteady state. Due to the nonlinearities involved in tires, suspension, weight load transfer as well as the traction and braking characteristics, the vehicle may oversteer during severe dynamic maneuvers. For such operating conditions, the vehicle dynamics can be linearized so that the equations are still applicable. The vehicle remains stable only if its speed is below the critical speed. The following example shows behaviors of an oversteered vehicle at different operating conditions when the vehicle speed is below and above the critical speed.

Example 9.4

This example shows the frequency response of the yaw rate and sideslip angle of an oversteered vehicle, which speed is below and above the critical speed. The same simulation program and vehicle parameters were used in Example 9.3. The calculated critical speed is at $v_{crit} = 39 \text{ m/s}$. In Figure 9.13a the vehicle remains stable at $v = 22.2 \text{ m/s}$ while Figure 9.13b shows the unstable vehicle at $v = 48.2 \text{ m/s}$. One can see that the phase angle of the yaw rate is at $-180°$ when $\omega \to 0$ indicating the second-order system has a pole in right-half s-plane.

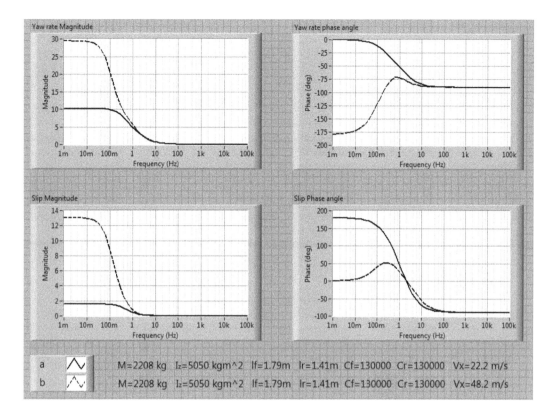

FIGURE 9.13 Frequency response of an oversteered vehicle at different speed.

❏

9.7 INFLUENCE OF 4WS CONFIGURATION

From the steering geometry introduced in the previous chapter, we understand the relation of the steering angles between the front and rear. The rear steer angle has a significant influence on steering performance measured by turnability, dynamic response and stability of vehicles. These phenomena will be examined in this section.

In the previous chapter, the four-wheel steering geometry was outlined. We continue the analysis based on the bicycle model. As mentioned before, the rear steer angle is limited due to safety reasons, typically $\delta_r < 4°$. In order to ease the analysis and control design, the rear steer angle can be defined as a function of the front one, while the ratio is

$$\kappa_s = \frac{\delta_r}{\delta_f} \tag{9.52}$$

where factor κ_s is not necessarily constant. This factor can also be considered as a function of controller parameters or some vehicle parameters. If $\kappa_s = 0$, it corresponds to the front-wheel steer. For $\kappa_s > 0$, the rear wheels are steered in-phase with the front wheels. For the bicycle model, $\delta_A = \delta_f$. Substituting (9.52) into (9.4), the Ankermann angle results in

$$\delta_A = \frac{l_f + l_r}{R_T} + \kappa_s \delta_A \tag{9.53}$$

One can see that the turn radius is enlarging for the vehicle with the front and rear wheels steered in phase as compared to the same vehicle with the front steered wheels only. This follows from the following expression:

$$R_T = \frac{l_f + l_r}{\delta_A (1 - \kappa_s)} \tag{9.54}$$

In order to keep the turn radius constant, the front steer needs to increase to compensate the factor of $(1 - \kappa_s)$. The above results of the in-phase steering are reflected in the geometry shown in Figure 9.14a, where one also can see that a larger turn radius worsen the vehicle turnability.

For $\kappa_s < 0$, the rear wheels are steered out-of-phase with the front wheels. Equations (9.53) and (9.54) conclude the opposite result that the turn radius is decreasing for a constant front steer. Figure 9.14b presents out of-phase geometry and illustrates that the out-of-phase steering makes vehicle more manoeuvrable.

Based on the analysis in Chapter 8 and the result given by (8.40), the rear steering introduces an impact on vehicle dynamics and thus can influence the dynamic response as the front steering vehicle does. Merging (9.52) into (9.19), we have

$$\begin{bmatrix} \dot{\beta} \\ \ddot{\psi} \end{bmatrix} = \begin{bmatrix} -\dfrac{C_{\alpha f} + C_{\alpha r}}{Mv} & -1 - \dfrac{C_{\alpha f} l_f - C_{\alpha r} l_r}{Mv^2} \\ -\dfrac{C_{\alpha f} l_f - C_{\alpha r} l_r}{I_z} & -\dfrac{C_{\alpha f} l_f^2 + C_{\alpha r} l_r^2}{I_z v} \end{bmatrix} \begin{bmatrix} \beta \\ \dot{\psi} \end{bmatrix} + \begin{bmatrix} \dfrac{C_{\alpha f} + \kappa_s C_{\alpha r}}{Mv} \\ \dfrac{C_{\alpha f} l_f - \kappa_s C_{\alpha r} l_r}{I_z} \end{bmatrix} \delta_f \mathfrak{A}$$

$$\tag{9.55}$$

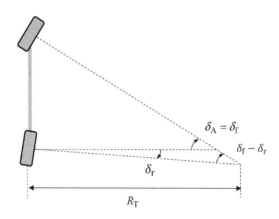

a. in-phase turning geometry

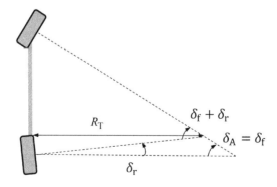

b. out-of-phase turning geometry

FIGURE 9.14   Turning geometry in four-wheel steering vehicles.

Depending on the value of factor κ_s, the behavior of the yaw rate as well as and the sideslip angle is influential as seen from (9.55).

At low-speed turning of vehicles with the four-wheel steering, the transient response of the side-slip angle and the yaw rate is not an essential factor. Vehicle dynamics is still in its linear zone according to the model given by (9.55), and the lateral acceleration reaches the same level as compared to the front-wheel steering. The out-of-phase rear steer enhances maneuverability by reducing the turn radius.

At high-speed turning, where the vehicle lateral acceleration normally beyond $4 \ \text{m}/\text{s}^2$, the out-of-phase rear steer can no longer be used. This is because of the increase of the yaw rate; the vehicle tends to move outward, and possibly constitutes an oversteer influence. However, in high-speed situations, the in-phase steering can improve vehicle performance in turning. The transient response plays here an essential role. Figure 9.15 shows the transient response of the yaw rate, the sideslip angle and the lateral acceleration, respectively. The vehicle model simulated by CarSim® software product was then exported to the LabVIEW® environment. Three responses are simulated based on the variation of the rear steer angle by $\kappa_s = 0, 0.2$ and 0.3. With the rear-wheel steer, the yaw rate response time is longer, but it results in a stronger damping behavior of the yaw rate as well as the sideslip angle. As geometrically displayed in Figure 9.14a, the yaw response is lesser when the rear wheels are steered in-phase with the front wheels because the lateral force at the rear is contrary to the yaw moment. However, the rear wheels contribute more lateral force to the turn center, hence the lateral acceleration response will be stronger. Also Figure 9.15 shows a faster buildup of the lateral acceleration with the use of 4WS. This property of the four-wheel steering would be favorable

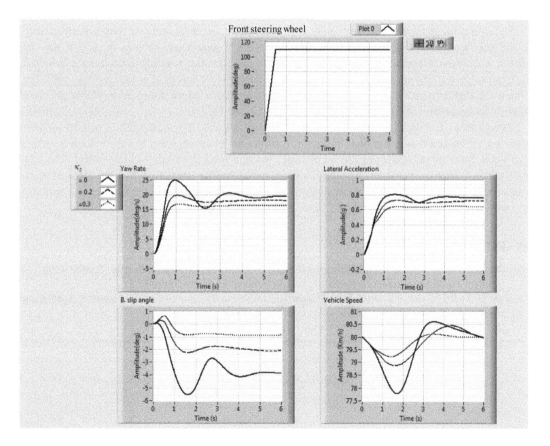

FIGURE 9.15 Yaw rate and lateral acceleration response in 4WS.

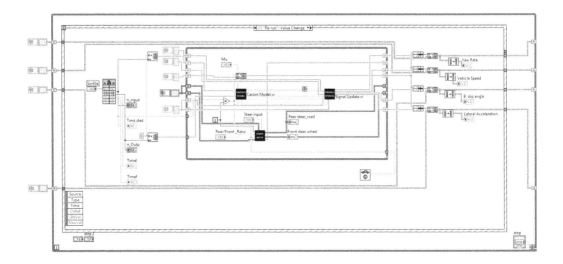

FIGURE 9.16   Block diagram function of 4WS vehicle system.

to vehicles at high speed, i.e., when the lateral acceleration builds up faster and the yaw rate reacts slower with a smaller amplitude. In this case, a vehicle is easier to be controlled and stabilized. Comparing to the front only steering, the sideslip angle is getting smaller with 4WS. This effect will benefit the stabilization when the vehicle is in cornering. In overall, the additional rear steer is able to arbitrarily soften the oscillation and better control vehicle stability during transient maneuvers.

So far, it is recognizable that the rear steer is available resource to improve vehicle performance. If the rear steer angle is selected depending on the vehicle parameters such as the speed, the yaw rate as well as the lateral acceleration, the vehicle can be controlled towards its understeer, which in turn improves stability in turning situations. That is basically a concept of a rear steering controller, which automatically reacts upon a change in driving conditions. More information will be provided later on in the control design described in Chapters 13 and 14.

The simulation shown in Figure 9.15 was conducted in LabVIEW with a vehicle model inputted from a CarSim simulation package. Figure 9.16 shows the block diagram of the simulation, where the block "steer input" shown as sub VI provides the steering input to the vehicle model. The "signals update" function manages the output signals for each time interval so that the measurements can be saved for the plots. The vehicle dynamics model is actually running in CarSim and interfacing with LabVIEW through the block "CarSim model", which is a library function block.

REFERENCES

1. M. Mitschke, Dynamik der Kraftfahrzeuge, Band C: Fahrverhalten, Berlin: Springer-Verlag, 1990.
2. A. Zomotor, Fahrwerktechnik: Fahrverhalten, Würzburg: Vogel Verlag, 1991.
3. A. F. Andreev, V. I. Kabanau, V. V. Vantsevich, Driveline Systems of Ground Vehicles: Theory and Design, Boca Raton, London, New York: CRC Press, 2010.
4. *Road vehicles - Lateral transient response test methods*, ISO Standard 7401, 2011.

10 Normal and Roll Dynamics

In preceding Chapters, we studied the modeling of vehicle longitudinal and lateral dynamics, where the vehicle motion in the normal direction was included in the models in the form of the wheel normal reactions, but has not been analyzed yet. The normal dynamics of a vehicle mostly refers to its response properties to inputs from road roughness. The normal motion is influenced through *suspension* behavior as well as tire/wheel excitation (considering the vehicle body rigid). The suspension is employed to minimize this movement, which affects essentially the normal dynamics and plays an important role for occupants' comfort and the safety. Roll dynamics extends beyond normal dynamics to include lateral forces and moments that influence vehicle motion about the vehicle longitudinal axis and thus can lead to vehicle rollover. The major purposes of this chapter are to describe

- Normal and roll motions
- Transient behavior following a road displacement input
- Relationship between normal dynamics and lateral dynamics.

Vehicles travel on road and experience a broad spectrum of vibrations. From the mechanical construction point of view, the suspension system divides the vehicle body into two parts: *sprung mass* and *unsprung mass*. The sprung mass is the integral unit of the vehicle body supported by the suspension while the axles and associated wheel hardware forms the unsprung mass. Elements of the suspension and the steering system are usually split between the sprung and unsprung masses.

The vibration environment is one of the most important criteria by which consumers judge the design and quality of vehicles. As defined in SAE Terminology [1], the spectrum of vibrations of the sprung mass is usually divided up to frequency and classified as *ride* (0–5 Hz), *shake* (5–25 Hz), *harshness* (25–100 Hz), which is vibrations of the structure and/or components that is perceived tactually and/or audible, and boom (25–100 Hz), which is perceived audibly and characterized as sensation of pressure by the ear. The 25 Hz boundary is approximately the lower frequency threshold of human hearing. The term "ride" is commonly used in reference to low frequency vibrations of vehicles. The ride dynamics basically describes three components of ride vibration in three-dimensional space: *bounce*, *pitch* and *roll*. "Bounce" is a translational component of ride vibration in the direction of the vehicle z-axis. "Pitch" and "roll" refer to the angular components of the ride vibrations about the vehicle y-axis and x-axis, respectively. The modeling described in this chapter focuses on bounce and roll components that actually characterize vehicle normal and roll dynamics.

The vibrations of the unsprung masses are known as *hop*, which is the oscillatory motion between the road and the sprung mass in the direction perpendicular to the road surface. If a pair of wheels hop in phase, this is parallel hop. Tramp is hop of a pair of wheels in opposite phase. Power hop is an oscillatory hopping motion of a single wheel of a pair of wheels when the traction forces are developed by the wheels. Braking hop occurs when the brakes are in action.

There are multiple sources, from which vehicle ride vibrations may be excited. These generally fall into two classes – road roughness and on-board sources. The on-board sources arise from rotating components, and thus include tire/wheel assemblies and powertrain. Without undermining the importance of studying powertrain vibrations, only the road roughness and the tire/wheel as influence sources are considered in this chapter.

A corresponding physical-level scheme of the suspension will be utilized for the modeling of vehicle normal dynamics. Due to the physical properties of design elements and their mechanical connection, the suspension linkages and its other elements possess various nonlinearities. A complete model, which considers all nonlinear effects, will increase the complexity that is not desired

DOI: 10.1201/9781003134305-10

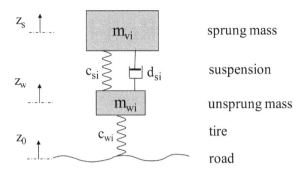

FIGURE 10.1 Quarter-car diagram.

for control design. For this reason, a linear system is used to approach the normal force transfers between the road, the tire, the suspension and the vehicle body. So, the linear model simplifies the dynamics to a system with spring-damper elements. The model has proved in practice as an effective approach to examine the vehicle vibration resulted from the previously-listed normal inputs and established to be fundamental for normal dynamics analysis.

In the following, we start with a simplified suspension approach leading to a quarter-car model. The quarter-car model is then extended to describe the vehicle roll dynamics in the two-dimensional normal plane. Using the roll dynamics model, the steady-state roll behavior will be analyzed to determine the system characteristics of the steady-state response. Finally, a vehicle three-dimensional model will be obtained by integrating the roll dynamics with the lateral dynamics.

10.1 QUARTER-CAR MODEL

The first step is to obtain a quarter-car model, which approaches the physical chain from the partial sprung mass through suspension over the unsprung mass down to the tire and the road. Then we can simply imagine that the entire car model results from four parallel quarter-car models, which sprung masses are combined together in the vehicle sprung mass.

The quarter-car model is widely used in the suspension modeling and control [2–5]. Figure 10.1 illustrates the physical elements of the model, where $i = 1,4$ indicating ith of quarter-car. The suspension is idealized as a spring-damper system that supports the sprung mass. The tire is represented as a simple spring, which usually has higher stiffness then the suspension. That is a further reason why the unsprung mass is treated separately. The road displacement, denoted by z_0, is the system input.

Now, we derive the motion equations of m_{vi} and m_{wi}. The suspension and tire stiffness are denoted by c_{si} and c_{wi}, respectively. d_{si} is the damping ratio of the suspension. The zero position is chosen at the static *equilibrium*[1], where the weight is counterbalanced by a spring and damper forces. Denoting, respectively, by z_s and z_w the displacements of m_{vi} and m_{wi} from their equilibrium positions, we observe the elongation of the spring c_{wi} (measured from the equilibrium position) is equal to the relative displacement $z_w - z_0$, while the elongation of the spring c_{si} is equal to the relative displacement $z_s - z_w$ of m_{vi} with respect to m_{wi}. Figure 10.2 shows the free-body diagram of each mass. The forces from the spring c_{si} on the two masses are equal in magnitude but occur in opposite direction, and likewise for damper d_{si}. The motion equations of m_{wi} and m_{vi} are therefore

$$m_{wi}\ddot{z}_w - d_{si}(\dot{z}_s - \dot{z}_w) - c_{si}(z_s - z_w) + c_{wi}z_w = c_{wi}z_0$$
$$m_{vi}\ddot{z}_s + d_{si}(\dot{z}_s - \dot{z}_w) + c_{si}(z_s - z_w) = 0$$

(10.1)

---

[1] Considering the masses as particles, a particle is said to be in equilibrium when the resultant of all external forces acting on it is zero.

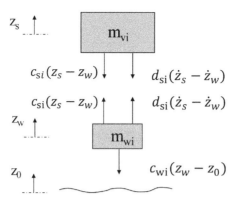

FIGURE 10.2   Free-body diagram of the quarter-car model.

As mentioned before, all the nonlinearities are neglected in the modeling. In the reality, suspensions have continuous distributed stiffness and mass as well as frictions in the struts and bushings. The damper represents the hydraulic shock absorber, which exhibits hysteretic behavior in jounce and rebound directions. Additionally, frictions are also given between piston and cylinder. Therefore, for a comprehensive treatment of suspension in ride analysis, the nonlinearities should be taken into account. More details can be found in literature [2]. The quarter-car model presents the essential dynamics of the vehicle at the most basic level and can be used for the analysis of vehicle vibrations and resonance.

Using the state space form, (10.1) can be written as

$$\mathbf{M}\,\ddot{z}(t) + \mathbf{D}\dot{z}(t) + \mathbf{C}z(t) = \mathbf{h}(t) \tag{10.2}$$

where

$$\mathbf{M} = \begin{bmatrix} m_{wi} & 0 \\ 0 & m_{vi} \end{bmatrix}, \quad \mathbf{D} = \begin{bmatrix} d_{si} & -d_{si} \\ -d_{si} & d_{si} \end{bmatrix}, \quad \mathbf{C} = \begin{bmatrix} c_{wi} + c_{si} & -c_{si} \\ -c_{si} & c_{si} \end{bmatrix},$$

$$z(t) = \begin{bmatrix} z_w \\ z_s \end{bmatrix}, \quad \mathbf{h}(t) = \begin{bmatrix} c_{wi} z_0 \\ 0 \end{bmatrix}$$

VIBRATION FREQUENCY

For the mechanical vibration system given by (10.2), the natural frequency is able to deliver a quantitative view of the vibration behavior. In the absence of damping, i.e., $\mathbf{D} = 0$, the *natural frequency* of system (10.2) is delivered by the eigenvalues determined by the characteristic equation of the system. Following (11.16) in Chapter 11, the characteristic equation is given then

$$p(\lambda) = \lambda^4 + \left(\frac{c_{si}}{m_{vi}} + \frac{c_{wi} + c_{si}}{m_{wi}} \right)\lambda^2 + \frac{c_{wi}c_{si}}{m_{wi}m_{vi}} = 0 \tag{10.3}$$

which leads to the eigenvalues

$$\lambda_{1,2,3,4} = \pm j\omega_{1,2} \tag{10.4}$$

with two natural frequencies

$$\omega_{1,2} = \sqrt{\frac{m_{wi}c_{si} + m_{vi}\left(c_{wi} + c_{si}\right) \pm \sqrt{\left(m_{wi}c_{si} + m_{vi}\left(c_{wi} + c_{si}\right)\right)^2 - 4m_{wi}m_{vi}c_{wi}c_{si}}}{2m_{wi}m_{vi}}} \qquad (10.5)$$

On the other hand, vehicle practical design shows the following typical relationship:

$$m_{vi} \approx 10m_{wi}, \quad c_{wi} \approx 10c_{si} \qquad (10.6)$$

Based on this relation, the natural frequencies can be computed:

$$\omega_1 \approx \sqrt{\frac{c_{si}}{m_{vi}}}, \quad \omega_2 \approx \sqrt{\frac{\left(c_{wi} + c_{si}\right)}{m_{wi}}} \qquad (10.7)$$

The first natural frequency, ω_1, indicates the natural frequency of the isolated sprung mass while ω_2 is the one of unsprung mass.

In a forced vibration, the amplitude may have extreme high peaks at one or more frequencies, so-called *resonance*. When damping is present, the resonance occurs at the "damped natural frequency" of a system. However, due to practical mechanical design of shock absorbers, the damping ratio exhibits less significant influence to the resonant frequency, so that the natural frequency can be approximately considered as the resonant frequency. Since the resonance occurrence is undesirable for the ride comfort, the best isolation is achieved by keeping the natural frequency as low as possible. In practice, the natural frequency is chosen around 1 Hz as the design optimum. For the vehicle roll motion the natural frequency is also one of essential characteristics.

An important criterion of the ride quality is the amplitude frequency response to road acceleration rather than road displacement. The analysis in reference [2] shows that for a given vehicle size the suspension constrains the natural frequency for most cars to a minimum in the 1–1.5 Hz range. Performance cars with stiff suspensions, in which ride is sacrificed for handling benefits, have natural frequencies up to 2 or 2.5 Hz. The natural frequency of the wheel hop is usually chosen near 10 Hz allowing more acceleration transmission in the high-frequency range.

As the shock absorber is an essential element in the suspension, the system shown in Figure 10.1 is a damped forced vibration system. To optimize the vibration behavior of its sprung mass, the suspension stiffness and damping ratio as well as the tire stiffness have to be determined first. They often do not have fixed values but vary within a certain range. Then, a cost function should be selected for the optimization process. Referring to reference [6] there are various cost criteria for optimization can be applied. Given that $D \neq 0$ and $\lambda_j, j = 1...4$ are the eigenvalues of the characteristic equation of (10.2), the *degree of stability* is defined here as the cost function

$$d_s = -\max_{j=1..4}\left\{Re\,\lambda_j\right\} \qquad (10.8)$$

which characterizes the decaying rate of the time transient response. The second criterion is the *degree of damping:*

$$d_d = -\max_{j=1..4}\left\{\frac{Re\,\lambda_j}{|\lambda_j|}\right\} \qquad (10.9)$$

The degree of damping characterizes the damped oscillations before they reach the steady state. For $d_d = 1$, the transient response is aperiodic. Both criteria above can be used to determine the suspension damping ratio.

Another criterion of the ride quality is the *integral of squared error*

$$J = \int_0^\infty \mathbf{x}^T(\tau) \mathbf{S} \mathbf{x}(\tau) d\tau \tag{10.10}$$

with $\mathbf{S} = \mathbf{S}^T \geq 0$ as a weighting matrix and $\mathbf{x}(t) = (z(t), \dot{z}(t))^T$. This cost criterion evaluates the amplitudes. From above, one can see that the minimized cost criteria imply small vibration amplitudes, and they all will yield different results of the optimized parameters. In practice, the parameters are often subject to constraints and the optimization will be adjusted based on the design requirements.

10.2 ROLL MOVEMENT

Vehicle body roll is mainly due to suspension deflection and is considered as the rotational movement of the sprung mass about a longitudinal axis. Besides providing normal compliance for the ride comfort, which was discussed in the previous section, other primary functions of the suspension include to resist to the roll of the chassis and also to react two other forces produced by the tires, i.e., the tire longitudinal and lateral forces. Thus, suspension works, i.e., generate forces, in three-dimensional space. A full kinematic analysis of the required generality to cope with the 3D-operation of suspensions for control design purposes is highly complex. Therefore, the presentation here is based on an appeal to relatively simple notions in common engineering usage for the vehicle control design. We first describe the roll movement in three spatial dimensions and then derive the two-dimensional model of the roll dynamics.

We start with a description of the roll movement in three-dimensional space to explain of the physics of the rollover. Under cornering conditions, the centrifugal force, Ma_y, tends to rollover the vehicle (see Figure 10.3). This force is taken by the tires in the lateral directions. Thus, the moment created by the pair of the centrifugal and lateral tire forces generates a load transfer between the inside and outside wheels. Figure 10.3 shows the vehicle in a left turn, where it is assumed that the steer angle for the turn is small. While the normal loads ΔF_{Nf} and ΔF_{Nr} on the outside wheels increase, the inside wheels have the same amount of the load reduced. In its turn, the varying of the normal wheel load results in a change of the cornering stiffness of the tires, and hence the wheel slip angles. This mechanism is at work on both axles of the vehicle. Greater slip angles on the front

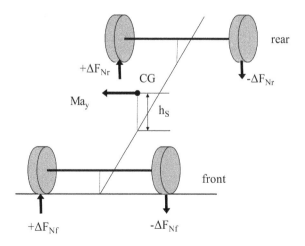

FIGURE 10.3   Lateral load transfer on two-axle vehicle.

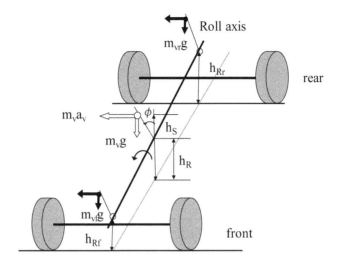

FIGURE 10.4   Roll centers on two-axle vehicle.

wheels contribute to vehicle understeer, whereas greater slip angles on the rear wheels contribute to oversteer.

Since the roll movement affects the individual normal tire forces, it is important to consider the roll angular motion of the sprung mass and the load transfer at the front and rear axles. This will help reach a better understanding in vehicle lateral dynamics with respect to the suspension influence. As indicated in Figure 10.3, however, the lateral load transfers of the front and rear wheels are two unknown parameters. The calculation of these two parameters needs two related equations. For this reason, the suspension compliance has to be taken into account. Also, the *roll center, roll axis* and *roll angle* will be introduced to describe the vehicle roll movement.

Figure 10.4 illustrates a schematic two-axle vehicle diagram, which distinguishes the vehicle sprung weight from the front and the rear, each with its own center of gravity. The masses of the weight that load the front and the rear wheels are m_{vf} and m_{vr}. The vehicle body roll is caused by a change of the suspension geometry known as the suspension roll and the axle roll due to the tire deflections, which presents the roll of the unsprung mass. The *suspension roll center* is defined as the point in the transverse normal plane through the left and right wheel centers, at which the lateral forces may be applied to the sprung mass without producing suspension roll [7]. It is to understand that the roll centers of the front and rear suspension are in the center plane of the vehicle and at center heights, at which the lateral forces developed by the wheels are transmitted to the sprung mass. The height of roll centers is denoted as h_{Rf} and h_{Rr} for the front and the rear, respectively. In the two-axle vehicle, one roll center is located above the front axle and the other one is above the rear axle. This means, the sprung mass rotates about the both roll centers in the front and the rear. It can be imagined that the entire vehicle sprung mass rotates about an axis, which is the line connecting two roll centers and called the *roll axis*. The position of the roll axis continuously changes during the vehicle motion due to the suspension travel and the tire deflections. To determine instantaneous positions of the roll axis is important since the vehicle rolls about this axis with respect to the ground. The significance of the roll axis is that it permits calculation of the roll moment about this axis and the load transfer between the wheels of the front and rear axles. The lateral force application at the roll centers, which belong to the roll axis, has no contribution to the roll moment.

The locations of the roll centers change depending on the suspension type and geometry. The roll center locations can be determined from the layouts of the suspension geometry in the plan and elevation views. General procedures for finding roll centers were presented in many literatures [3, 7, 8].

Another one of the key parameters joining the three-dimensional dynamics of vehicle roll is the *roll angle*, which is the angular displacement produced by the rotation of the sprung mass about the roll axis. It is positive for clockwise rotation viewed from the rear, as denoted by ϕ in Figure 10.4. As mentioned before, the introduction of these all roll parameters makes it possible to determine the relationship between the roll moment and load transfers between the left and right wheels of both the front and the rear axles. Apparently the calculation is best performed in the steady-state. Before getting there, we first derive the roll dynamics models in the following section.

10.3 VEHICLE TRANSVERSE MODEL

For the roll movement modeling, the vehicle with an *independent suspension* can be represented by a two-dimensional vehicle transverse diagram when the following assumptions are made:

- Front and rear suspensions have the same characteristics.
- Road condition is the same for both wheels on one side of the vehicle.

The mechanics governing the roll moment applied to an axle is functionally summarized as a combination of two quarter-car models, that is a half-car model. Considering the same half-car model at the front and the rear, the vehicle transverse model combines two half-car models and is used here to model the roll movement of the entire vehicle. Due to nonlinearities, the suspension parameters on two sides are actually different, greater parameters are on the lower side. For simplification, we assume that the suspension parameters on both sides of the vehicle are the same. It is to point out, that the roll angle range considered for the roll dynamics and control design is small. Indeed, the angle should not be big to provide vehicle safety. This will also reduce the complexity of the modeling of the dynamics. Figure 10.5 shows the rear view of a vehicle in left turn, where the following notation is used:

c_s effective stiffness of suspensions on each side of vehicle
d_s effective damping ratio of the suspensions on each side
c_w tire normal stiffness on each side
m_v vehicle sprung mass
m_w equivalent mass of two unsprung masses on each side of vehicle
m_u entire vehicle unsprung mass

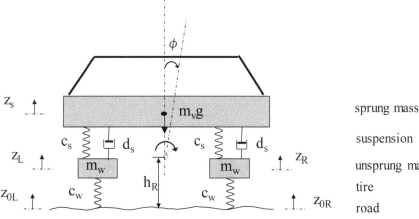

FIGURE 10.5   Vehicle transverse diagram.

a_v lateral acceleration of sprung mass
a_y lateral acceleration of unsprung mass (equal to vehicle lateral acceleration)
I_{x_v} moment of inertia of sprung mass about roll center
I_{x_u} moment of inertia of unsprung mass about roll center
h_R height of roll center
h_s CG height of entire vehicle (assumed equal to height of CG sprung mass)
h_u CG height of unsprung mass
d_t half wheel track
z_L left normal displacement of unsprung mass
z_R right normal displacement of unsprung mass
z_s normal displacement of sprung mass CG
z_{0L} left normal road displacement
z_{0R} right normal road displacement

The sprung mass is located much higher than the unsprung mass, so that the CG height of the sprung mass is assumed the same as the one of the entire vehicle. To find the force and moment relations, we use free-body diagrams with regard to the sprung and unsprung masses as shown in Figure 10.6 with a vehicle-fixed coordinate system, V_{xyz}, with origin on the roll axis below the vehicle CG. The force balances are given along the normal and the lateral directions. The moment balances can be then determined about the roll center.

One fact to be noted here is that the gravitational force contributes to the roll as the center of gravity of the sprung mass is slightly shifted to the right side due to the roll angle as shown in Figure 10.6. This lateral motion of CG could be ignored as it gives only a minor percentage of the roll motion comparing with other factors. For the complete understanding of the force relations, it can be kept in the consideration as long as it does not complicate the model applications.

From the view of the sprung mass, F_r is the force exerted by the unsprung mass acting at the roll center. In the opposite direction the sprung mass exerts the same force on the unsprung mass.

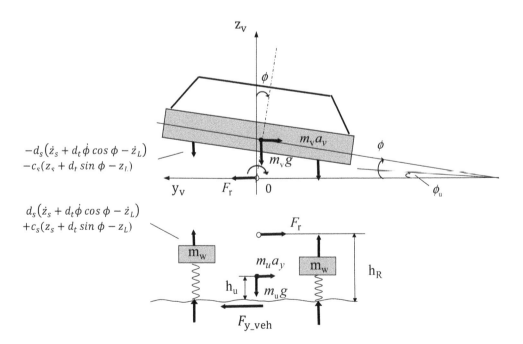

FIGURE 10.6 Transverse free-body diagrams.

Observe that when both masses together are considered to be a single system, then both F_r forces become internal forces and cancel each other out. The unsprung mass centrifugal compensation force $m_u a_y$ is applied at the unsprung center of gravity while the sprung mass centrifugal compensation force $m_v a_v$ is applied at the sprung center of gravity (assumed as vehicle CG). F_{y_veh} is introduced here to present the sum of the lateral forces developed at the tires to balance the centrifugal force of the unsprung mass. Each force balance of the sprung and unsprung masses along the lateral axis gives:

$$F_r = m_v a_v + m_v g \tan \phi$$
$$F_{y\_veh} - F_r = m_u a_y \qquad (10.11)$$

Note that a_v and a_y are the lateral accelerations of the sprung and unsprung masses, respectively. Based on the definition in Figure 10.6, the vehicle mass is the sum of the sprung and unsprung masses, i.e.,

$$M = m_v + m_u \qquad (10.12)$$

In the normal direction, the forces resulted from the suspension spring and damper acting between the sprung mass and unsprung masses are opposite in their directions as shown in the free-body diagram in Figure 10.6. By applying Newton's equation to unsprung masses on the left and right sides, the equations of motion for the wheels are

$$m_w \ddot{z}_L - d_s \left(\dot{z}_s + d_t \dot{\phi} \cos \phi - \dot{z}_L \right) - c_s \left(z_s + d_t \sin \phi - z_L \right) + c_w z_L = c_w z_{0L} \qquad (10.13)$$

$$m_w \ddot{z}_R - d_s \left(\dot{z}_s - d_t \dot{\phi} \cos \phi - \dot{z}_R \right) - c_s \left(z_s - d_t \sin \phi - z_R \right) + c_w z_R = c_w z_{0R} \qquad (10.14)$$

The spring and damper also impose forces on the sprung masses, so that the equation of motion for the sprung mass is

$$m_v \ddot{z}_s + d_s \left(\dot{z}_s - \dot{z}_L \right) + c_s \left(z_s - z_L \right) + d_s \left(\dot{z}_s - \dot{z}_R \right) + c_s \left(z_s - z_R \right) = 0 \qquad (10.15)$$

The normal shift of the CG due to the small roll angle is neglected. Also, the centripetal acceleration, which arises due to the roll of the sprung mass, is negligible comparing to the gravitational acceleration.

Consider the moment balance, applying Euler's equation (2.70) to the sprung mass yields

$$I_{x\_v} \ddot{\phi} - d_t d_s \left(\dot{z}_L - \dot{z}_R \right) + 2 d_s d_t^2 \dot{\phi} \cos \phi - d_t c_s \left(z_L - z_R \right) + 2 c_s d_t^2 \sin \phi$$
$$= m_v a_v \left(h_s - h_R \right) + m_v g \left(h_s - h_R \right) \tan \phi \qquad (10.16)$$

The first term on the right-hand side of the above equation is the moment due to the centrifugal force acting on the CG. The second term is the moment contributed by the gravitational force due to the roll angle. The unsprung mass does not have the same rotation as the sprung mass, and hence it has another roll angle ϕ_u. Taking the roll movement of the unsprung mass about the roll center and using the force relation given in (10.11) yields

$$I_{x\_u} \ddot{\phi}_u + d_t d_s \left(\dot{z}_L - \dot{z}_R \right) - 2 d_s d_t^2 \dot{\phi} \cos \phi + d_t c_s \left(z_L - z_R \right) - 2 c_s d_t^2 \sin \phi + 2 \Delta F_N d_t$$
$$= m_v a_v h_R + m_u a_y h_u + m_v g h_R \tan \phi - c_w d_t \left(z_{0L} - z_{0R} \right) \qquad (8.17)$$

where the load transfer represents

$$\Delta F_N = \frac{1}{2} c_w \left(z_R - z_L \right) \qquad (10.18)$$

Equation (10.17) may not be necessarily used for controller design because of the small roll angle of the unsprung mass. However, this equation will be used later on to relate load transfer ΔF_N to the roll moment. The load transfer will be examined later in the steady-state to understand the relation of the understeer gradient to the suspension system parameters.

The roll dynamics of the vehicle transverse model is given now by (10.13)–(10.16) presenting four state variables $z_L, z_R, z_s,$ and ϕ. Placing the state variables in a vector, these equations can be formulated in a state space form. On the other hand, the linear system is preferred in practical applications. This is achieved as the roll angle is assumed to be small so that $\cos\phi = 0$, $\sin\phi = \phi$ and $\tan\phi = \phi$. Hence, the system is written in a linear state-space form as follows:

$$
\begin{bmatrix} m_w & & & \\ & m_w & & \\ & & m_v & \\ & & & I_{x\_v} \end{bmatrix}
\begin{bmatrix} \ddot{z}_L \\ \ddot{z}_R \\ \ddot{z}_s \\ \ddot{\phi} \end{bmatrix}
+
\begin{bmatrix} d_s & 0 & -d_s & -d_s d_t \\ 0 & d_s & -d_s & d_s d_t \\ -d_s & -d_s & 2d_s & 0 \\ -d_t d_s & d_t d_s & 0 & 2d_s d_t^2 \end{bmatrix}
\begin{bmatrix} \dot{z}_L \\ \dot{z}_R \\ \dot{z}_s \\ \dot{\phi} \end{bmatrix}
+
$$

$$
\begin{bmatrix} c_s + c_w & 0 & -c_s & -d_t c_s \\ 0 & c_s + c_w & -c_s & d_t c_s \\ -c_s & -c_s & 2c_s & 0 \\ -d_t c_s & d_t c_s & 0 & 2c_s d_t^2 - m_v g(h_s - h_R) \end{bmatrix}
\begin{bmatrix} z_L \\ z_R \\ z_s \\ \phi \end{bmatrix}
=
\begin{bmatrix} c_w z_{0L} \\ c_w z_{0R} \\ 0 \\ m_v a_v(h_s - h_R) \end{bmatrix}
\tag{10.19}
$$

Suspension Roll vs. Body Roll

The vehicle body roll consists of suspension roll and axle roll from tire deflection which presents the roll of the unsprung mass, so the roll angle is the sum of both angles:

$$
\phi = \phi_v + \phi_u
\tag{10.20}
$$

Suspension roll ϕ_v is the body roll relative to the axle, i.e., to a line joining the wheel centers while axle roll ϕ_u is between the axle and the level ground where the tire deflection is zero. As seen in Figure 10.6, the axle roll angle resulted from the tire deflection is given by

$$
sin\,\phi_u = \frac{z_L - z_R}{2d_t}
\tag{10.21}
$$

In fact, axle roll is very small since tire spring is relatively stiff to the suspension spring. Typically axle roll is one-eighth of suspension roll, reaching 1–2° [7]. Therefore, vehicle roll angle is usually set to be equal to the suspension roll angle:

$$
\phi \approx \phi_v
\tag{10.22}
$$

Based on this simplification, and by using linear approximation $\sin\phi = \phi$, (10.16) and (10.17) are linearized to

$$
I_{x\_v}\ddot{\phi} + 2d_s d_t^2 \dot{\phi} + \left(2c_s d_t^2 - m_v g(h_s - h_R)\right)\phi = m_v a_v(h_s - h_R)
\tag{10.23}
$$

$$
I_{x\_u}\ddot{\phi}_u - 2d_s d_t^2 \dot{\phi} - 2c_s d_t^2 \phi + 2\Delta F_N d_t = m_v a_v h_R + m_u a_y h_u + m_v g h_R \phi - c_w d_t(z_{L0} - z_{R0})
\tag{10.24}
$$

To characterize the roll moment, *roll damping* and *roll stiffness* are introduced. It is defined as the product of the suspension damping/stiffness and the lateral separation of the springs, i.e.,

$$d_\phi = 2d_s d_t^2$$
$$c_\phi = 2c_s d_t^2$$
(10.25)

where d_ϕ and c_ϕ are roll damping and roll stiffness, respectively. Hence, (10.23) becomes

$$I_{x\_v}\ddot{\phi} + d_\phi\dot{\phi} + \left(c_\phi - m_v g\left(h_s - h_R\right)\right)\phi = m_v a_v \left(h_s - h_R\right)$$
(10.26)

which describes the roll dynamics of the sprung mass about the *x*-axis.

10.4   VEHICLE TWO-AXLE MODEL

Considering the case, where the vehicle front and rear dynamics are different, such as the suspension characteristics and the tire forces, the transverse model does not work anymore, and the entire vehicle should be modeled differently. A two-axle model is applied for this purpose. The other reason for the use of the two-axle model is the longitudinal torsional compliance of the vehicle frame. The torsional flexibility of the frame influences the distribution of roll moments between the front and rear, and a significant frame flexibility might be expected to affect vehicle roll and handling performance noticeably. Especially for heavy duty vehicles it is essential to include the torsional flexibility of the frame in the vehicle model [9]. Since we focus on passenger cars and the light trucks, the torsional compliance will not be considered here, i.e., the front and rear are torsional stiff and thus have equal roll angles.

A two-axle model can be easily obtained combining two vehicle transverse models, and each of them represents a half-car model in the front and rear respectively. Figure 10.7 presents a diagram of the two-axle model, where the physics of notations is similar to the notations shown in Figure 10.5. Here in Figure 10.7, the indices *f* and *r* are added to distinguish the front and the rear.

Let $z_{LF}, z_{RF}, z_{LR}, z_{RR}$ be the normal displacements of the unsprung masses and $z_{0\_LF}, z_{0\_RF}, z_{0\_LR}, z_{0\_RR}$ be the displacements of the road at four wheels, respectively. z_f and z_r are

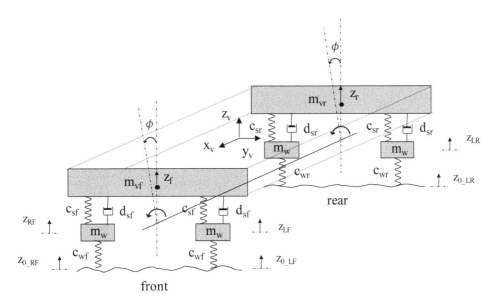

FIGURE 10.7   Vehicle two-axle diagram of the roll movement.

the displacements of the CG of the sprung mass at the front and rear, respectively. The equations of motion for the unsprung masses are

$$m_w \ddot{z}_{LF} - d_{sf}\left(\dot{z}_f + d_t\dot{\phi}\cos\phi - \dot{z}_{LF}\right) - c_{sf}\left(z_f + d_t\sin\phi - z_{LF}\right) + c_w z_{LF} = c_w z_{0\_LF} \tag{10.27}$$

$$m_w \ddot{z}_{RF} - d_{sf}\left(\dot{z}_f - d_t\dot{\phi}\cos\phi - \dot{z}_{RF}\right) - c_{sf}\left(z_f - d_t\sin\phi - z_{RF}\right) + c_w z_{RF} = c_w z_{0\_RF} \tag{10.28}$$

$$m_w \ddot{z}_{LR} - d_{sr}\left(\dot{z}_r + d_t\dot{\phi}\cos\phi - \dot{z}_{LR}\right) - c_{sr}\left(z_r + d_t\sin\phi - z_{LR}\right) + c_w z_{LR} = c_w z_{0\_LR} \tag{10.29}$$

$$m_w \ddot{z}_{RR} - d_{sr}\left(\dot{z}_r - d_t\dot{\phi}\cos\phi - \dot{z}_{RR}\right) - c_{sr}\left(z_r - d_t\sin\phi - z_{RR}\right) + c_w z_{RR} = c_w z_{0\_RR} \tag{10.30}$$

The spring and damper also impose forces on the sprung masses, so that the equations of motion for the front and rear sprung masses are

$$m_{vf}\ddot{z}_f + d_{sf}\left(\dot{z}_f - \dot{z}_{LF}\right) + c_{sf}\left(z_f - z_{LF}\right) + d_{sf}\left(\dot{z}_f - \dot{z}_{RF}\right) + c_{sf}\left(z_f - z_{RF}\right) = 0 \tag{10.31}$$

$$m_{vr}\ddot{z}_r + d_{sr}\left(\dot{z}_r - \dot{z}_{LR}\right) + c_{sr}\left(z_r - z_{LR}\right) + d_{sr}\left(\dot{z}_r - \dot{z}_{RR}\right) + c_{sr}\left(z_r - z_{RR}\right) = 0 \tag{10.32}$$

Adding up both the front and rear rotations of the sprung masses, the roll movement is determined by Euler's equation, that yields

$$I_{x\_v}\ddot{\phi} - d_t d_{sf}\left(\dot{z}_{LF} - \dot{z}_{RF}\right) + 2 d_{sf} d_t^2 \dot{\phi}\cos\phi - d_t c_{sf}\left(z_{LF} - z_{RF}\right) \tag{10.33}$$

$$+ 2 c_{sf} d_t^2 \sin\phi - d_t d_{sr}\left(\dot{z}_{LR} - \dot{z}_{RR}\right) + 2 d_{sr} d_t^2 \dot{\phi}\cos\phi - d_t c_{sr}\left(z_{LR} - z_{RR}\right)$$

$$+ 2 c_{sr} d_t^2 \sin\phi = m_v a_v\left(h_s - h_R\right) + m_v g\left(h_s - h_R\right)\tan\phi$$

The first term on the right-hand side of the above equation is the moment due to the centrifugal force acting on the vehicle center of gravity. The second term is the moment contributed by the gravitational force due to the roll angle. Hence, the roll dynamics of the entire vehicle two-axle system is given now by (10.27) to (10.33) presenting seven state variables. Similar to (10.19), the system can be formulated in a linear state space form:

$$
\begin{bmatrix}
m_w & & & & & & \\
& m_w & & & & & \\
& & m_w & & & & \\
& & & m_w & & & \\
& & & & m_f & & \\
& & & & & m_r & \\
& & & & & & I_{x\_v}
\end{bmatrix}
\begin{bmatrix}
\ddot{z}_{LF} \\ \ddot{z}_{RF} \\ \ddot{z}_{LR} \\ \ddot{z}_{RR} \\ \ddot{z}_f \\ \ddot{z}_r \\ \ddot{\phi}
\end{bmatrix}
+
\begin{bmatrix}
d_{sf} & 0 & 0 & 0 & -d_{sf} & 0 & -d_{sf}d_t \\
0 & d_{sf} & 0 & 0 & -d_{sf} & 0 & d_{sf}d_t \\
0 & 0 & d_{sr} & 0 & 0 & -d_{sr} & -d_{sr}d_t \\
0 & 0 & 0 & d_{sr} & 0 & -d_{sr} & d_{sr}d_t \\
-d_{sf} & -d_{sf} & 0 & 0 & 2d_{sf} & 0 & 0 \\
0 & 0 & -d_{sr} & -d_{sr} & 0 & 2d_{sr} & 0 \\
-d_t d_{sf} & d_t d_{sf} & -d_t d_{sr} & d_t d_{sr} & 0 & 0 & 2\left(d_{sf}+d_{sr}\right)d_t^2
\end{bmatrix}
\begin{bmatrix}
\dot{z}_{LF} \\ \dot{z}_{RF} \\ \dot{z}_{LR} \\ \dot{z}_{RR} \\ \dot{z}_f \\ \dot{z}_r \\ \dot{\phi}
\end{bmatrix}
$$

$$
\begin{bmatrix}
c_{sf}+c_w & 0 & 0 & 0 & -c_{sf} & 0 & -d_t c_{sf} \\
0 & c_{sf}+c_w & 0 & 0 & -c_{sf} & 0 & d_t c_{sf} \\
0 & 0 & c_{sr}+c_w & 0 & 0 & -c_{sr} & -d_t c_{sr} \\
0 & 0 & 0 & c_{sr}+c_w & 0 & -c_{sr} & d_t c_{sr} \\
-c_{sf} & -c_{sf} & 0 & 0 & 2c_{sf} & 0 & 0 \\
0 & 0 & c_{sr} & -c_{sr} & 0 & 2c_{sr} & 0 \\
-d_t c_{sf} & d_t c_{sf} & -d_t c_{sr} & d_t c_{sr} & 0 & 0 & 2\left(c_{sf}+c_{sr}\right)d_t^2 - m_v g(h_s-h_R)
\end{bmatrix}
\begin{bmatrix}
z_{LF} \\ z_{RF} \\ z_{LR} \\ z_{RR} \\ z_f \\ z_r \\ \phi
\end{bmatrix}
=
\begin{bmatrix}
c_w z_{0\_LF} \\ c_w z_{0\_RF} \\ c_w z_{0\_LR} \\ c_w z_{0\_RR} \\ 0 \\ 0 \\ m_v a_v\left(h_s - h_R\right)
\end{bmatrix}
$$

$$\tag{10.34}$$

In order to distinguish the load transfer between the left and right wheels at the front and rear respectively, the roll moment of the unsprung mass can be determined for each of the front and the rear axles:

$$I_{x\_uf}\ddot{\phi}_u - 2d_{sf}d_t^2\dot{\phi} - 2c_{sf}d_t^2\phi + 2\Delta F_{Nf}d_t = m_{vf}a_vh_r + m_{uf}a_yh_u + m_{vf}gh_r\phi - c_wd_t\left(z_{0\_LF} - z_{0\_RF}\right) \quad (10.35)$$

$$I_{x\_ur}\ddot{\phi}_u - 2d_{sr}d_t^2\dot{\phi} - 2c_{sr}d_t^2\phi + 2\Delta F_{Nr}d_t = m_{vr}a_vh_r + m_{ur}a_yh_u + m_{vr}gh_r\phi - c_wd_t\left(z_{0\_RF} - z_{0\_RR}\right) \quad (10.36)$$

For simplification in this place, it is assumed that the lateral accelerations at the front and rear are equal since the purpose is to indicate the relation of the load transfers. Otherwise the acceleration can be calculated as given by Example 2.3 of Chapter 2. Note that the weight distributions on the front and rear can be readily computed using the following relationships of the masses assuming that the gravity center of the vehicle and the CG of the sprung mass coincide

$$m_f = \frac{Ml_r}{l_f + l_r}, \quad m_r = \frac{Ml_f}{l_f + l_r}$$
$$m_{vf} = \frac{m_vl_r}{l_f + l_r}, \quad m_{vr} = \frac{m_vl_f}{l_f + l_r} \quad (10.37)$$

likewise, for the front and rear unsprung masses:

$$m_{uf} = \frac{m_ul_r}{l_f + l_r} = m_f - m_{vf}, \quad m_{ur} = \frac{m_ul_f}{l_f + l_r} = m_r - m_{vr} \quad (10.38)$$

10.5 STEADY-STATE

As we already learned in vehicle lateral dynamics, the consideration of steady-state behavior due to driver command inputs and external disturbances has been an important component in control design. To understand the roll angle response to steady-state inputs, such as a constant lateral acceleration, will help to quantify the impact of the load transfer between the vehicle axles on the understeer gradient. In this manner, the preferred characteristics of vehicle lateral dynamics can be determined to achieve their optimum by considering the roll dynamics.

For simplification, it is assumed that the four wheels have equal road displacements. In the steady state, the roll angle and normal displacements of the suspensions remain constant, i.e., $\dot{\phi} = 0$ and $\dot{z}_{LF} = \dot{z}_{LR} = \dot{z}_{RF} = \dot{z}_{RR} = 0$. From (10.23) the roll angle can be computed:

$$\phi = \frac{m_va_v\left(h_s - h_r\right)}{2\left(c_{sf} + c_{sr}\right)d_t^2 - m_vg\left(h_s - h_r\right)} \quad (10.39)$$

Using (10.25) for each front and rear roll stiffness respectively, the roll angle is given then

$$\phi = \frac{m_va_v\left(h_s - h_r\right)}{\left(c_{\phi f} + c_{\phi r}\right) - m_vg\left(h_s - h_r\right)} \quad (10.40)$$

LOAD TRANSFER

Once the roll angle is determined, the load transfer can be calculated from (10.35) and (10.36) for the steady-state. The load transfer on the wheels of the front axle is obtained

$$\Delta F_{Nf} = \frac{\left(c_{\phi f} + m_{vf}gh_r\right)m_va_v\left(h_s - h_r\right)}{2d_t\left(\left(c_{\phi f} + c_{\phi r}\right) - m_vg\left(h_s - h_r\right)\right)} + \frac{m_{vf}a_vh_r + m_{uf}a_yh_u}{2d_t} \quad (10.41)$$

and likewise, for the rear axle wheels:

$$\Delta F_{\mathrm{Nr}} = \frac{\left(c_{\phi r} + m_{\mathrm{vr}}gh_r\right)m_{\mathrm{v}}a_{\mathrm{v}}\left(h_s - h_r\right)}{2d_t\left(\left(c_{\phi f} + c_{\phi r}\right) - m_{\mathrm{v}}g\left(h_s - h_r\right)\right)} + \frac{m_{\mathrm{vr}}a_{\mathrm{v}}h_r + m_{\mathrm{ur}}a_yh_u}{2d_t} \tag{10.42}$$

Under the practical assumption that lateral contributions of the unsprung masses is negligible comparing to roll stiffness $C_{\phi f}$ or $C_{\phi r}$, (10.41) and (10.42) lead to

$$\Delta F_{\mathrm{Nf}} = \frac{\left(c_{\phi f} + m_{\mathrm{vf}}gh_r\right)m_{\mathrm{v}}a_{\mathrm{v}}\left(h_s - h_r\right)}{2d_t\left(\left(c_{\phi f} + c_{\phi r}\right) - m_{\mathrm{v}}g\left(h_s - h_r\right)\right)} + \frac{m_{\mathrm{vf}}a_{\mathrm{v}}h_r}{2d_t} \tag{10.43}$$

$$\Delta F_{\mathrm{Nr}} = \frac{\left(c_{\phi r} + m_{\mathrm{vr}}gh_r\right)m_{\mathrm{v}}a_{\mathrm{v}}\left(h_s - h_r\right)}{2d_t\left(\left(c_{\phi f} + c_{\phi r}\right) - m_{\mathrm{v}}g\left(h_s - h_r\right)\right)} + \frac{m_{\mathrm{vr}}a_{\mathrm{v}}h_r}{2d_t} \tag{10.44}$$

From the above equations, it is clear to see the dependence of the load transfer on the lateral acceleration. In the steady-state with a constant lateral acceleration, a bigger roll moment will increase the load transfer between the wheels of each axle. The roll stiffness is the key design parameter of the roll moment. The weight distribution depending on l_f and l_r, the heights of the CG as well as the roll center are the other design parameters to influence the load transfer.

Now, we take a close look at the lateral tire forces. In particular, we analyze the force changes when the wheel load varies. As shown in Figure 3.4, the lateral tire force depends on the normal load in a nonlinear relation. Basically, the lateral tire force can be considered "proportional" to the load; however, the proportion is getting smaller while the normal load increases. That is the key property that influences the significance of lateral tire force alteration due to the wheel load transfer. An example in Figure 10.8, which is zoomed from Figure 3.4, will illustrate the relation between the lateral force and the normal load. For a vehicle at 3000 N load on each wheel, about 2700 N of the lateral force will be developed by each wheel at 4-degree slip angle. When making a turn, the normal load changes to 2000 N on the inside wheel and 4000 N on the outside wheel. Then the average lateral force from both tires will be reduced to about 2500 N. In order to maintain the lateral force 2700 N for the turn, the slip angle changes from 4-degree to about 5-degree as shown in Figure 10.8. As a conclusion, the load transfer on the axle wheels implies an increase of the slip angle. Hence the roll moment on the front wheels will contribute to understeer whereas the roll moment on the rear contributes to oversteer.

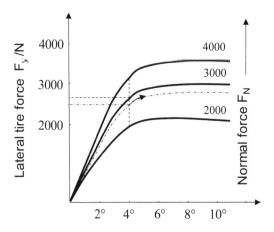

FIGURE 10.8   Lateral force alteration due to normal load.

Vehicles are desired to have understeer propensity satisfying the stable condition. That can be achieved by obtaining higher wheel load difference on the front axle than that on the rear, for example by enhancing the front roll stiffness. However, a higher spring rate will produce a higher normal acceleration and hence degraded the driving comfort. To avoid this, an additional spring bar called *stabilizer* is often used on an axle to vary the spring stiffness under the roll condition. While the roll, the stabilizer creates a force on the inside wheel through its torsion stiffness and decreases the same amount of the force on the outside wheel. The stabilizer is designed to increase the roll stiffness and may be installed on the front and rear axles. Mathematically, the front roll stiffness will be changed to $c_{\phi f} + c_{st}$ and that of the rear to $c_{\phi r} + c_{st}$, where c_{st} is the additional roll stiffness of the stabilizer. More details about stabilizer can be found in reference [8].

In the same way, the load transfer for the vehicle traverse model can be obtained from (10.24):

$$\Delta F_N = \frac{\left(c_\phi + m_v g h_r\right) m_v a_v \left(h_s - h_r\right)}{2d_t \left(c_\phi - m_v g \left(h_s - h_r\right)\right)} + \frac{m_v a_v h_r}{2d_t} \tag{10.45}$$

which will be used in the rollover analysis further in Chapter 13.

10.6   THREE-DIMENSIONAL DYNAMICS MODEL

This section presents a three-dimensional dynamics model combining the vehicle's lateral and normal dynamics. For the purpose of control design, the vehicle modeling should be simple and have reduced complexity as mentioned before. Therefore, a compromise between the contents of the modeling and the accuracy of model is essential. Due to the strong couplings of various aspects of vehicle dynamics, such as lateral acceleration, yaw rate and roll motion, each one interferes with other aspects of vehicle dynamics and adversely influences vehicle performance. In order to complement each other and optimize overall vehicle performance, some levels of integration of subsystems of modeling are of importance when they are present in the vehicle model simultaneously.

Some studies concluded the necessity to incorporate the roll dynamics into the lateral motion [2]. Yaw motions produce lateral accelerations causing roll motions, and a roll motion in turn alters a yaw response through the lateral tire force arising from the lateral load transfer and suspension action. In reference [10], an example showed the effects of the suspension roll on the frequency responses caused by steering inputs to the lateral and yaw motions. Without suspension roll dynamics, the model in reference [10] exhibits lowered gain characteristics over the frequency range around 1–2 Hz, which attributed to the natural frequency of the suspension system. Considering the suspension roll dynamics will improve the model accuracy and benefit the performance of a controller based on that model.

The three-dimensional model can be realized as a combination of the single-track and the vehicle traverse models, which were derived in the previous sections. The lateral dynamics were detailed and fully described in Chapter 8. The movement in y-direction and the yaw moment around z-direction made the base to derive the key model elements as given by (8.23) and (8.24). In the end the linear bicycle model came through, where the assumption was made that the sideslip angle is small and the longitudinal forces are negligible. The linear bicycle model is a useful platform for the control design and analysis, and has been proven to be practical and effective in vehicle dynamics control [11, 12]. From this background, the bicycle model can be based and extended by the roll motion to obtain a three-dimensional model. Comparing to longitudinal and lateral dynamics, the following features of the description need to be taken into account as they were already introduced in vehicle normal dynamics:

- Vehicle fixed coordinate system originates on the roll axis under vehicle CG;
- Sprung and unsprung masses are treated separately.

In the previous chapters, the vehicle fixed coordinate system originated at vehicle's center of gravity. Now, we use the roll center as the reference point for the roll motion. The origin is shifted down to the roll center along the z-axis, where the orientation remains unchanged. The sprung mass and the unsprung mass are considered separately since their roll movements behave differently due to the stiffness characteristics of the suspension and the tires.

Let $\mathbf{v}_R = (v_x, v_y, 0)^T$ present the velocity of the origin of the vehicle-fixed coordinate system, the vehicle velocity at the CG can be determined using equation (2.44):

$$\mathbf{v} = \mathbf{v}_R + \begin{bmatrix} \dot{\phi} \\ 0 \\ \dot{\psi} \end{bmatrix} \times \begin{bmatrix} 0 \\ 0 \\ h_s - h_R \end{bmatrix} = \begin{bmatrix} v_x \\ v_y - \dot{\phi}(h_s - h_R) \\ 0 \end{bmatrix} \tag{10.46}$$

Using (2.41) the acceleration of the gravity center is computed:

$$\mathbf{a} = \dot{\mathbf{v}} + \begin{bmatrix} \dot{\phi} \\ 0 \\ \dot{\psi} \end{bmatrix} \times \begin{bmatrix} v_x \\ v_y - \dot{\phi}(h_s - h_r) \\ 0 \end{bmatrix} = \begin{bmatrix} \dot{v}_x - \left(v_y - \dot{\phi}(h_s - h_R)\right)\dot{\psi} \\ \dot{v}_y + v_x\dot{\psi} - \ddot{\phi}(h_s - h_R) \\ \left(v_y - \dot{\phi}(h_s - h_R)\right)\dot{\phi} \end{bmatrix} \tag{10.47}$$

$\mathbf{v}_R$ is still identical to the velocity of the vehicle, $\mathbf{v}$, without considering the roll movement. Comparing with (2.51), the roll motion affects all three components of the acceleration, where term $\ddot{\phi}(h_s - h_R)$ is added to the lateral acceleration a_y. Hence, we have

$$a_v = a_y - \ddot{\phi}(h_s - h_R) \tag{10.48}$$

Since the sprung mass is rolling about the x-axis, vehicle lateral dynamics is now featured with the roll angle as an element of the roll movement. Recalling Figure 8.6, F_{yf} and F_{yr} are the lateral forces acting on the tires. Now we look at (10.11) again, which considers both the sprung and unsprung masses. By summing all lateral forces, they are balanced at

$$F_{yf} + F_{yr} - m_v g\phi = m_v a_v + m_u a_y$$
$$= M a_y - m_v \ddot{\phi}(h_s - h_R) \tag{10.49}$$

by considering (10.47).

To determine the rotational moment, Euler's equations are applied. The roll center is assumed to be below the CG, where the z-axis goes through both CG and the vehicle roll center. In this case, the coordinate system axes coincide with the principle axes. By summing the moments about the z-axis, the yaw moment is balanced at

$$\sum T_{Gz} = I_{z\_v}\ddot{\psi} = F_{yf}l_f - F_{yr}l_r \tag{10.50}$$

where F_{yf} and F_{yr} are the lateral forces specified by (8.21) and (8.31).

For the roll motion about the x-axis, the total moments are balanced at

$$\sum T_{Gx} = I_{x\_v}\ddot{\phi} = m_v a_v (h_s - h_r) + m_v g(h_s - h_r)\phi - d_\phi\dot{\phi} - c_\phi\phi \tag{10.51}$$

which is a representation of (10.26). In this moment balance, the roll damping and the roll stiffness are explicitly considered. The lateral acceleration of the sprung mass, a_v, is the y-component in (10.47).

Equations (10.49)–(10.51) build the base of the three-dimensional dynamics model for the vehicle roll motion. For the further evaluation, the same assumptions made to the linear bicycle model, such as small sideslip angle β and negligible longitudinal tire forces. Substituting the lateral tire forces by (8.21), then (10.49) through (10.51) can be written as

$$C_{\alpha f}\alpha_f + C_{\alpha r}\alpha_r - m_v g\phi = M\left(\dot{v}_y + v_x\dot{\psi}\right) - m_v\ddot{\phi}\left(h_s - h_r\right) \tag{10.52}$$

$$I_{z\_v}\ddot{\psi} = C_{\alpha f}\alpha_f l_f - C_{\alpha r}\alpha_r l_r \tag{10.53}$$

$$I_{x\_v}\ddot{\phi} = \left[m_v a_y - m_v\ddot{\phi}\left(h_s - h_r\right)\right]\left(h_s - h_r\right) + m_v g\left(h_s - h_r\right)\phi - d_\phi\dot{\phi} - c_\phi\phi \tag{10.54}$$

Subsequently, using (8.29) through (8.30) and y-component in (10.47) yields

$$\dot{\beta} - \frac{m_v\left(h_s - h_r\right)}{Mv}\ddot{\phi} = -\left(1 + \frac{C_{\alpha f}l_f - C_{\alpha r}l_r}{Mv^2}\right)\dot{\psi} - \frac{C_{\alpha f} + C_{\alpha r}}{Mv}\beta - \frac{m_v g}{Mv}\phi + \frac{C_{\alpha f}}{Mv}\delta_f \tag{10.55}$$

$$I_{z\_v}\ddot{\psi} = -\frac{C_{\alpha f}l_f^2 + C_{\alpha r}l_r^2}{v}\dot{\psi} - \left(C_{\alpha f}l_f - C_{\alpha r}l_r\right)\beta + C_{\alpha f}l_f\delta_f \tag{10.56}$$

$$I_{x\_v}\ddot{\phi} + m_v\left(h_s - h_r\right)^2\ddot{\phi} = m_v v\left(\dot{\beta} + \dot{\psi}\right)\left(h_s - h_r\right) + m_v g\left(h_s - h_r\right)\phi - d_\phi\dot{\phi} - c_\phi\phi \tag{10.57}$$

Those can also be formulated in a state space form:

$$\mathbf{M}\dot{\mathbf{x}} = \mathbf{A}\mathbf{x} + \mathbf{b}u \tag{10.58}$$

with

$$\mathbf{x} = \begin{pmatrix} \beta & \dot{\psi} & \dot{\phi} & \phi \end{pmatrix}^T; \quad u = \delta_f;$$

$$\mathbf{M} = \begin{bmatrix} Mv & 0 & -m_v\left(h_s - h_r\right) & 0 \\ 0 & I_{z\_v} & 0 & 0 \\ -m_v v\left(h_s - h_r\right) & 0 & I_{x\_v} + m_v\left(h_s - h_r\right)^2 & 0 \\ 0 & 0 & 0 & 1 \end{bmatrix}$$

$$\mathbf{A} = \begin{bmatrix} -C_{\alpha f} - C_{\alpha r} & -Mv - \dfrac{C_{\alpha f}l_f - C_{\alpha r}l_r}{v} & 0 & m_v g \\ -C_{\alpha f}l_f + C_{\alpha r}l_r & -\dfrac{C_{\alpha f}l_f^2 + C_{\alpha r}l_r^2}{v} & 0 & 0 \\ 0 & -m_v v\left(h_s - h_r\right) & -d_\phi & m_v g\left(h_s - h_r\right) - c_\phi \\ 0 & 0 & 1 & 0 \end{bmatrix}$$

$$\mathbf{b} = \begin{bmatrix} C_{\alpha f} & C_{\alpha f}l_f & 0 & 0 \end{bmatrix}^T$$

REFERENCES

1. *Vehicle Dynamics* Terminology, SAE Standard J670, revised in 2008.
2. T. D. Gillespie, Fundamentals of Vehicle Dynamics, Warrendale: SAE, 1992.
3. A. Alleyne and J. K. Hedrick, "Nonlinear Adaptive Control of Active Suspensions," *IEEE Transactions on Control Systems Technology*, vol. 3, no. 1, 1995.

4. R. Rajamani and J. K. Hedrick, "Adaptive Observers for Automotive Suspensions: Theory and Experiment," *IEEE Transactions. on Control Systems Technology*, vol. 3, no. 1, 1995.

5. Georg Rill, Abel Arrieta Castro, Road Vehicle Dynamics: Fundamentals and Modeling, Second Edition, Boca Raton, London, New York: CRS Press, Taylor & Francis Group, 2020.

6. P. C. Müller and W. O. Schiehlen, Linear Vibrations, Dordrecht: Martinus Nijhoff Publishers, 1985.

7. J. C. Dixon, Tires, Suspension and Handling, Warrendale: SAE, 1996.

8. J. Reimpell, Fahrwerktechnik: Grundlagen, Würzburg: Vogel Verlag, 1995.

9. D. J. M. Sampson, Active Roll Control of Articulated Heavy Vehicles, Ph.D. thesis, University of Cambridge, UK, 2000.

10. K. T. Feng, H. S. Tan and M. Tomizuka, "Automatic Steering Control of Vehicle Lateral Motion with the Effect of Roll Dynamics," in *Proceedings of the American Control Conference*, Philadelphia, Pennsylvania, 1998.

11. H. Fennel, R. Gutwein, A. Kohl, M. Latarnik and G. Roll, "Das modulare Regler- und Regelkonzept beim ESP von ITT Automotive," in *7th Achener Kolloquium Fahrzeug- und Motortechnik*, Achen, Okt. 1998.

12. A. van Zanten, R. Erhardt and G. Pfaff, "FDR-Die Fahrdynamikregelung von Bosch," *ATZ Automobiltechnische Zeitschrift*, vol. 96, no. 11, pp. 674–689, 1994.

Part II

Control Design

So far, we have seen that the vehicle dynamic system is a high-degree-of-freedom system. It is highly complex and nonlinear. However, in our description, the system has been taken apart by the approximations into various individual such as wheel dynamics, vehicle straight-line and longitudinal motion, vehicle lateral and normal dynamics. Based on the assumptions, the dynamic models can be considered individually and sometimes independently, and they can be described by linear or nonlinear differential equations. In order to use linear control methods, the system should be modeled by linear differential equations. The dynamic models given in the previous chapters are more properly described by nonlinear differential equations. In some circumstances, the nonlinear dynamics can be approximated by a linear model and make the use of the linear control methods possible. It is often reasonable and practicable to make such approximations.

We will see now how effective the control can be. Part II is dedicated to control design that is based on the models of vehicle dynamics developed in the previous chapters. Prior to the study of the control design methods, the reader is referred to Chapter 11, in which we review the concepts of control theory needed for the control designs described in the book. After Chapter 11, we have the fundamental control techniques to influence the vehicle dynamics following the desired motions. In further chapters from 12 to 14, we design the controllers which can actually cause the vehicle or the vehicle's systems to perform desired dynamics. Although self-driving technology is not the primary focus in the book, some controllers are specifically designed for the use on autonomous vehicles.

In Chapter 11, it is shown how to use state-space form to analyze the behavior of a dynamic system. Throughout this chapter, the state-space formulation is repeatedly used for control design, either in nonlinear form or in linear form.

We first introduce the wheel slip control in Chapter 12, which primarily deals with the wheel dynamics. There are three types of controllers designed, where the linear approximation is made for the nonlinear wheel dynamics in order to utilize the linear observer. The *brake slip controller* can be used in *anti-lock braking system* (ABS) by preventing the wheels from getting locked during braking. The *tractive slip* controller and the *speed differential* controller can be a part of traction control system (TCS) preventing the wheels from excessive slipping during acceleration.

In Chapter 13, we design the vehicle speed control by using the linear quadratic optimal control method. In addition, the path-following control is presented based on the extended linear quadratic optimal method, which is a very useful control for autonomous vehicles. This design involves the decomposition of the control problem into inner-loop and outer-loop design.

Chapter 14 is concerned with vehicle stability controls. In Section 14.1 we design yaw stability control, where three types of controllers are shown – the state feedback control, robust control and adaptive yaw stability controllers. We begin with the application of the state feedback control technique including a feedforward compensation to the single-track model of lateral dynamics. This control technique requires the exact knowledge of the model. In real engineering, the model parameters are not completely known and some of the parameters vary in time. Robust control is an advanced control method, which does not need the assumption for the knowledge of the model parameters. In a similar way, the adaptive control algorithm is designed to deal with model uncertainties and adapt the model parameters during the control. At the end of the section, simulation results are plotted and discussed. A rollover control scheme is shown in Section 14.2, which involves the estimation of the roll angle based on a linear state observer. Finally, we present a method for the semi-trailer stabilization by using rear steering in Section 14.3. There, an approximation is made so that the vehicle-trailer system behaves as a linear time-varying system. The Kalman filter is used to estimate the hitch angle, which is the input to the rear steering to mitigate the trailer's oscillation, and thus to maintain stability of the vehicle-trailer system.

11 Introduction to Control Theory and Methods

After vehicle dynamics is modeled by differential equations, controllers and control systems are then based and designed on those models. In general, there are two categories of the control systems – linear and nonlinear systems. The control methods that are discussed in this book fall into both categories. Since the field of control theory is large, we can only focus on few methods which are suited to the vehicle dynamic models presented in the previous chapters.

In this chapter, we review the concepts of control theory that are needed for our controller designs. We begin with a review of second-order linear systems and the state-space description of linear systems in Sections 11.1 and 11.2. Then in the following Sections 11.3 and 11.4, the state observer as well as Kalman filter will be presented. Section 11.5 will introduce the method of the stability analysis known as Lyapunov's method. Finally, the linear quadratic optimal control is presented. All methods are reviewed from the application point of view, and references are made to literatures where the proofs are provided.

11.1 SECOND-ORDER LINEAR SYSTEMS

Systems can often be described by second-order differential equations, and the second-order linear system characteristics are considered dominant in their dynamic nature. For this reason, we will review some basic characteristics of the second-order linear systems starting with a simple mechanical system as an example. It is assumed that the readers are familiar with simple differential equations since those equations do not require to know complete linear control theory in first place.

Figure 11.1 shows a spring-mass-damper system, in which a body of mass m is attached to a spring of stiffness c and a viscous damper with coefficient k in parallel. The vibration can be initiated by a vertical movement of the mass from its zero position, and the mass keeps moving up and down across its zero position. The zero position is chosen at the static equilibrium, where the weight is counterbalanced by a spring force. Considering both an external force $u(t)$ and the spring force exerted on the mass body and its viscous damping leads directly to the following equation of motion

$$m\ddot{x}(t) + c\dot{x}(t) + kx(t) = u(t) \tag{11.1}$$

Hence, the open loop dynamics of this one degree of freedom system is described by a second-order linear differential equation with constant coefficients.

QUADRATIC LAG

In the following, we consider (11.1) a general form of the second-order linear systems, in which $u(t)$ is the system input and m, k, c are constant coefficients. Note that the coefficients are not necessarily positive. For the study of control systems, we are interested in determining the dynamic behavior of the output $x(t)$. In control theory, linear systems are commonly described by means of transfer functions. For the linear, time-invariant system is given by (11.1), the transfer function is the ratio of the Laplace transform of the output X(s) to the Laplace transform of the corresponding input U(s) with all initial conditions assumed to be zero. The response of the dynamic system to a step input is essential, and provides information on the stability of the system as well as the transient behavior

DOI: 10.1201/9781003134305-11

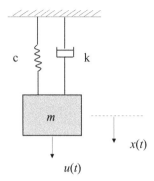

FIGURE 11.1   Spring-mass-damper system.

to reach a steady state. For the purpose of control system analysis, it suffices to consider $u(t)$ to be a unit step function. Introducing $H(s)$ as the transform of the output due to the unit step input, we have for the spring-mass-damper system

$$H(s) = \frac{1}{\left(ms^2 + cs + k\right)} \cdot \frac{1}{s} \tag{11.2}$$

as the result of the transfer function multiplied by function $1/s$, which is the Laplace transform of the unit step function. The transfer function given by the first fraction on the right-hand side of (11.2) forms a *quadratic lag* system.

Inverse *Laplace transform* of this transfer function will give a corresponding response in the time domain. Exactly speaking, inverse Laplace transform provides the system response $x(t)$ to a given input $u(t)$ with specified initial conditions.

On the other side, the system (11.1) might exhibit several different characteristic motions depending upon the system parameters. From *Laplace transform* (11.2), we know that the form of the transient response of the second-order system depends on the roots of its *characteristic equation*,

$$ms^2 + cs + k = 0 \tag{11.3}$$

The roots are as follows:

$$s_1 = -\frac{c}{2m} + \frac{\sqrt{c^2 - 4mk}}{2m}$$
$$s_2 = -\frac{c}{2m} - \frac{\sqrt{c^2 - 4mk}}{2m} \tag{11.4}$$

The location of s_1 and s_2 in the *s-plane*, which is called the *poles* of the system, depends on parameters m, k, c and dictates the nature of the motions of the system. A common way to describe and analyze characteristic equations is to do so by using the *damping ratio* and the *natural frequency* of the system. These terms are defined by the parameterization of the characteristic equation as

$$s^2 + 2\zeta\omega_o s + \omega_o^2 = 0 \tag{11.5}$$

where ζ is the *damping ratio*, and ω_0 is the *natural frequency*. Hence, the system poles given by (11.5) can be expressed in the form

$$s_1 = -\zeta\omega_o + \omega_o\sqrt{\zeta^2 - 1}$$
$$s_2 = -\zeta\omega_o - \omega_o\sqrt{\zeta^2 - 1} \tag{11.6}$$

with

$$\zeta = \frac{c}{2\sqrt{km}}$$
$$\omega_o = \sqrt{\frac{k}{m}} \tag{11.7}$$

The imaginary part of the poles, $\omega_o\sqrt{1-\zeta^2}$, is called the *damped natural frequency*.

In the following, we specify some typical set of poles and, for each pole, give the system's response to the unit step function. The system response in the time domain results from the inverse Laplace transform that is obtained by using (11.2). The examples below are demonstrated by using LabVIEW®.

First, Laplace transform (11.2) is implemented in the block diagram shown in Figure 11.2. The transfer function is specified by numerator and denominator, and the transient response is defined by the block of unit step response, in which the sampling time and length of time period are specified. The third block maps the poles and zeros of the transfer function. The parameters for the transfer function are to insert in the front panel.

Second, we consider special cases of the damping ratio.

$0 < \zeta < 1$

For this condition, (11.6) yields a complex conjugate pair of poles. Given initial conditions with $\dot{x}(0) = x(0) = 0$, the unit step system response in the time domain, which is the inverse Laplace transformation of (11.2), results in

$$h(t) = \frac{1}{k}\left\{1 - e^{-\zeta\omega_0 t}\left[\cos\left(\omega_0\sqrt{1-\zeta^2}t\right) + \frac{\zeta}{\sqrt{1-\zeta^2}}\sin\left(\omega_0\sqrt{1-\zeta^2}t\right)\right]\right\} \tag{11.8}$$

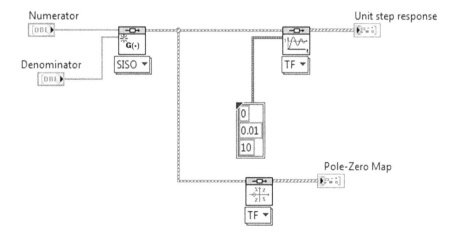

FIGURE 11.2   Block diagram of the spring-mass-damper system.

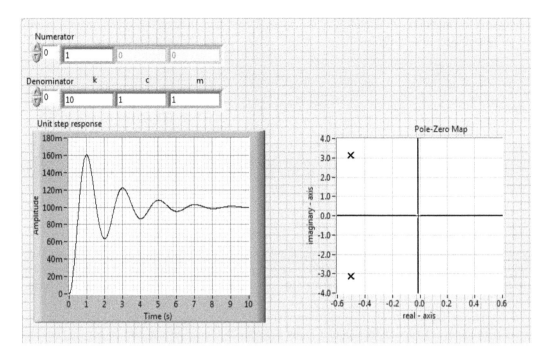

FIGURE 11.3   Root location and unit step response for the example system with $m = 1$, $c = 1$, $k = 10$.

In this form, called *underdamped*, the resulting motion is an oscillation whose amplitude is exponentially decreasing toward zero as plotted in Figure 11.3. On the right side of Figure 11.3 shows the system poles in the *s*-plane, which is built on a real axis and an imaginary axis.

$\zeta > 1$

In this case, both system poles are negative real. The unit step response is then given by

$$h(t) = \frac{1}{k}\left\{1 - \frac{\zeta + \sqrt{\zeta^2 - 1}}{2\sqrt{\zeta^2 - 1}}\, e^{-\omega_0\left(\zeta - \sqrt{\zeta^2 - 1}\right)t} + \frac{\zeta - \sqrt{\zeta^2 - 1}}{2\sqrt{\zeta^2 - 1}}\, e^{-\omega_0\left(\zeta + \sqrt{\zeta^2 - 1}\right)t}\right\} \tag{11.9}$$

which is overdamped. The motion of the system is sluggish and nonoscillatory, as shown in Figure 11.4.

$\zeta = 1$

Under this condition, the characteristic equation yields real and equal roots. The unit step response is then given by

$$h(t) = \frac{1}{k}\left\{1 - e^{-\zeta\omega_0 t}\left(1 + \omega_0 t\right)\right\} \tag{11.10}$$

The system is critically damped and gains its final position in the shortest possible time without oscillation. Figure 11.5 shows an example of the data plot.

$\zeta = 0$

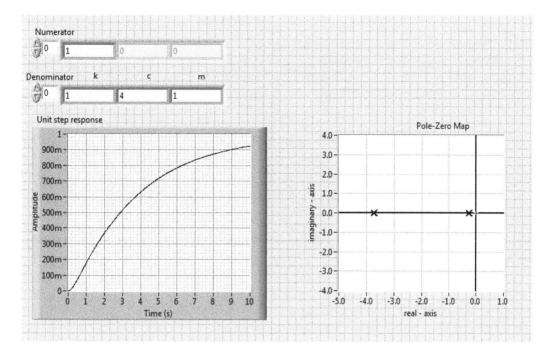

FIGURE 11.4 Root location and unit step response for the example system with $m = 1$, $c = 4$, $k = 1$.

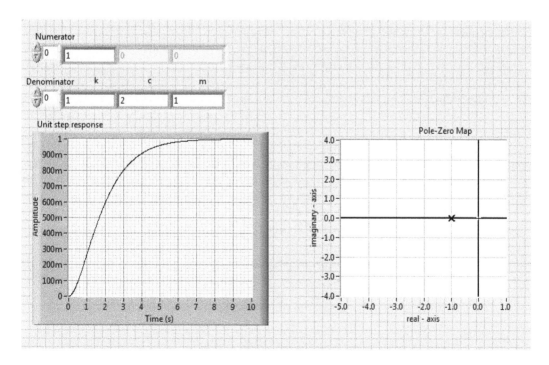

FIGURE 11.5 Root location and unit step response for the example system with $m = 1$, $c = 2$, $k = 1$.

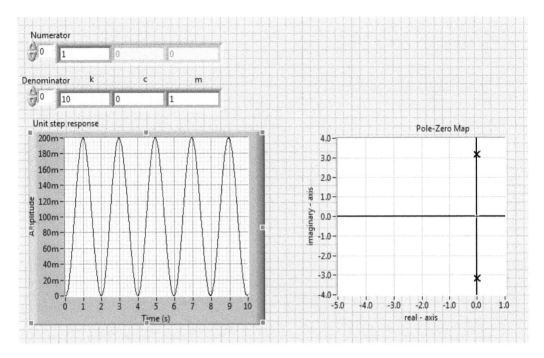

FIGURE 11.6 Root location and unit step response for the example system with $m = 1$, $c = 0$, $k = 10$.

This happens when $c = 0$ in (11.3). Both system poles are on the imaginary $j\omega$-axis of the s-plane. Oscillation without damping will result in the system's motion, which can be described by the following equation for the unit step input:

$$h(t) = \frac{1}{k}(1 - \cos\omega_0 t) \tag{11.11}$$

An example of this kind of motion is illustrated in Figure 11.6.

$\underline{\zeta < 0}$

In this parameter category, the system poles always lie in the right half of the s-plane. They may be real or complex. In these cases, (11.8) to (11.10) still apply. However, the exponential function is no longer decaying, and the function value tends to the infinite when $t \to \infty$, as shown in Figure 11.7. Such kind of behavior is defined as *unstable*. The poles with a positive real part are called *unstable poles*.

$\underline{\omega_0^2 < 0}$

In some systems, the third term, ω_0^2, in (11.5) can be negative, for example, oversteered vehicles as illustrated before in (9.26). One of the poles given by (11.6) is positive real while the other one is negative. The response of the system to the unit step input is then given by

$$h(t) = \frac{1}{k}\left\{1 - \frac{\omega_0\zeta + \sqrt{\omega_0^2\zeta^2 - \omega_0^2}}{2\sqrt{\omega_0^2\zeta^2 - \omega_0^2}}e^{-\left(\omega_0\zeta - \sqrt{\omega_0^2\zeta^2 - \omega_0^2}\right)t} + \frac{\omega_0\zeta - \sqrt{\omega_0^2\zeta^2 - \omega_0^2}}{2\sqrt{\omega_0^2\zeta^2 - \omega_0^2}}e^{-\left(\omega_0\zeta + \sqrt{\omega_0^2\zeta^2 - \omega_0^2}\right)t}\right\} \tag{11.12}$$

Due to the negative ω_0^2, the second term in the equation above tends to the infinity when $t \to \infty$, The transient response of $h(t)$ is shown in Figure 11.8.

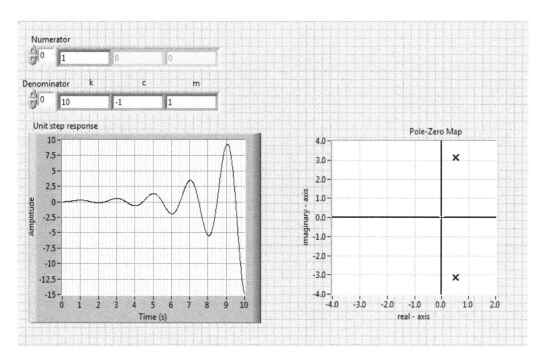

FIGURE 11.7 Root location and unit step response for the example system with $m = 1$, $c = -1$, $k = 10$.

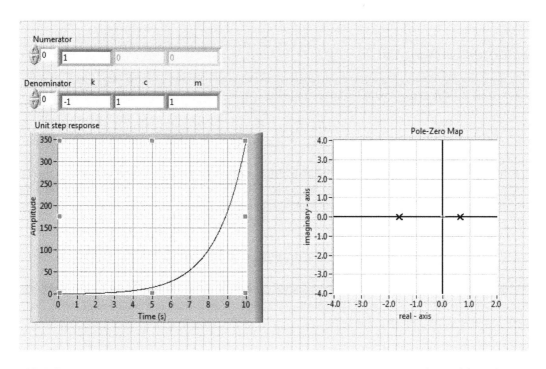

FIGURE 11.8 Root location and unit step response for the example system with $m = 1$, $c = 1$, $k = -1$.

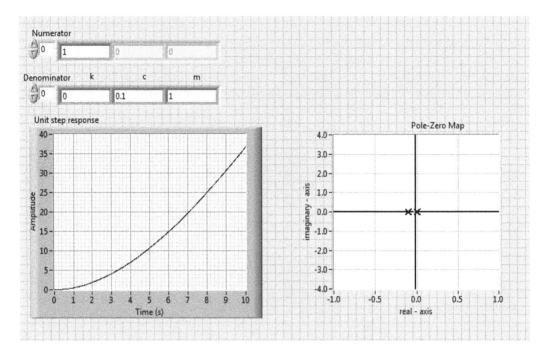

FIGURE 11.9   Root location and unit step response for the example system with $m = 1$, $c = 0.1$, $k = 0$.

INTEGRATOR AND SIMPLE-LAG

The quadratic lag becomes a serial connection of an integrator $1/s$ and a simple-lag $1/(ms + c)$ when $k = 0$ in (11.3). The corresponding transfer function will be

$$\frac{X(s)}{U(s)} = \frac{1}{s(ms + c)} \tag{11.13}$$

It is readily to see that one pole of the system lies on the origin of the s-plane. The zero pole actually corresponds to the integrator behavior of the system. The transient response of the system to the unit step input is obtained by inverse Laplace transformation:

$$h(t) = \frac{t}{c} + \frac{m}{c^2}\left(e^{-\frac{c}{m}t} - 1\right) \tag{11.14}$$

The first term on the right-hand side of the above equation is resulted from the integrator and tends to the infinity when $t \rightarrow \infty$. Figure 11.9 shows an example of pole locations and corresponding transient response to the unit step function.

LINEAR TIME-INVARIANT VIBRATING SYSTEMS

One class of mechanical systems is the linear time-invariant vibrating systems, which can be described by second-order differential equations in the matrix form as

$$\mathbf{M}\ddot{\mathbf{x}}(t) + \mathbf{D}\dot{\mathbf{x}}(t) + \mathbf{C}\mathbf{x}(t) = \boldsymbol{h}(t) \tag{11.15}$$

where

   $\mathbf{x}(t)$ $n \times 1$ output position vector

   $\mathbf{h}(t)$ $n \times 1$ input vector

   $\mathbf{M}$ $n \times n$ inertia matrix

   $\mathbf{D}$ $n \times n$ velocity dependent forces

   $\mathbf{C}$ $n \times n$ position dependent forces

The characteristic equation is determined by

$$p(s) = \frac{1}{det\mathbf{M}} det\left[\mathbf{M}\,s^2 + \mathbf{D}s + \mathbf{C}\right] = 0 \tag{11.16}$$

As often happens in practice, engineer models vibrating systems within a small range of the equilibrium points in order to obtain a linear or linearized model. In this case, the matrix form given by (11.15) will play an important role. As an example of such systems, we can refer to the quarter-car model described in Chapter 10. The characteristic equation can be derived from (11.16). Hence, the natural frequencies of the system are determined based on the poles.

Example 11.1

system matrices are given:

$$\mathbf{M} = \begin{bmatrix} m_1 & 0 \\ 0 & m_2 \end{bmatrix}, \quad \mathbf{D} = \begin{bmatrix} d_2 & -d_2 \\ -d_2 & d_2 \end{bmatrix}$$

$$\mathbf{C} = \begin{bmatrix} c_1 + c_2 & -c_2 \\ -c_2 & c_2 \end{bmatrix}$$

The characteristic equation is calculated then

$$p(s) = \frac{1}{det\mathbf{M}} det\begin{bmatrix} m_1 s^2 + d_2 s + c_1 + c_2 & -(d_2 s + c_2) \\ -(d_2 s + c_2) & m_2 s^2 + d_2 s + c_2 \end{bmatrix}$$

$$= s^4 + \left(\frac{d_2}{m_1} + \frac{d_2}{m_2}\right) s^3 + \left(\frac{c_1}{m_1} + \frac{c_2}{m_1} + \frac{c_2}{m_2}\right) s^2 + \frac{c_1 d_2}{m_1 m_2} s + \frac{c_1 c_2}{m_1 m_2}$$

❏

11.2 STATE-SPACE MODEL

While conventional control theory is based on the single-input single-output relationship, or Laplace transfer functions, modern control theory deals with more complex systems, such as those with multiple inputs and multiple outputs. The approach to the analysis and control design is based on the use of n-number of first-order differential equations to describe a system's motion, which is called a state-space model of the system. The state-space approach is best suited for system analysis and control design, even for very large systems such as multiple-input-multiple-output systems.

For an n-order linear system with r inputs, m outputs, the general form of the state-space model is as follows:

$$\begin{aligned} \dot{\mathbf{x}}(t) &= \mathbf{A}\mathbf{x}(t) + \mathbf{B}\mathbf{u}(t) \\ \mathbf{y}(t) &= \mathbf{C}\mathbf{x}(t) \end{aligned} \tag{11.17}$$

where the $n \times 1$ vector x of the state variables is the *state vector*, $\mathbf{y}$ is $m \times 1$ *output vector* while $\mathbf{u}$ is $r \times 1$ *input vector*. $\mathbf{A}$ is $n \times n$ *system matrix*, and $\mathbf{B}, \mathbf{C}$ are the $n \times r$ *input matrix* and $m \times n$ *output matrix*, respectively.

The Laplace transformation can be applied to the state-space model

$$s\mathbf{x}(s) - \mathbf{x}_0 = \mathbf{A}\mathbf{x}(s) + \mathbf{B}\mathbf{u}(s)$$
$$\mathbf{y}(s) = \mathbf{C}\mathbf{x}(s) \tag{11.18}$$

hence

$$\mathbf{y}(s) = \mathbf{G}(s)\mathbf{u}(s) + \mathbf{C}(s\mathbf{I} - \mathbf{A})^{-1}\mathbf{x}_0 \tag{11.19}$$

where $\mathbf{x}_0$ is the initial conditions, $\mathbf{G}(s)$ is called the *transfer function matrix*:

$$\mathbf{G}(s) = \frac{\mathbf{C} \; adj(s\mathbf{I} - \mathbf{A})\mathbf{B}}{|s\mathbf{I} - \mathbf{A}|} \tag{11.20}$$

For zero initial conditions, the state response becomes

$$\mathbf{x}(s) = \frac{adj(s\mathbf{I} - \mathbf{A})}{|s\mathbf{I} - \mathbf{A}|}\mathbf{B}\mathbf{u}(s) \tag{11.21}$$

The state-space model is used to describe multi-input multi-output systems in the time domain. The systems can be linear or nonlinear, time invariant or time varying. One can see now that the state-space model is applicable to all models formulated in Chapter 8, and provide a basis for control design that will be presented in further chapters.

11.3   STATE OBSERVER

A state observer determines an estimate $\hat{\mathbf{x}}$ of the state $\mathbf{x}$ from measured output $\mathbf{y}$ and control variable $\mathbf{u}$. The reason for computing estimates of the state by an observer is that in most cases, not all state variables are measurable. The cost of required sensors may be prohibitive, or it may be physically impossible to measure all state variables. At the same time, the state variables can be needed for control design, for example, a feedback control requires a complete measurement of the state variables. In many applications the state variables are also used for the monitoring of the system behavior, or for the calculating of other system signals.

Consider the following system:

$$\dot{\mathbf{x}}(t) - \mathbf{A}\mathbf{x}(t) + \mathbf{B}\mathbf{u}(t)$$
$$\mathbf{y}(t) = \mathbf{C}\mathbf{x}(t) \tag{11.22}$$

For which, the state observer is given by

$$\dot{\hat{\mathbf{x}}}(t) = \mathbf{A}\hat{\mathbf{x}}(t) + \mathbf{B}\mathbf{u}(t) + \mathbf{K}_o\left(\mathbf{y}(t) - \mathbf{C}\hat{\mathbf{x}}(t)\right) \tag{11.23}$$

Note that the observer has $\mathbf{y}$ and $\mathbf{u}$ as inputs and $\hat{\mathbf{x}}$ as output. The basic idea of observer design is to construct a model of the system dynamics to be estimated. The last term on the right-hand side of the model given by (11.23) is a correction term that involves the differences between the measured output $\mathbf{y}$ and estimated output $\mathbf{C}\hat{\mathbf{x}}$, i.e., the last term in (11.23) corrects the model continuously with the error signal. Gain matrix $\mathbf{K}_o$ serves as a weighting matrix. The addition of the correction term

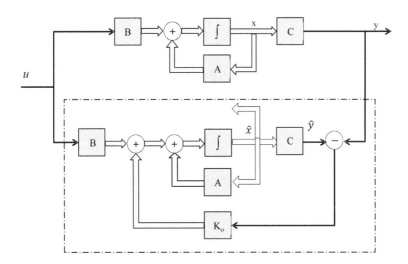

FIGURE 11.10 System model and state observer structure.

will help to influence the model dynamics of the observer by choosing appropriate gain matrix $\mathbf{K}_o$. The target is to let $\hat{\mathbf{x}}$ converge to $\mathbf{x}$. Figure 11.10 shows a block diagram of the system model and the state observer.

Let the error be $\mathbf{e} = \mathbf{x} - \hat{\mathbf{x}}$, and then subtracting (11.23) from (11.22) we obtain

$$\dot{\mathbf{e}} = (A - \mathbf{K}_o\mathbf{C})\mathbf{e} \tag{11.24}$$

We see the dynamic behavior of the error vector is determined by the eigenvalues of matrix $A - \mathbf{K}_o\mathbf{C}$. If $\mathbf{K}_o$ is chosen so that $A - \mathbf{K}_o\mathbf{C}$ is stable, the error vector will converge to zero, regardless of the value of $\mathbf{e}(0)$. Usually, the eigenvalues of the observer matrix $A - \mathbf{K}_o\mathbf{C}$ are chosen in such a way that the dynamic behaviour of the error vector is asymptotically stable and is adequately fast.

One of the methods to determine the $\mathbf{K}_o$ matrix is the *pole placement*, which means, selecting the desired location of the poles as $s_1, s_2 \ldots s_n$ and taking the same s_i's as the desired poles of the observer matrix:

$$\left| s\mathbf{I} - (A - \mathbf{K}_o\mathbf{C}) \right| = (s - s_1)(s - s_2)\ldots(s - s_n) \tag{11.25}$$

Hence, the required elements of $\mathbf{K}_o$ are obtained by matching coefficients of the like powers of s on both sides of (11.25). This approach is convenient if the system order of n is low. There are other methods in practice that can be found in reference [1].

Now the question is: How it can be sure that $\mathbf{K}_o$ always exists? What is the condition for the observer exitance presented by (11.23)? The condition can be shown that the system is observable. The *observability* refers to the ability of an observer to deduce information about all the modes of the system by monitoring the sensed outputs. *Unobservability* results from some modes or subsystems being disconnected physically from the output and therefore not appearing in the measurements. The mathematical formulation of the observability is that the system, or the pair $(\mathbf{A}, \mathbf{C})$, is observable if the matrix

$$\left[\mathbf{C}^T \vdots \quad \mathbf{A}^T\mathbf{C}^T \vdots \quad (\mathbf{A}^T)^2\mathbf{C}^T \vdots \quad \ldots (\mathbf{A}^T)^{n-1}\mathbf{C}^T \right] \tag{11.26}$$

has the rank n, which means that its determinant is not equal to zero. The matrix is called the *observability matrix*. An example should help understand the observer design process.

Example 11.2

Consider a system defined by

$$\dot{\mathbf{x}} = \mathbf{A}\mathbf{x} + \mathbf{B}u$$
$$y = \mathbf{C}\mathbf{x}$$

where

$$\mathbf{A} = \begin{bmatrix} -1 & 1 \\ 1 & -2 \end{bmatrix}, \mathbf{B} = \begin{bmatrix} 1 \\ 0 \end{bmatrix}, \mathbf{C} = \begin{bmatrix} 1 & 0 \end{bmatrix}$$

Design a state observer. The desired poles for the observer are to be placed at $s_1 = -5$, $s_2 = -6$.

Due to $\mathbf{C} = [1 \quad 0]$, the only measurable output is $y = x_1$, so the observer is supposed to estimate x_2. To check the observability first, we have

$$det\begin{bmatrix} \mathbf{C}^T & \mathbf{A}^T\mathbf{C}^T \end{bmatrix} = \begin{vmatrix} 1 & -1 \\ 0 & 1 \end{vmatrix} = 1$$

From (11.25),

$$|s\mathbf{I} - \mathbf{A} + \mathbf{K}_0\mathbf{C}| = \begin{vmatrix} s + k_{01} + 1 & -1 \\ k_{02} - 1 & s + 2 \end{vmatrix} = s^2 + (k_{01} + 3)s + (2k_{01} + k_{02} + 1)$$

Comparing the coefficients in the above equation to the coefficients in the desired characteristic equation,

$$(s - s_1)(s - s_2) = (s + 5)(s + 6) = s^2 + 11s + 30,$$

we find

$$\mathbf{K}_0 = \begin{bmatrix} 8 \\ 13 \end{bmatrix}$$

Following (11.23), the observer is designed as

$$\begin{bmatrix} \dot{\hat{x}}_1 \\ \dot{\hat{x}}_2 \end{bmatrix} = \begin{bmatrix} -1 & 1 \\ 1 & -2 \end{bmatrix}\begin{bmatrix} \hat{x}_1 \\ \hat{x}_2 \end{bmatrix} + \begin{bmatrix} 1 \\ 0 \end{bmatrix}u + \begin{bmatrix} 8 \\ 13 \end{bmatrix}\begin{bmatrix} y - \hat{x}_1 \end{bmatrix} \qquad (11.27)$$

❏

Example 11.3

The observer in Example 11.2 above is now implemented by using LabVIEW. Figure 11.11 shows the observer implementation in the block diagram, which reflects the observer structure illustrated before in Figure 11.10. The input data is implemented on the left side of the block diagram. Also, the observability is determined by the CD observability matrix VI returning the result in the front panel. The system model is programmed on the top while the observer is placed below the model. On the bottom, the observer gain is calculated by CD Ackerman VI based on the system model matrices. The observer poles are to be entered in the front panel. Both VIs are in the module "Control Design and Simulation Module" of LabVIEW.

In the simulation, the state initial conditions of the system model are set to zero while the observer states are initialized with $\hat{x}_1(0) = 1$ and $\hat{x}_2(0) = 0$. Figure 11.12 shows a plot of the system response to a step unit input, and the estimated state vector, $\hat{\mathbf{x}}$, follows exactly the system model after 1.3 sec. Depending on the location of poles chosen, the observer dynamics can be faster or slower. The farther the poles located on the left side of the *s-plane*, the faster the estimation will be.

❏

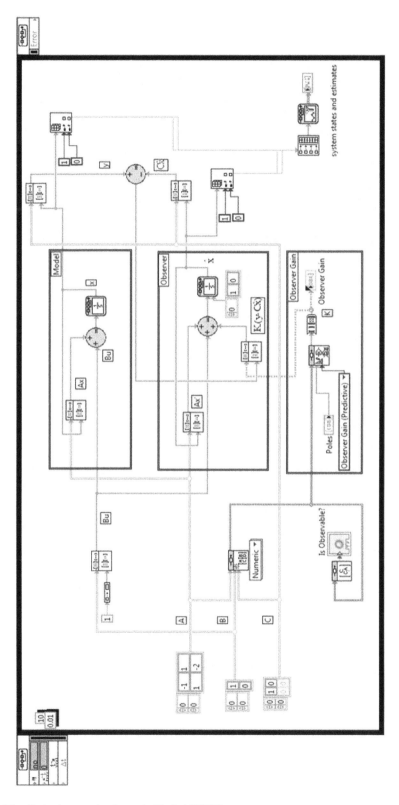

FIGURE 11.11 State observer implemented in LabVIEW.

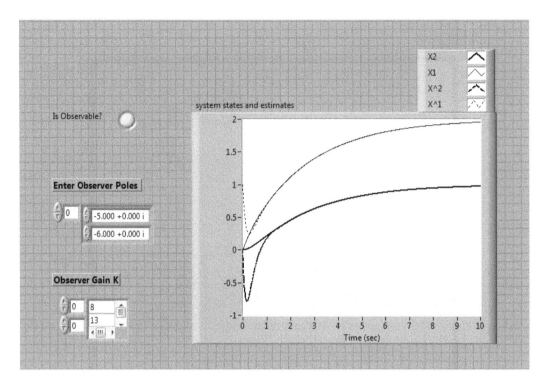

FIGURE 11.12   State estimates.

11.4   KALMAN FILTER

This section attempts to present important results of Kalman filter in a simple way omitting some details of mathematical rigor. The purpose is to focus on how to apply the Kalman filter to control problems from a practical point of view. The theoretical background as well as the derivation of the method are here not in the foreground. At the end, the differences to the standard observer described in the preceding section will be highlighted.

In the original work of 1961, Kalman and Bucy proved that this filter is the minimum mean-square linear estimator. The name "optimal filter" came from the key property that the mean-square estimation error can achieve the minimum under any reasonable performance criterion, provided the random processes are white and Gaussian. Here we view the Kalman filter from an of observer point of view, going this way is more transparent in understanding the control applications and is also a straightforward extension of the previous section.

Basically, the Kalman filter is an observer, but the Kalman filter gives an optimal estimate if the system is contaminated by noise. Let us consider a linear system

$$\dot{\mathbf{x}}(t) = \mathbf{A}(t)\mathbf{x}(t) + \mathbf{B}(t)\mathbf{u}(t) + \mathbf{v}(t)$$
$$\mathbf{y}(t) = \mathbf{C}(t)\mathbf{x}(t) + \mathbf{w}(t)$$

(11.28)

where $\mathbf{x}$ is $n \times 1$ state vector. $\mathbf{y}$ is $m \times 1$ output vector while $\mathbf{u}$ is $r \times 1$ input vector; $\mathbf{A}$ is $n \times n$ system matrix; and $\mathbf{B}, \mathbf{C}$ are the $n \times r$ *input* and $m \times n$ output matrix, respectively; $\mathbf{v}$ is $n \times 1$ state excitation noise vector; and $\mathbf{w}$ is $m \times 1$ output noise vector. Note that $\mathbf{A}(t)$, $\mathbf{B}(t)$ and $\mathbf{C}(t)$ are time varying, hence it is a *linear time varying* system.

It is to assume, that both random processes $\mathbf{v}(t)$ and $\mathbf{w}(t)$ are white-noise processes for which the *mean-value*

$$E\{\mathbf{v}(t)\} = 0$$

$$E\{\mathbf{w}(t)\} = 0 \tag{11.29}$$

and *covariance matrix*

$$E\{\mathbf{v}(t)\mathbf{v}^T(\tau)\} = \mathbf{V}(t)\delta(t-\tau)$$

$$E\{\mathbf{w}(t)\mathbf{w}^T(\tau)\} = \mathbf{W}(t)\delta(t-\tau) \tag{11.30}$$

are defined, where $\mathbf{V}, \mathbf{W}$ are non-negative definite $n \times n$ and $m \times m$ matrix, respectively and referred to as the intensity of the process at time t. The two stochastic processes, $\mathbf{v}(t)$ and $\mathbf{w}(t)$, as it often occurs in practice, are assumed to be independent of each other:

$$E\{\mathbf{v}(t)\mathbf{w}^T(\tau)\} = 0 \tag{11.31}$$

Further it is to assume, that the initial state $\mathbf{x}(t_0)$ is a random vector with mean

$$E\{\mathbf{x}(t_0)\} = \mathbf{x}_{0m} \tag{11.32}$$

and $n \times n$ is a positive semi-definite *covariance* matrix

$$E\{(\mathbf{x}(t_0) - \mathbf{x}_{0m})(\mathbf{x}(t_0) - \mathbf{x}_{0m})^T\} = \mathbf{P}_0 \tag{11.33}$$

Also, $\mathbf{x}(t_0)$ is independent of $\mathbf{v}(t)$ and $\mathbf{w}(t)$

$$E\{\mathbf{x}(t_0)\mathbf{w}^T(\tau)\} = E\{\mathbf{x}(t_0)\mathbf{v}^T(\tau)\} = 0 \tag{11.34}$$

Now for the system given by (11.28) with all assumptions presented by (11.29) and (11.30), it is to find an optimum estimator of the state $\mathbf{x}(t)$. The solution of the problem, as given by Kalman and Bucy, has the form of an observer, as shown in Figure 11.13. In other words, it can be expressed by the differential equation

$$\dot{\hat{\mathbf{x}}}(t) = \mathbf{A}(t)\hat{\mathbf{x}}(t) + \mathbf{B}(t)\mathbf{u}(t) + \mathbf{K}(t)(\mathbf{y}(t) - \mathbf{C}(t)\hat{\mathbf{x}}(t)) \tag{11.35}$$

provided that the gain matrix, $\mathbf{K}(t)$, is optimally chosen. The question is how to determine gain matrix $\mathbf{K}(t)$, which makes the covariance matrix of the error least.

Let the estimation error be $\mathbf{e}(t) = \mathbf{x}(t) - \hat{\mathbf{x}}(t)$. By subtracting (11.35) from (11.28), the differential equation for the error is obtained:

$$\dot{\mathbf{e}}(t) = (\mathbf{A}(t) - \mathbf{K}(t)\mathbf{C}(t))\mathbf{e}(t) + \mathbf{v}(t) - \mathbf{K}(t)\mathbf{w}(t) \tag{11.36}$$

Denote the covariance matrix of $\mathbf{e}(t)$ by $\mathbf{P}(t)$ and the mean of $\mathbf{e}(t)$ by $\mathbf{e}_m(t)$, i.e.,

$$E\{\mathbf{e}(t)\} = \mathbf{e}_m(t) \tag{11.37}$$

$$E\{(\mathbf{e}(t) - \mathbf{e}_m)(\mathbf{e}(t) - \mathbf{e}_m)^T\} = P(t) \tag{11.38}$$

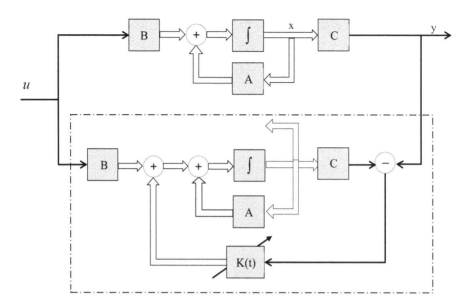

FIGURE 11.13   Kalman Filter structure.

it leads to

$$E\left\{\mathbf{e}(t)\mathbf{e}^{T}(t)\right\} = \mathbf{e}_{m}(t)\mathbf{e}_{m}^{T}(t) + \boldsymbol{P}(\mathrm{t}) \tag{11.39}$$

We are now dealing with an optimization problem, and wish to find the particular gain matrix of $\mathbf{K}(t)$ that minimizes the individual terms along the major diagonal of $\mathbf{P}(t)$, because these terms represent the estimation error variances for the elements of the state vector being estimated. Hence, the mean-square estimation error can be then formulated as

$$E\left\{\mathbf{e}^{T}(t)\mathbf{e}(t)\right\} = \mathbf{e}_{m}^{T}(t)\mathbf{e}_{m}(t) + tr\boldsymbol{P}(\mathrm{t}) \tag{11.40}$$

The optimization can be done in a way that the two terms at the right-hand side of the above equation are minimized individually. For the second term $tr\mathbf{P}(t)$, it first involves (11.36) to obtain a recursive algorithm for calculating $\mathbf{P}(t)$ and then uses the well-known result of optimal linear regulator. The result and its proof can be found in references [2, 3]. The results show that the mean-square estimation error given by (11.40) reaches the minimum if $\mathbf{K}(t)$, called *gain matrix*, is chosen as:

$$\mathbf{K}(t) = \mathbf{P}(t)\ \mathbf{C}^{\mathrm{T}}(\mathrm{t})\ \mathbf{W}^{-1}(t) \tag{11.41}$$

where the optimizing *covariance matrix* satisfies the *matrix Riccati equation*:

$$\dot{\mathbf{P}}(t) = \mathbf{A}(t)\mathbf{P}(t) + \mathbf{P}(t)\mathbf{A}^{T}(t) - \mathbf{P}(t)\mathbf{C}^{T}(t)W^{-1}(\mathrm{t})\mathbf{C}(t)\mathbf{P}(t) + \mathbf{V}(\mathrm{t}) \tag{11.42}$$

The algorithm described by (11.35), (11.41) and (11.42) is known as the *Kalman filter*[1], it computes estimates of the state vector. Since the original work of Kalman and Bucy, there were many research work published on the same topics, where the different methods were used to derive the solution. An interesting point was that further correlations with other estimation methods were noticed during that time, e.g., maximum-likelihood method, minimum variance with least-squares, etc. It is to note that

---

[1] Strictly speaking, the continuous-time filter is the Kalman-Bucy filter. The discrete-time one is called Kalman filter, published by Kalman in 1960.

the notion of the observer came several years after Kalman filter theory. As seen through the above-given results, the observer provided a perceptive viewpoint to understand the Kalman filter theory. Comparing to the standard observer, we can conclude on the following features of the Kalman filter:

- It has the same structure as the observer.
- It applies to linear time varying systems subjected to white noise.
- Its gain matrix is not constant and calculated through the covariance matrix satisfying the minimum variance estimation condition.

If the random processes, $\mathbf{v}(t)$ and $\mathbf{w}(t)$, are not only white but also Gaussian, it can be proved the Kalman filter achieves the best performance with respect to the minimum mean-square error, within linear or nonlinear systems.

VALIDITY OF ASSUMPTIONS

In analysis, one frequently makes simplifying assumptions in order to make the problem mathematically tractable. With regard to the assumptions specified in the above subsection for $\mathbf{v}(t)$ and $\mathbf{w}(t)$, the white noise seems to be the most difficult one to be justified on a practical basis. The question is whether the end result in the presence of white noise still makes sense given the conditions of the physical system.

White noise is defined to be a stationary random process having a constant spectral density function. Taking a scalar variable V as an example, we have

$$S_{vv}(j\omega) = V_s \tag{11.43}$$

with the spectral amplitude as V_s. The corresponding autocorrelation function is then

$$R_{vv}(\tau) = V_s \delta(\tau) \tag{11.44}$$

Here $\delta(\tau)$ is a delta function. One can see that no actual signal can ever satisfy the white noise assumption. Instead, the bandlimited white noise is more realistic. The spectral amplitude of the bandlimited white noise is constant over a finite range of frequencies, and zero outside that range. But, how different are the outputs if the input of each is given to the same linear system? The analysis shows that the difference is relatively small if the input spectrum is flat considerably out beyond the frequency range of the system response one or more decades [4]. Even so, the resulting simplification in the analysis is important by assuming the pure white noise. Many control applications operate in low frequency range, for example the bandwidth of vehicle dynamics responding to the steering input is around 1.0 Hz (see Figure 9.9). Therefore, the white noise assumption can be justified for the typical low frequency control systems.

Another property of the random variable is called *normal* or *Gaussian*. Taking the same example as before, the random variable v is Gaussian if its *probability density function* is given by

$$pdf(\mathrm{v}) = \frac{1}{\sqrt{2\pi V_s}} \exp\left\{ -\frac{1}{2V_s}(\mathrm{v} - \mathrm{v}_m)^2 \right\} \tag{11.45}$$

where V_s is the variance, v_m is the mean-value. The *normal distribution function* is the integral of the density function

$$ndf(\mathrm{v}) = \int_{-\infty}^{\mathrm{v}} \frac{1}{\sqrt{2\pi V_s}} \exp\left\{ -\frac{1}{2V_s}(\mathrm{u} - \mathrm{v}_m)^2 \right\} du \tag{11.46}$$

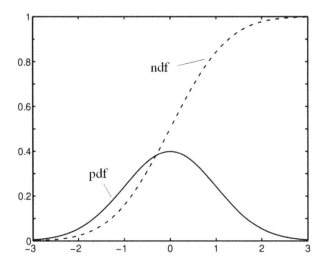

FIGURE 11.14   Probability density function and normal distribution function.

Both functions are sketched in Figure 11.14 for $v_m = 0$ and $V_s = 1$. In the normal distribution function, the portion around zero mean value is kind of linear and reflects equivalent events in the random variable space, i.e., v occurs equally likely. From the probability density function, the mean is seen to be the most likely value, with values on both sides of the mean gradually becoming less and less likely. The normal distribution is frequently applied in the analysis since many natural random phenomena seem to approximately exhibit this central-tendency property.

It should be mentioned that the random vector is a collection of random variables, such as $\mathbf{v} = [v_1, v_2 \ldots v_n]^T$, and the random process is the collection of random vectors. The term "random process" is used due to the nature of the random signals. Unlike the deterministic signal, the random signal is not predictable in a deterministic sense. More details about analysis of the random signals can be found in reference [4].

STEADY-STATE KALMAN FILTER

For a linear time-invariant system, i.e., $\mathbf{A}$, $\mathbf{B}$ and $\mathbf{C}$ in (11.28) are constant matrices, if the random processes, $\mathbf{v}(t)$ and $\mathbf{w}(t)$, are stationary so that $\mathbf{V}$ and $\mathbf{W}$ are constant as well, the covariance matrix $\mathbf{P}$(t) may become steady. Under these conditions, Kalman filter becomes

$$\dot{\hat{\mathbf{x}}}(t) = \mathbf{A}\hat{\mathbf{x}}(t) + \mathbf{B}u(t) + \mathbf{K}\left(y(t) - \mathbf{C}\hat{\mathbf{x}}(t)\right) \tag{11.47}$$

$$\mathbf{K} = \mathbf{P}\mathbf{C}^T\mathbf{W}^{-1} \tag{11.48}$$

$$0 = \mathbf{A}\mathbf{P} + \mathbf{P}\mathbf{A}^T - \mathbf{P}\mathbf{C}^T\mathbf{W}^{-1}\mathbf{C}\mathbf{P} + \mathbf{V} \tag{11.49}$$

In this steady state case, (11.49) is also called the *Algebraic Riccati Equation*. The conditions for the existence of the solution of (11.49) are similar to those for the *optimal linear regulator*:

*   The pair $(\mathbf{A}, \mathbf{C})$ is completely observable.
*   The filter (11.47) is asymptotically stable if the pair $\left(\mathbf{A}, \mathbf{V}^{1/2}\right)$ is controllable.

The first condition is sufficient for the existence of a steady-state solution of the algebraic Riccati equation while the second one ensures the filter to be asymptotically stable. In the case that solution exists, there are two ways to lead to the same solution of (11.49). One way is to solve the equation directly, where matrices $\mathbf{A}$, $\mathbf{C}$, $\mathbf{W}$ and $\mathbf{V}$ are constant. Here $\mathbf{P}$ may have more than one solutions, but the correct one must be positive definite. The alternative way is to integrate the equation by setting any arbitrary nonnegative definite matrix $\mathbf{P}_0$. The integration will reach a steady-state value of $\mathbf{P}$. However, that the initial matrix $\mathbf{P}_0$ has no influence on the steady-state value.

The following example shows an application of steady-state Kalman filter and the implementation in the LabVIEW:

Example 11.4

Consider the system defined by

$$\dot{\mathbf{x}} = \mathbf{A}\mathbf{x} + \mathbf{B}\mathbf{u} + \mathbf{F}\mathbf{v}$$
$$\mathbf{y} = \mathbf{C}\mathbf{x} + \mathbf{w}$$

where $\mathbf{A}$, $\mathbf{B}$ and $\mathbf{C}$ are the same as given in Example 11.2

$$\mathbf{A} = \begin{bmatrix} -1 & 1 \\ 1 & -2 \end{bmatrix}, \mathbf{B} = \begin{bmatrix} 1 \\ 0 \end{bmatrix}, \mathbf{C} = \begin{bmatrix} 1 & 0 \end{bmatrix}, \mathbf{F} = \begin{bmatrix} 0 \\ 1 \end{bmatrix},$$

v and w are uncorrelated limited-band white noise characterized by their scalar intensity $V = 1$ and $W = 0.1$, respectively.

Define

$$\mathbf{P} = \begin{bmatrix} p_{11} & p_{12} \\ p_{21} & p_{22} \end{bmatrix}.$$

According to (11.28), $\mathbf{v}(t) = v\mathbf{F}$ and hence $\mathbf{V} = \mathbf{F}V\mathbf{F}^T$.
Applying the Algebraic Riccati Equation (11.49)

$$\begin{bmatrix} -1 & 1 \\ 1 & -2 \end{bmatrix} \begin{bmatrix} p_{11} & p_{12} \\ p_{21} & p_{22} \end{bmatrix} + \begin{bmatrix} p_{11} & p_{12} \\ p_{21} & p_{22} \end{bmatrix} \begin{bmatrix} -1 & 1 \\ 1 & -2 \end{bmatrix}$$

$$-\frac{1}{0.1} \begin{bmatrix} p_{11} & p_{12} \\ p_{21} & p_{22} \end{bmatrix} \begin{bmatrix} 1 \\ 0 \end{bmatrix} [1\ 0] \begin{bmatrix} p_{11} & p_{12} \\ p_{21} & p_{22} \end{bmatrix} + \begin{bmatrix} 0 \\ 1 \end{bmatrix} [0\ 1] = 0,$$

we obtain the steady-state solution

$$\mathbf{P} = \begin{bmatrix} 0.069 & 0.093 \\ 0.093 & 0.275 \end{bmatrix} \text{and } \mathbf{K} = \frac{1}{0.1} \begin{bmatrix} 0.069 & 0.093 \\ 0.093 & 0.275 \end{bmatrix} \begin{bmatrix} 1 \\ 0 \end{bmatrix} = \begin{bmatrix} 0.69 \\ 0.93 \end{bmatrix}$$

The designed Kalman filter can be implemented in LabVIEW, Figure 11.15 shows the implementation in the block diagram, which reflects the Kalman filter structure illustrated before in Figure 11.13. The input data is implemented on the left side of the block diagram. The difference to Example 10.4 is that the system model and Kalman filter in block diagram are built as simulation subsystem VI, respectively. The system model VI is shown in Figure 11.16a corresponding to the system equation while the Kalman filter VI is in displayed in Figure 11.16b reflecting (11.47). Figure 11.16c shows the calculation of Riccati equation (11.49) using CD Continuous Algebraic Riccati Equations VI in LabVIEW.

In the simulation, the state initial conditions of the system model are set to zero while the estimated states are initialized with $\hat{x}_1(0) = 1$ and $\hat{x}_2(0) = 0$. Figure 11.17 displays a step-response plot of the system, and the estimated state vector $\hat{\mathbf{x}}$ follows exactly the model after 1 sec. System state x_2 is excited by white noise with variance of 1.0, and system output x_1 is disturbed by white noise with variance of 0.1. Both states are filtered as expected.

❏

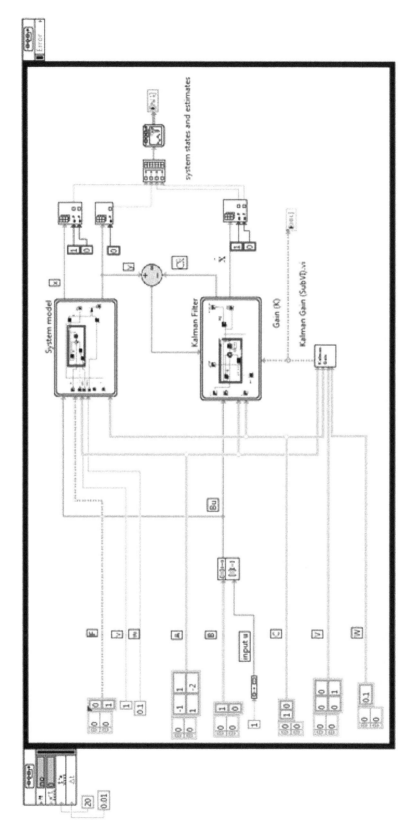

FIGURE 11.15   Steady-state Kalman filter implemented in LabVIEW.

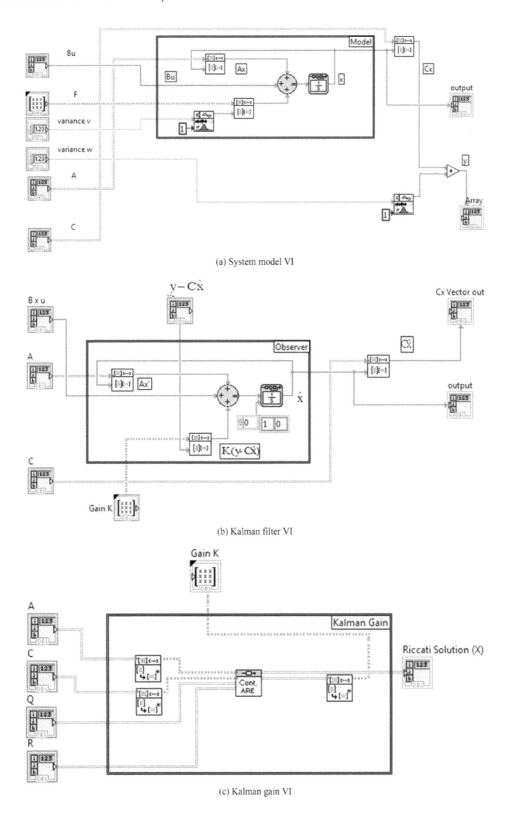

(a) System model VI

(b) Kalman filter VI

(c) Kalman gain VI

FIGURE 11.16   System model and Kalman filter illustrated as subsystems in LabVIEW.

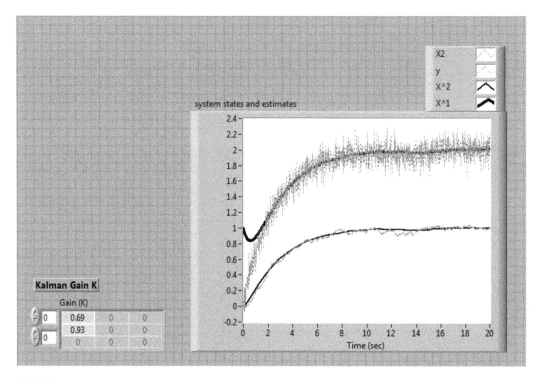

FIGURE 11.17   System states and estimates.

11.5   LYAPUNOV STABILITY THEORY

Stability plays a very important role in the design of control systems. Normally, a control system is first designed and then its stability is analyzed. So far for nonlinear systems and time-varying systems, stability analysis can be difficult or even impossible in many cases. Lyapunov stability theory has been proved to be a very useful technique for dealing with stability problems of nonlinear dynamic systems. The advantage of Lyapunov theory is that it allows for analyzing stability without explicitly solving the differential equations since solving nonlinear differential equations could be extremely difficult.

In this section we present first some basic stability definitions, and then the second method of Lyapunov, which we apply to design stable controllers for nonlinear dynamic systems. For nonlinear systems, there are different aspects in regard to the scope of systems, stability concepts and theorems, which go beyond this book. The main focus here is to apply Lyapunov stability theorem to vehicle control applications, thus only relevant topics are addressed. For more details and contents, the interested reader is referred to references [1, 5, 6].

STABILITY DEFINITIONS

Unlike linear systems, a nonlinear system can demonstrate behavior that is considered stable in some region of state space and unstable in other regions. Therefore, the question of stability should refer to the *equilibrium points* of a system rather than to the system itself.

The state-space model is described for linear systems in Section 11.2. For nonlinear systems, the system dynamics can also be formulated in a state-space model as

$$\dot{\mathbf{x}}(t) = f(\mathbf{x}, t) \tag{11.50}$$

where $n \times 1$ vector $\mathbf{x}$ of the state variables is the *state vector* and $f(\mathbf{x},t)$ is an $n \times 1$ vector whose elements are functions of x_1, $x_2 \ldots x_n$ and t. A state $\mathbf{x}_e$ is defined as an *equilibrium point* where

$$f(\mathbf{x}_e,t) = \mathbf{0} \text{ for all } t \tag{11.51}$$

Basically, the equilibrium point is a point in the state space, at which $\dot{\mathbf{x}}$ is zero. That means, the system remain in that state once it is placed there. In general, any isolated equilibrium point can be shifted to the origin of the coordinates by translating the coordinates. Therefore, the stability analysis of an equilibrium point can be performed under the assumption $\mathbf{x}_e = \mathbf{0}$.

Stability

An equilibrium point $\mathbf{x}_e$ is said to be *stable* if for a given nonzero initial condition the state $\mathbf{x}(t)$ will remain inside a certain region. More precisely, the system described by (11.50) is stable at $\mathbf{x}_e$ if for every ε there is a δ such that if

$$\|\mathbf{x}_0 - \mathbf{x}_e\| \leq \delta$$

then

$$\|\mathbf{x}(t) - \mathbf{x}_e\| \leq \varepsilon \text{ for all } t$$

where $\mathbf{x}_0$ is the initial state, and $\|\mathbf{x}\|$ is the Euclidean norm. The definition above is also called *stability in the sense of Lyapunov*.

Asymptotic Stability

The equilibrium point, $\mathbf{x}_e$, is said to be *asymptotically stable* if it is stable in the sense of Lyapunov, and furthermore state $\mathbf{x}(t)$ in fact approaches to $\mathbf{x}_e$ as time goes to infinity.

The asymptotic stability has a local character, simply to establish asymptotic stability may not mean that the system will operate properly. In this case, it is necessary to know the size of region that guarantees the stability.

Global Asymptotic Stability

The equilibrium point, $\mathbf{x}_e$, is said to be *globally asymptotically stable* if it is asymptotically stable for all states from which trajectories originate.

In control design, it is always desirable to achieve the global asymptotic stability. The common approach is to ensure the asymptotic stability first, and then to verify that it is applicable for all states in the defined state space.

Instability

The equilibrium point, $\mathbf{x}_e$, is said to be un*stable* if the condition for stability cannot be met, i.e., there is always a state, $\mathbf{x}_0$, within the region $S(\delta)$ defined by δ, so that the trajectory will leave the region of $S(\varepsilon)$ defined by ε at a later time.

The foregoing definitions are graphically represented in Figure 11.18. In a two-dimensional case, the equilibrium points and the associated trajectories are shown corresponding to the above-given stability definitions, respectively – stability, asymptotic stability and instability.

SECOND METHOD OF LYAPUNOV

The core of the *second method of Lyapunov* is to utilize the energy concept from physics to interpret the stability concept in a mathematical form. The energy concept is widely and well understood in the physics world. We can look at the same typical example, the mass-spring oscillator, given by

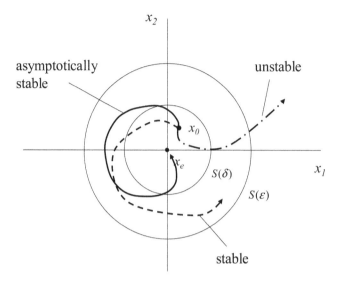

FIGURE 11.18   Stability definitions of equilibrium points.

Figure 11.1 in the beginning of this chapter. If the mass is displaced and released, it will slow down and stop as a result of action of the damping mechanism. The fact is that the stored energy will continually decrease while the time derivative of the energy is negative, until the system becomes stable at the equilibrium position, i.e., at the equilibrium point of the system.

Lyapunov generalized the energy concept and introduced a scalar function $V(\mathbf{x})$ known as *Lyapunov function*. Mathematically function $V(\mathbf{x})$ is kind of "energy" function. Provided that a Lyapunov function can be found for a given system, Lyapunov proved that if the function is always decreasing, the system will come to rest, thus stable.

To match Lyapunov stability method, the following definitions are used for positiveness of scalar functions. From now on, we treat an equilibrium point as the origin of the state space.

- Scalar function $V(\mathbf{x})$ is *positive (negative) definite* in a region containing the origin if $V(\mathbf{x}) > 0$ ($V(\mathbf{x}) < 0$) for $\mathbf{x} \neq \mathbf{0}$ and $V(\mathbf{0}) = 0$.
- Scalar function $V(\mathbf{x})$ is *positive (negative) semidefinite* in a region containing the origin if $V(\mathbf{x}) \geq 0$ ($V(\mathbf{x}) \leq 0$) while $V(\mathbf{x}) = 0$ not only for $\mathbf{x} = \mathbf{0}$.
- Scalar function $V(\mathbf{x})$ is *indefinite* if it can have both positive and negative signs in the region.

A very often used class of scalar functions has quadratic form, for example,

$$V(\mathbf{x}) = \mathbf{x}^T \mathbf{P} \mathbf{x},$$

where $\mathbf{P}$ is a symmetrical and constant matrix and defined as

$$\mathbf{P} = \begin{vmatrix} p_{11} & p_{12} & \cdots & p_{1n} \\ p_{12} & p_{22} & \cdots & \\ \vdots & & & \\ p_{1n} & \cdots & \cdots & p_{nn} \end{vmatrix} \tag{11.52}$$

For this form, the following Sylvester's theorem is used to determine the definiteness.

Theorem 11.1 (Sylvester)
$V(\mathbf{x})$ is positive definite if and only if all principal minors of determinant $|\mathbf{P}|$ are positive:

$$p_{11} > 0, \quad \begin{vmatrix} p_{11} & p_{12} \\ p_{12} & p_{22} \end{vmatrix} > 0, \ ..., \ |\boldsymbol{P}| > 0 \tag{11.53}$$

If one or more are zero, $V(\mathbf{x})$ is semidefinite. $V(\mathbf{x})$ is negative definite if $-V(\mathbf{x})$ is positive definite.

Example 11.5

Determine if the given quadratic form is positive or negative definite:

$$V(\mathbf{x}) = 7x_1^2 + 2x_2^2 + 4x_1x_2$$

The quadratic form can be written

$$V(\mathbf{x}) = \mathbf{x}^T \mathbf{P} \mathbf{x} = \begin{bmatrix} x_1 & x_2 \end{bmatrix} \begin{bmatrix} 7 & 1 \\ 3 & 2 \end{bmatrix} \begin{bmatrix} x_1 \\ x_2 \end{bmatrix}$$

Using (11.53), we have

$$7 > 0, \quad \begin{vmatrix} 7 & 1 \\ 3 & 2 \end{vmatrix} > 0.$$

Thus, $V(\mathbf{x})$ is positive definite.

❏

In vehicle control systems, which this book is dealing with, the open-loop systems are autonomous, i.e., they have no explicit time dependence. For this reason, Lyapunov function can be a time independent function. Now, we review Lyapunov stability theorem in the following.

Theorem 11.2 (Lyapunov)
Consider an autonomous nonlinear system given by

$$\dot{\mathbf{x}}(t) = f(\mathbf{x}) \tag{11.54}$$

where

$$f(\mathbf{0}) = \mathbf{0} \text{ for all } t$$

If there exists a scalar function $V(\mathbf{x})$ having continuous, first partial derivatives and

- $V(\mathbf{x})$ is positive definite.
- $\dot{V}(\mathbf{x})$ is negative definite, then the equilibrium at the origin is *asymptotically stable*.
- The equilibrium at the origin is *asymptotically stable*, and $V(\mathbf{x})$ is radially unbounded, i.e., $V(\mathbf{x}) \to \infty$ as $\|\mathbf{x}\| \to \infty$, then it is *globally asymptotically stable*.

The first derivative $\dot{V}(\mathbf{x})$ is not guaranteed to be negative definite. In the case that it is semidefinite, the LaSalle's theorem can be used.

Theorem 11.3 (LaSalle)
Consider an autonomous nonlinear system given by

$$\dot{\mathbf{x}}(t) = f(\mathbf{x}) \tag{11.55}$$

where

$$f(0) = 0 \text{ for } \quad \text{all } t$$

If there exists a scalar function $V(\mathbf{x})$ having continuous, first partial derivatives and

- $V(\mathbf{x})$ is positive definite
- $\dot{V}(\mathbf{x})$ is negative semidefinite, then the equilibrium at the origin is *asymptotically stable* if and only if $\dot{V}(\mathbf{x}) = 0$ at $\mathbf{x} = 0$
- The equilibrium at the origin is *asymptotically stable*, and $V(\mathbf{x})$ is radially unbounded, i.e., $V(\mathbf{x}) \to \infty$ as $\|\mathbf{x}\| \to \infty$, then it is *globally asymptotically stable*

In the following, Example 11.6 shows how to determine the Lyapunov stability while Example 11.7 describes the basic idea to specify a control input based on Lyapunov stability theorem.

Example 11.6

Determine the stability at the origin of the following system:

$$\dot{x}_1 = -x_1 + 3x_2 + x_1(x_1^2 + x_2^2)$$

$$\dot{x}_2 = -3x_1 - x_2 + x_2(x_1^2 + x_2^2)$$

Using the following quadratic functions as a Lyapunov function:

$$V(\mathbf{x}) = x_1^2 + x_2^2$$

The time derivative of $V(\mathbf{x})$ is given by

$$\dot{V}(\mathbf{x}) = 2x_1\dot{x}_1 + 2x_2\dot{x}_2$$

$$= 2(x_1^2 + x_2^2)(x_1^2 + x_2^2 - 1)$$

$\dot{V}(\mathbf{x})$ is negative definite in the region where $x_1^2 + x_2^2 < 1$. Therefore, the system at origin is asymptotically stable within $x_1^2 + x_2^2 - 1$, but it is not globally asymptotically stable.

❏

Example 11.7

Consider a linear system described by

$$\begin{bmatrix} \dot{x}_1 \\ \dot{x}_2 \end{bmatrix} = \begin{bmatrix} 0 & 1 \\ -2 & a \end{bmatrix} \begin{bmatrix} x_1 \\ x_2 \end{bmatrix} + \begin{bmatrix} 0 \\ 1 \end{bmatrix} u \qquad (11.56)$$

where a is an unknown parameter in the range $0 < a < a_m$. Specify a control input to stabilize the equilibrium point at the origin.
Define a Lyapunov function by

$$V(\mathbf{x}) = \frac{1}{2}\mathbf{x}^T\mathbf{x},$$

thus, the first derivative becomes

$$\dot{V}(\mathbf{x}) = \mathbf{x}^T \dot{\mathbf{x}}$$
$$= x_2(-x_1 + ax_2 + u)$$

If we choose u so that

$$u = -x_2 + x_1 - \text{sgn}(x_2)a_m|x_2|$$

then

$$\dot{V}(\mathbf{x}) \leq -x_2^2 - (a_m - a)x_2^2$$

So $\dot{V}(\mathbf{x})$ is negative semidefinite. In case of $x_2 = 0$, then $x_1 = 0$. Hence, $\dot{V}(\mathbf{x})$ is zero only for $\mathbf{x} = 0$. Therefore, the equilibrium point $\mathbf{x} = 0$ is asymptotically stable.

❏

Example 11.7 shows the basic principle for designing a controller. One defines a Lyapunov function, then can calculate its total derivative which exhibits an explicit dependence on a control signal. Afterward the control signal is to be determined so that the derivative of the Lyapunov function is negative definite, then the closed-loop control system is ensured to be stable.

Lyapunov's method establishes only sufficient conditions for the stability. Lyapunov function $V(\mathbf{x})$ is not unique. Different choices may lead to different stability regions. Generally, Lyapunov's method is very useful and powerful, the most difficult step is to search for Lyapunov function because there is no universal method to find it.

11.6   LINEAR QUADRATIC OPTIMAL CONTROL

There are different optimization techniques are discussed in many literatures. In regard to vehicle control applications, we are interested to find out how to use the minimum energy to achieve an acceptable control performance. In order to perform optimization, a mathematical scale must established for quantifying an optimization criterion or a "cost function". Then a control law needs to be determined to satisfy the optimal value of the criterion.

In this section, we describe the linear-quadratic (LQ) optimal control, which is a design method of stable control systems. The LQ method is basically to provide a control input, which is optimized based on quadratic performance criteria. The LQ optimal control is structured as the linear feedback control. But due to the optimal design, it provides improved robustness in the presence of disturbances and unmodeled uncertain dynamics.

We consider here a linear time-invariant system again

$$\dot{\mathbf{x}}(t) = \mathbf{A}\mathbf{x}(t) + \mathbf{B}\mathbf{u}(t) \tag{11.57}$$

where $n \times 1$ vector $\mathbf{x}$ is the state vector and $\mathbf{u}$ is the $n \times 1$ control input vector. $\mathbf{A}$ is the $n \times n$ system matrix, and $\mathbf{B}$ is the $n \times r$ input matrix. In control applications, the control input can be related to energy, for example engine torque or braking force which can be thought as a part of performance criterion. At the same time, the deviation of the state from the control target should be taken into consideration while choosing the control input to minimize expenditure of energy. For this purpose, a quadratic performance criterion is used:

$$J = \int_0^\infty \left(\mathbf{x}^T \mathbf{Q}\mathbf{x} + \mathbf{u}^T \mathbf{R}\mathbf{u}\right) dt \tag{11.58}$$

where $\mathbf{Q}$ is a $n \times n$ symmetric, positive definite (or positive semidefinite) matrix, and $\mathbf{R}$ is a $r \times r$ symmetric, positive definite matrix. The first term on the right side of the above equation can be seen as is a measure of the system transient response while the second term represents a measure of instantaneous rate of energy expenditure. The first term can also be interpreted as a penalty term for physical constraints on $\mathbf{u}$. Both terms are weighted by $\mathbf{Q}$ and $\mathbf{R}$, respectively. The state *weighting matrix*, $\mathbf{Q}$, specifies the importance of the various components of the state vector relative to each other. The control *weighting matrix, $\mathbf{R}$*, often has the purpose to limit the magnitude of the control signal $\mathbf{u}$, so that it will not saturate at the maximum which can be produced. For example, smaller value of $\mathbf{R}$ leads to faster response, but at the cost of a larger input amplitude. These matrices basically contain design parameters that are selected by engineer depending on the desired form of the closed-loop transient responses and the control input.

The LQ optimal control is to design control input $\mathbf{u}(t)$ in the form

$$\mathbf{u}(t) = -\mathbf{K}\mathbf{x}(t) \tag{11.59}$$

so that the performance index in (11.58) is minimized, where $\mathbf{K}$ is the feedback gain matrix. In a standard feedback control, the gain matrix can be chosen to achieve specified closed-loop pole locations. The optimal control is to seek a gain matrix to minimize the specified performance criterion. Some approaches to find the optimal solution of this problem are using the dynamic programming method [3, 6].

In order to show a direct relationship between Lyapunov function and the above-performance criterion, we present here the approach based on the second method of Lyapunov [1]. Applying (11.59) to (11.57), the closed-loop dynamics is given by

$$\dot{\mathbf{x}}(t) = (\mathbf{A} - \mathbf{B}\mathbf{K})\mathbf{x}(t)$$

For now, it is assumed that the closed-loop dynamics matrix $\mathbf{A} - \mathbf{B}\mathbf{K}$ is stable. In the same way, the performance criterion (11.58) becomes

$$J = \int_0^\infty \mathbf{x}^T \left(\mathbf{Q} + \mathbf{K}^T \mathbf{R} \mathbf{K} \right) \mathbf{x} dt \tag{11.60}$$

Now we chose a Lyapunov function in the form

$$\mathbf{x}^T \left(\mathbf{Q} + \mathbf{K}^T \mathbf{R} \mathbf{K} \right) \mathbf{x} = \frac{d}{dt} \mathbf{x}^T \mathbf{P} \mathbf{x} \tag{11.61}$$

where $\mathbf{P}$ is a symmetric, positive definite matrix. Following (11.61), we obtain

$$(\mathbf{A} - \mathbf{B}\mathbf{K})^T \mathbf{P} + \mathbf{P}(\mathbf{A} - \mathbf{B}\mathbf{K}) + \mathbf{Q} + \mathbf{K}^T \mathbf{R} \mathbf{K} = 0 \tag{11.62}$$

and the performance criterion becomes

$$J = \mathbf{x}^T(0) \mathbf{R} \mathbf{x}(0) \tag{11.63}$$

It shows that the performance criterion depends on the initial condition, $\mathbf{x}(0)$, and matrix $\mathbf{P}$.

With further evaluation of (11.62), as detailed in reference [1], the feedback gain matrix is determined as follows:

$$\mathbf{K} = \mathbf{R}^{-1} \mathbf{B}^T \mathbf{P} \tag{11.64}$$

and the matrix $\mathbf{P}$ is the solution of Riccati equation

$$\mathbf{A}^T \mathbf{P} + \mathbf{P} \mathbf{A} - \mathbf{P} \mathbf{B} \mathbf{R}^{-1} \mathbf{B}^T \mathbf{P} + \mathbf{Q} = 0 \tag{11.65}$$

Based on Lyapunov theorem, the system is stable if there is a positive-definite matrix $\mathbf{P}$ exists. Therefore, the approach is to ensure that P results in a positive-definite matrix, then the feedback gain matrix $\mathbf{K}$ of the controller is the optimal matrix. Since $\mathbf{A} - \mathbf{BK}$ is a stable matrix in this case, we have $\mathbf{x}(\infty) \to 0$.

In the calculation of the Riccati equation given by (11.65), quadratics will have multiple solutions. A stronger and numerically safer set of requirements for one unique positive definite solution of matrix $\mathbf{P}$ is given [6]: the algebraic Riccati equation has a unique, positive definite solution if

- $\mathbf{A}$ is asymptotically stable, or
- $(\mathbf{A}, \mathbf{B})$ is controllable and $\left(A, \sqrt{Q}\right)$ is observable.

Observability was discussed in Section 11.3. System (11.57), or the pair $(\mathbf{A}, \mathbf{B})$, is said to be controllable if it is possible, by means of the input, to transfer the system from any initial state $\mathbf{x}(t_0)$ to any other state in a infinite interval of time. The algebraic criterion for complete state controllability is that $n \times n$ matrix

$$\left[\mathbf{B} \ \vdots \ \mathbf{AB} \ \vdots \ (\mathbf{A})^2\mathbf{B} \ \vdots \ ... \left(\mathbf{A}\right)^{n-1}\mathbf{B}\right] \tag{11.66}$$

has the rank n, which means that its determinant is not equal to zero. The matrix is called the *controllability matrix*.

Furthermore, the robustness of the LQ design was studied in different literatures [6]. It can be shown that the designed system guarantees infinite gain margin and $60°$ of phase margin. This means that the LQ system will remain stable even though the gain has any upward increase or downward reduction to half the nominal design value.

The following example shows how to calculate the LQ optimal controller. The same example is also implemented in LabVIEW so that the reader can practice the design using the available modules in LabVIEW.

Example 11.8

Consider a linear system described by

$$\begin{bmatrix} \dot{x}_1 \\ \dot{x}_2 \end{bmatrix} = \begin{bmatrix} 0 & 1 \\ 0 & 0 \end{bmatrix} \begin{bmatrix} x_1 \\ x_2 \end{bmatrix} + \begin{bmatrix} 0 \\ 1 \end{bmatrix} u$$

It is to find the optimal control signal u(t) such that the performance criterion

$$J = \int_0^\infty \mathbf{x}^T \mathbf{Q} \mathbf{x} + u^2 dt, \quad Q = \begin{bmatrix} 9 & 0 \\ 0 & 2 \end{bmatrix}$$

is minimized.

In this case *R=1*. Using the Riccati equation (11.65), it follows

$$\begin{bmatrix} 0 & 0 \\ 1 & 0 \end{bmatrix} \begin{bmatrix} p_{11} & p_{12} \\ p_{21} & p_{22} \end{bmatrix} + \begin{bmatrix} p_{11} & p_{12} \\ p_{21} & p_{22} \end{bmatrix} \begin{bmatrix} 0 & 1 \\ 0 & 0 \end{bmatrix}$$

$$- \begin{bmatrix} p_{11} & p_{12} \\ p_{21} & p_{22} \end{bmatrix} \begin{bmatrix} 0 \\ 1 \end{bmatrix} \begin{bmatrix} 0 & 1 \end{bmatrix} \begin{bmatrix} p_{11} & p_{12} \\ p_{21} & p_{22} \end{bmatrix} = \begin{bmatrix} -9 & 0 \\ 0 & -2 \end{bmatrix}$$

which can be simplified to

$$\begin{bmatrix} -p_{12}^2 & p_{11} - p_{22}p_{12} \\ p_{11} - p_{22}p_{12} & 2p_{12} - p_{22}^2 \end{bmatrix} = \begin{bmatrix} -9 & 0 \\ 0 & -2 \end{bmatrix}$$

Among the multiple solutions, $\mathbf{P}$ has a positive-definite matrix

$$\mathbf{P} = \begin{bmatrix} 6\sqrt{2} & 3 \\ 3 & 2\sqrt{2} \end{bmatrix}$$

Thus,

$$\mathbf{K} = \mathbf{R}^{-1}\mathbf{B}^{\mathsf{T}}\mathbf{P} = \begin{bmatrix} 3 & 2\sqrt{2} \end{bmatrix}$$

So, the optimal controller is determined by

$$u = -3x_1 - 2\sqrt{2}x_2$$

In LabVIEW, a VI named LQR solves the associated Riccati equation and calculates the feedback gain matrix $\mathbf{K}$. Figure 11.19 shows the LQR used in a block diagram. By entering $\mathbf{A}$, $\mathbf{B}$, $\mathbf{Q}$ and $\mathbf{R}$ the calculated results show up in the front panel. In the case that matrix $\mathbf{A}\text{-}\mathbf{BK}$ cannot be made stable for any $\mathbf{K}$, then a positive-definite of $\mathbf{P}$ does not exist. LQR gives a solution which is not positive definite. ❑

11.7   LINEAR QUADRATIC OPTIMAL CONTROL WITH OUTPUT TARGET

From the LQ optimal control presented in the previous section, it is to see that the use of controller (11.59) results in system state $\mathbf{x}$ to become zero in the steady state, i.e., $\mathbf{x}(\infty) = 0$. In practical applications, however, a target or desired value for the system output is required, and additionally there

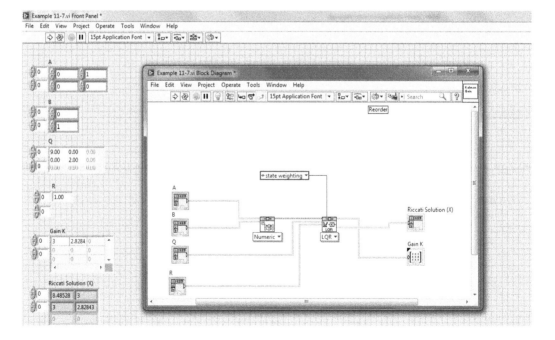

FIGURE 11.19   An example of using LQR in LabVIEW.

will be disturbances that need to be considered in the system as well. This controller does not offer the possibility, to force the output to follow the target value and to compensate disturbances. Let us consider a system with defined output **y** and disturbance **z**

$$\dot{\mathbf{x}}(t) = \mathbf{A}\mathbf{x}(t) + \mathbf{B}\mathbf{u}(t) + \mathbf{B}_z\mathbf{z}(t) \tag{11.67}$$

$$\mathbf{y}(t) = \mathbf{C}\mathbf{x}(t) \tag{11.68}$$

where $n \times 1$ vector **x** is the state vector, **u** is the $r \times 1$ input vector, **y** is the $m \times 1$ output vector, and **z** is the $q \times 1$ disturbance vector. The dimensions of **A**, **B** and **C** are matched accordingly. The output target is defined as $\mathbf{y}_d(t)$. We assume here that output target $\mathbf{y}_d(t)$ and disturbance $z(t)$ are step functions, respectively, i.e.,

$$\begin{aligned} \mathbf{y}_d(t) &= \mathbf{y}_0 \quad \text{for} \quad t \geq 0 \\ z(t) &= z_0 \quad \text{for} \quad t \geq 0 \end{aligned} \tag{11.69}$$

For $t \geq 0$, both are constant. Hence, the control error is defined as

$$\mathbf{e}(t) = \mathbf{y}_d(t) - \mathbf{y}(t) \tag{11.70}$$

Applying a feedback controller such as (11.59), the system given by (11.67) becomes

$$\dot{\mathbf{x}}(t) = (\mathbf{A} - \mathbf{BK})\mathbf{x}(t) + \mathbf{B}_z\mathbf{z}(t) \tag{11.71}$$

Even though **u**(t) provides stable poles for the closed-loop system, the steady state will not be zero, i.e., $\mathbf{x}(\infty) \neq 0$. In this case, the performance criterion in (11.58) will approach infinite.

In order to use the linear optimal controller, we introduce a controller as

$$\mathbf{u}(t) = -\mathbf{K}\,\mathbf{x}(t) + \mathbf{u}_0 \tag{11.72}$$

where $\mathbf{u}_0$ is an auxiliary control input which has the same dimension as **u**(t). The target is to specify $\mathbf{u}_0$, hence **u**(t), so that the control error diminished in the steady state, i.e.,

$$\lim_{t \to \infty} \mathbf{e}(t) = 0 \tag{11.73}$$

Substituting (11.72) into system (11.67), we have

$$\dot{\mathbf{x}}(t) = (\mathbf{A} - \mathbf{BK})\mathbf{x}(t) + \mathbf{B}\mathbf{u}_0 + \mathbf{B}_z\mathbf{z}(t) \tag{11.74}$$

and the steady state will be

$$\mathbf{x}(\infty) = -(\mathbf{A} - \mathbf{BK})^{-1}(\mathbf{B}\mathbf{u}_0 + \mathbf{B}_z z_0) \tag{11.75}$$

From (11.68), the output target results in

$$\mathbf{y}_0 = \mathbf{C}\mathbf{x}(\infty) \tag{11.76}$$

Combining (11.75) and (11.76), we obtain

$$\mathbf{y}_0 = -\mathbf{C}(\mathbf{A} - \mathbf{BK})^{-1}(\mathbf{B}\mathbf{u}_0 + \mathbf{B}_z z_0) \tag{11.77}$$

Hence, the auxiliary control input, $\mathbf{u}_0$, can be solved from the above-presented equation. Each component of vector $\mathbf{u}_0$ can be calculated if the dimensions of $\mathbf{u}_0$ and $\mathbf{y}_0$ matche to each other, i.e., the unique solution exists when $m = r$. If the dimension of $\mathbf{y}_0$ is lower than the one of $\mathbf{u}_0$, i.e., $m < r$, then the solution of $\mathbf{u}_0$ is not unique. The solution does not exist if $m > r$. In the case of a full dimension $n = m = r$, the result can be given by

$$\mathbf{u}_0 = -(\mathbf{C}(\mathbf{A} - \mathbf{BK})^{-1}\mathbf{B})^{-1}(\mathbf{y}_0 + \mathbf{C}(\mathbf{A} - \mathbf{BK})^{-1}\mathbf{B}_z\mathbf{z}_0) \tag{11.78}$$

if $\mathbf{C}(\mathbf{A} - \mathbf{BK})^{-1}\mathbf{B}$ has the full rank n.

The equation given by (11.72) is basically an extension of the LQ optimal controller, for which the performance criterion has a modified version of (11.58), i.e., $\mathbf{u}(t)$ and $\mathbf{x}(t)$ are replaced by $\mathbf{u}(t) - \mathbf{u}(\infty)$ and $\mathbf{x}(t) - \mathbf{x}(\infty)$, respectively. In this sense, the linear feedback gain, $\mathbf{K}$, is optimized with regard to $\mathbf{u}(t) - \mathbf{u}(\infty)$ and $\mathbf{x}(t) - \mathbf{x}(\infty)$, which indeed corresponds to a coordination transformation.

As a conclusion for the system defined by (11.67) and (11.68), the LQ optimal controller is given by (11.72), (11.64) and (11.65), which are summarized here:

$$\mathbf{u}(t) = -\mathbf{K}\,\mathbf{x}(t) + \mathbf{u}_0 \tag{11.79}$$

with

$$\mathbf{K} = \mathbf{R}^{-1}\,\mathbf{B}^{\mathrm{T}}\,\mathbf{P}, \tag{11.80}$$

$$\mathbf{A}^{\mathrm{T}}\mathbf{P} + \mathbf{P}\mathbf{A} - \mathbf{P}\mathbf{B}\mathbf{R}^{-1}\mathbf{B}^{\mathrm{T}}\mathbf{P} + \mathbf{Q} = 0 \tag{11.81}$$

where $\mathbf{u}_0$ is calculated by (11.77).

REFERENCES

1. K. Ogata, Modern Control Engineering, 3rd ed., New Jersey: Prentice Hall, 1997.
2. B. Friedland, Control System Design, New York: Dover Publications, 2005.
3. M. Gopal, Modern Control System Theory, New Delhi: New Age International Ltd., 1993.
4. R. G. Brown, Introduction to Random Signal Analysis and Kalman Filtering, New Jersey: Wiley, 1983.
5. F. L. Lewis, C. T. Abdallah and D. M. Dawson, Control of Robot Manipulators, New York: Macmillan Publishing Company, 1993.
6. W. L. Brogan, Modern Control Theory, Upper Saddle River, NJ: Prentice Hall, 1991.

12 Wheel Slip Control

Essentially excessive wheel slip occurs due to the fact that the output torque of the actuator generates the longitudinal tire force that reaches the maximum friction force between the tire and the road surface. The current friction coefficient, hence the longitudinal force is a nonlinear function of wheel slip as exhibited by the μ-curve. However, the relationship between the longitudinal forces and slip varies depending on tire-road surface conditions. Also, that the longitudinal force cannot be measured directly makes the control design analysis more intricately.

The main goal of wheel slip control is to keep the operating point at a slip optimum, which corresponds to some user-defined strategy, including fuel saving, vehicle performance or other. Typically, the slip optimum for most road surfaces is established within the range $0.1 < \lambda < 0.3$. In this region, the tire has a stable gripping with the road surface and obtains sufficient longitudinal force. From control point of view, a control system must keep track of changes in road conditions and adjust the operating point at the desired slip target.

Wheel slip control can be applied during deceleration as well as acceleration of a vehicle. In the following, we refer it to as *brake slip control* while vehicle is braking and *tractive slip control* while vehicle is accelerating, respectively.

For control design, the model of wheel rotational dynamics from Chapter 3 is recalled. The crucial part is the nonlinear dynamics in the model. If the nonlinear dynamics can be measured, control design would be convenient and have more options for methods to apply. In the following, we first introduce an estimation algorithm for the nonlinearities of wheel rotational dynamics, and then design brake and tractive slip controls. The estimation method was introduced in reference [1] and uses a fictitious model in a linear form. Based on the linear model, the state observer can be applied as the effectiveness shown in references [1, 2].

ESTIMATION OF NONLINEARITIES

Wheel rotational dynamics is described by the moment balance equation presented in Chapter 3 as (3.25), and rewritten here as follows:

$$T_a - T_b - J_w \dot{\omega}_w = F_x r_w + F_R r_w + \tau_d \tag{12.1}$$

Here, T_a is the tractive torque and T_b is the braking torque applied through the axle shaft to the wheel. F_x is the longitudinal force and F_R is the rolling resistance force, which includes the effects of Camber and the alignment moment. In reality, any wheel is always affected by disturbances. Therefore, a *disturbance* term τ_d is added to represent internal factors, which are not included in the model of wheel dynamics, and external disturbances acting from the road.

On the right side of (12.1) are the nonlinear terms. The longitudinal force, F_x, is a nonlinear function of the wheel slip and time t as described before by (4.9) and (4.24). The rolling resistance is a difficult term to model for a given wheel, and indeed, may be the most contrary term to describe in the wheel dynamics. It depends on tire temperature, tire material, load and slip. For more details, see reference [3]. In order to make the control design easier, we combine the nonlinear terms and introduce a fictitious model as follows:

$$F_x r_w + F_R r_w + \tau_d \approx g z(t)$$

$$\dot{z}(t) = h z(t) \tag{12.2}$$

DOI: 10.1201/9781003134305-12

where $z(t)$ is a system state to describe approximately the time behavior of the nonlinearities. The model parameters, g and h, have to be chosen with respect to the nonlinear characteristics. In practice, they will be verified by simulation or experimental data. As the fictitious model, (12.2) can be placed into a state-space form to further leverage the control methods of the linear system.

Now we define a state $\mathbf{x}(t)$ by

$$\mathbf{x}(t) = \begin{vmatrix} \omega_w(t) \\ z(t) \end{vmatrix}$$

and the system input, $u(t)$, and output, $y(t)$, become, respectively

$$u(t) = T_a - T_b, \; y(t) = \omega_w(t)$$

The system can be controlled by the driving torque as well as the braking torque. Thus, (12.1) can be rewritten as

$$\dot{\mathbf{x}}(t) = \mathbf{A}\mathbf{x}(t) + \mathbf{B}(t)\mathbf{u}(t)$$
$$y(t) = \mathbf{C}\mathbf{x}(t) \tag{12.3}$$

with

$$\mathbf{A} = \begin{bmatrix} 0 & -\dfrac{g}{J_w} \\ 0 & h \end{bmatrix}, \; \mathbf{B} = \begin{bmatrix} \dfrac{1}{J_w} & 0 \end{bmatrix}^T, \quad \mathbf{C} = \begin{bmatrix} 1 & 0 \end{bmatrix}$$

It is to assume that the sensor, which is capable of measuring the angular speed of the wheel, is available. In Chapter 11 we considered the design of a state observer to estimate the state variables, which are not measurable. The state vector $\mathbf{x}(t)$ defined in (12.3) has $z(t)$ as the state variable. Since the system represented by (12.3) is completely observable, state $z(t)$ can be estimated. Hence, the nonlinear term is estimated by means of the state observer based on the fictitious model. With regard to (12.3), the state observer is designed as

$$\dot{\hat{\mathbf{x}}}(t) = \mathbf{A}\hat{\mathbf{x}}(t) + \mathbf{B}u(t) + \mathbf{K}_o\big(y(t) - \mathbf{C}\hat{\mathbf{x}}(t)\big) \tag{12.4}$$

where $\mathbf{K}_o$ is the gain matrix to be chosen. In reference [1], it was shown that the observer estimation system is asymptotically stable if the model error of the nonlinearities is relative bounded by the estimation error, i.e.,

$$F_x r_w + F_R r_w + \tau_d - gz(t) \leq \gamma\hat{\mathbf{x}}(t) - \mathbf{x}(t) \tag{12.5}$$

where $\cdot$ is Euclidean norm, γ is a small positive constant, and $\hat{\mathbf{x}}$ is the estimate of $\mathbf{x}$. As the result, the estimated state $\hat{z}$ converges to z. Thus,

$$F_x r_w + F_R r_w + \tau_d \approx g\hat{z}(t) \tag{12.6}$$

Figure 12.1 shows the design block diagram of the state observer. Equation (12.4) of the observer is exactly reflected in the block diagram, where the difference between the actual wheel speed as the model output and the estimate of the model output is feedbacked to the observer.

The effectiveness of the estimation will be shown by the simulation and the vehicle experiment in the following subsections of this chapter.

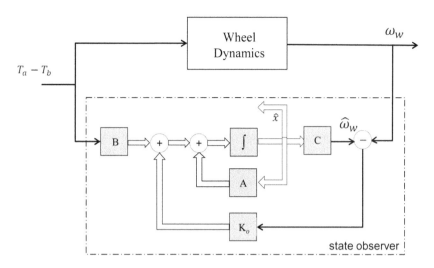

FIGURE 12.1   State observers for the nonlinearity estimation in the wheel dynamics.

12.1   BRAKE SLIP CONTROL

A typical application of the *brake slip control* is its implementation for *anti-lock brake control systems*. The anti-lock brake control is a part of the vehicle braking system and provides controlled braking torques to each wheel preventing the wheel from locking, i.e., from complete sliding.

In order to design the control algorithm, it is convenient to formulate wheel rotational dynamics in terms of the brake slip. For this purpose, the definition of the slip from Chapter 4 is recalled. Considering the wheel dynamics from (12.1) again, and combining it with (4.2) and (4.5), we obtain the following equation for the brake slip

$$\frac{J_w}{r_w}v\dot{\lambda} + \frac{J_w}{r_w}\dot{v}(\lambda - 1) = T_b - T_a + F_x r_w + F_R r_w + \tau_d \tag{12.7}$$

where index "B" of the slip is omitted for simplification. Note that the engine torque is equal to zero while braking, i.e., $T_a = 0$. The vehicle linear speed sensor and the sensor of the angular speed of the wheel are considered available, thus the wheel slip is to be calculated. Since the nonlinearities are estimated by the state observer, they can be directly compensated by the brake torque. Hence, the control law for the braking torque can be partitioned into three parts as follows

$$T_b = \frac{J_w}{r_w}v\,u_0 - g\hat{z} + \frac{J_w}{r_w}\dot{v}(\lambda - 1) \tag{12.8}$$

where u_0 is an auxiliary control input to be specified later while the second term, $g\hat{z}$, compensates the nonlinearities, and the third term compensates the dynamics that relates to the longitudinal acceleration, $\dot{v}$.

Combining (12.8) and (12.7) with (12.2) yields

$$\dot{\lambda} = u_0 \tag{12.9}$$

This is a linear system in the form of an integrator, in which input u_0 makes the use of feedback to modify the behaviour of the wheel dynamics. This feedback can be designed by using many approaches, including linear controller design methods, such as *P*, *PI* and *PID* controllers.

In order to detail the control implementation, we use a *PI* controller as an example. Suppose that the slip target, λ_d, has been selected at each time instance of the control, define a control error as follows

$$e(t) = \lambda_d(t) - \lambda(t) \tag{12.10}$$

The PI controller is given by

$$\dot{\epsilon}(t) = e(\mathrm{t})$$
$$u_0(t) = k_p e(t) + k_i \epsilon(t) \tag{12.11}$$

here, $\epsilon(t)$ is the integral of the control error $e(t)$. k_p, k_i are controller prameters to be chosen. Combining (12.11) and (12.9), yields

$$\ddot{e} + k_p \dot{e} + k_i e = \dot{\lambda}_d \tag{12.12}$$

The characteristic equation of the closed-loop system is then

$$s^2 + k_p s + k_i = 0 \tag{12.13}$$

Parameters k_p, k_i should be determined so that the system poles locate in the left-half s-plane. At the same time, these parameters should provide a quick dynamic control response of the system so that the wheel dynamics follows quickly to its desired behavior. For example, such quick response is needed to reduce the braking force when the vehicle is skidding in deceleration.

The goal of the control is to obtain the optimum slip. For the same road surface, it will have same value for the optimum. Thus, the slip target is ending up at a constant value, i.e., $\dot{\lambda}_d = 0$. The controller becomes a constant set-point controller to guarantee the error $e(t)$ converges to zero. Figure 12.2 shows the structure of the brake slip control. The controller compensates the known nonlinear part, the unknown nonlinearities estimated by the observer, and plus the input provided by the *PI* controller. The *PI* controller consists of the *P* controller and the *I* controller components. The *P* controller usually produces a steady state error and a long settling time. Due to the integrator behavior, the *I* controller can reduce the steady state error of the wheel slip, but will increase the overshoot. In the end, the *PI* controller may result in a longer settling time with no steady state error but a higher overshoot.

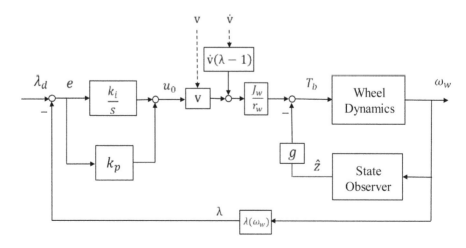

FIGURE 12.2 Structure of the brake slip control.

The above-presented method is implemented in a simulation program using the MATLAB software packet to demonstrate the performance of the brake slip control and to illustrate the estimation of the nonlinearities based on the state observer. A simulation of a one-wheel model control is considered while the vehicle speed is calculated based on vehicle model (3.35) assuming other three wheels have the same road condition and develop the same slippage as the controlled wheel. The parameters of a passenger car used in the simulation include: $M = 1400$ kg, $r_w = 0.31$ m, $J_w = 2.15$ kgm^2. For the drag force, the parameters are $\rho = 0.6$ kg $/$ m^3, $v_{Ad} = 0.3$ and $A = 3.2$ m^2.

In the one-wheel model given by (12.1), it is assumed that $f_R = 0$, $\tau_d = 0$ and $v_{ad} = v$. The μ − curve of the dry asphalt is simulated for the the slip calculation based on (4.16), in which $C_1 = 0.879$, $C_2 = 37.2$ and $C_3 = 0.179$ are selected. The controller parameters are chosen as $k_p = 50$ and $k_i = 0$.

In this example, the control is applied to reduce the speed of the car by achieving the slip target of $\lambda_d = 0.14$, Figure 12.3 shows the simulation results. As seen in Figure 12.3a, the wheel slip reaches the target value after 0.2 s. The estimation of the longitudinal force of the wheel is shown in Figure 12.3b, in which the observer-based estimate is able to approach the force, which is computed in the model, within 30 ms. Figure 12.3c shows the vehicle speed. The control starts at 72 km/h, and the speed decreases rapidly during the control.

12.2 TRACTIVE SLIP CONTROL

A typical application of the tractive slip control is the implementation for *traction control*, which reduces the engine torque and/or applies the brake mechanisms to eliminate the wheels spinning during acceleration. The braking of a spinning wheel allows the engine power to deliver to the other drive wheel or the wheels that have better road conditions. In such applications, a wheel speed sensor is used to detect when a drive wheel is spinning faster than the other wheels – meaning the vehicle is losing its traction.

Similar to the brake slip control, we start to formulate the model of wheel rotational dynamics in terms of tractive slip. Considering the wheel dynamics as (12.1) again, and combining it with (4.2) and (4.8), the dynamic model of rotational dynamics is given by the following equation:

$$J_w \omega_w \dot{\lambda} + \frac{J_w}{r_w} \dot{v} = (T_a - T_b - F_x r_w - F_R r_w - \tau_d)(1 - \lambda) \tag{12.14}$$

where index "T" of the slip is omitted for simplification. A vehicle linear speed sensor and a sensor of the angular speed of the wheel are considered available, thus the wheel slip can be to be calculated. Since the nonlinearities are estimated by the state observer, they can be directly compensated by the torque. Hence, the control law can be partitioned into two parts.

$$T_a - T_b = u_0 + g\hat{z} \tag{12.15}$$

where u_0 is an auxiliary control input that is specified later; the second term $g\hat{z}$ compensates the nonlinearities. The control input is made of both driving torque and brake torque.

Substituting (12.15) into (12.14) and (12.1), respectively, it yields

$$J_w \omega_w \dot{\lambda} + \frac{J_w}{r_w} \dot{v} = u_0(1 - \lambda) \tag{12.16}$$

$$J_w \dot{\omega}_w = u_0 \tag{12.17}$$

Equation (12.16) describes a first-order nonlinear system, in which input u_0 needs to be defined to stabilize system and achieve the slip target. For this purpose, a controller is designed using Lyapunov stability theory.

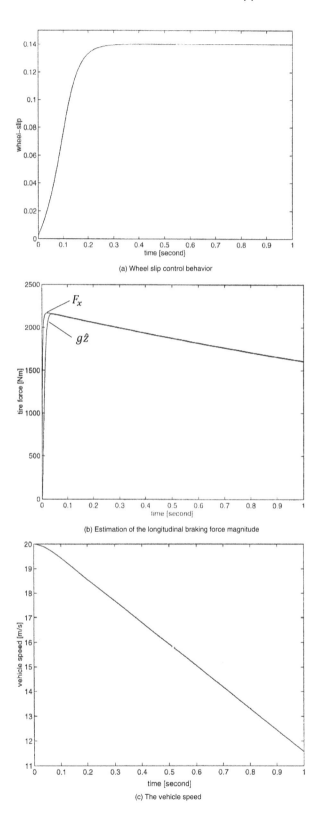

(a) Wheel slip control behavior

(b) Estimation of the longitudinal braking force magnitude

(c) The vehicle speed

FIGURE 12.3   Simulation of the brake slip control.

Suppose that a tractive slip target, $\lambda_d(t)$, has been selected for the control, define a control error as

$$e(t) = \lambda_d(t) - \lambda(t) \tag{12.18}$$

Since the optimum values of the tractive slip for different road surfaces are considered constant, i.e., we have $\dot{\lambda}_d = 0$ and $\dot{e} = -\dot{\lambda}$.

Now we select a Lyapunov function in terms of the rotational inertia and the control error

$$V(e) = \frac{1}{2} J_w \omega_w\, e^2(t) \tag{12.19}$$

Differentiating (12.19) with respect to time leads to

$$\dot{V}(e) = \frac{1}{2} J_w \dot{\omega}_w\, e^2 - J_w\, \omega_w \dot{\lambda} e \tag{12.20}$$

Substituting (12.16) and (12.17) into (12.20) gives

$$\dot{V}(e) = \frac{1}{2} u_0\, e^2 - \left(u_0(1-\lambda) - \frac{J_w}{r_w}\dot{v} \right) e \tag{12.21}$$

If we specify the auxiliary control input as

$$u_0 = \frac{-ke - \dfrac{J_w}{r_w}\dot{v}}{0.5e - (1-\lambda)} \tag{12.22}$$

where, k is a positive constant, then (12.21) becomes

$$\dot{V} = -ke^2 \tag{12.23}$$

This result shows Lyapunov function V decreases continuously. The only chance for $\dot{V} = 0$, would be at $e = 0$. Therefore, according to LaSalle's Theorem, the control error, e, is asymptotically stable. Note that $(0.5e - (1-\lambda)) \neq 0$ during the control due to the range of the wheel slip and the setting of the target slip, i.e., $0 \le \lambda \le 1$ and $0 < \lambda_d < 0.3$. Hence, the tractive slip controller is given by a nonlinear control law of (12.15) with the auxiliary control input given by (12.22), where the estimate of the nonlinearities is provided by a state observer. Figure 12.4 shows the structure of the tractive slip control.

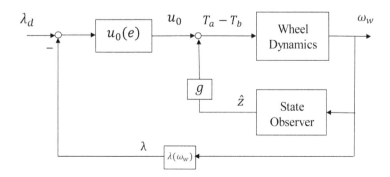

FIGURE 12.4   Structure of the tractive slip control.

The tractive slip control has two distinct types of actuation – driving and braking torques. Both of them can be chosen to provide the torque depending on the application, in which the vehicle control system is configured. The application can be, for example a stability control system, a traction control system. In the application of the *traction control system*, the tractive slip controller is implemented for each drive wheel or drive axle, and cooperates with *speed differential control* in order to increase vehicle traction as well as vehicle stability on *low-µ* road surfaces. The *speed differential controller* determines the braking torque needed for the wheel. The torque distribution between the drive wheels is managed through *torque vectoring*, which can be performed through the driving torque or the braking torque delivery. In the next section, we describe the speed differential controller between the drive wheels using *braking torque vectoring*, and afterwards its function in a traction control system.

12.3 SPEED DIFFERENTIAL CONTROL BY TOQUE VECTORING

The term of the speed differential refers to the difference of the rotational speeds of two wheels on one axle. The control of the speed differential intends to equalize the rotational speeds of two wheels of the axle. In regard to the actuation, there are two distinct types of torque vectoring. The first one is *driving torque vectoring*, which was introduced in Chapter 5. This type of torque vectoring uses an individual electric motor for each drive wheel or an active axle differential controlling the torque distribution between the drive wheels. The second type is called *braking torque vectoring* where a brake control system is utilized to enable the torque vectoring between two wheels connected by an open differential. Here, we consider the braking torque vectoring.

To increase the vehicle traction capability, dynamics and control of both wheels of the drive axle are considered. We continue to use the same driving scenario discussed in Section 5.2.6 for the explanation, in which the vehicle is moving on a *split-µ* surface. The left wheel is spinning on *low-µ* while the right wheel is driven on *high-µ*. With a standard open differential, the driving torque is equally distributed to both wheels. The goal of the control should be to vary the torque at each wheel, in particular to reduce the torque at the left wheel and increase the torque at the right wheel of the axle. This function can be achieved by controlling the brake of the spinning left wheel.

Applying the state observer result of (12.6) to the wheel dynamic model given by (12.1), we have

$$T_a - T_b - J_w \dot{\omega}_w = g\hat{z} \tag{12.24}$$

Extending (12.24) to the left and right wheels and remembering the open differential's property to split the driving torque equally between the wheels, the difference of the braking torques yields

$$J_w \left(\dot{\omega}_r - \dot{\omega}_l \right) + g_r \hat{z}_r - g_l \hat{z}_l = T_{bl} - T_{br} \tag{12.25}$$

where "l" and "r" are the index for the left and right wheels, respectively. Since term $g_r \hat{z}_r - g_l \hat{z}_l$ is the estimate by the observer, a controller can be defined by (12.26), which compensates the nonlinear estimate:

$$T_{bl} - T_{br} = -k \left(\omega_r - \omega_l \right) + g_r \hat{z}_r - g_l \hat{z}_l \tag{12.26}$$

where $k > 0$ is a controller parameter that needs to be chosen. After substituting (12.26) in (12.25), the closed-loop dynamics becomes

$$J_w \left(\dot{\omega}_r - \dot{\omega}_l \right) + k \left(\omega_r - \omega_l \right) = 0 \tag{12.27}$$

which implies

$$\lim_{t \to \infty} \left(\omega_r - \omega_l \right) = 0 \tag{12.28}$$

The rotational speed of the left wheel tends to equal the rotational speed of the right wheel by increasing the braking torque at the left wheel. Indeed, due to the open differential properties, the braking torque at the left wheel results in an increased torque supplied by the differential to the right wheel. The increased driving torque generates a bigger tractive force at this wheel with the *high-μ* condition, and thus improves vehicle traction performance. It should be emphasized that (12.26) determines the braking torque vectoring, where the braking torque on the *high-μ* side may be set to zero; in the above example we have $T_{br} = 0$.

The speed differential controller usually cooperates with the tractive slip controller, which controls the slip of the drive axle to the target slip. The next example explains such cooperation of the two controllers.

EXAMPLE OF A TRACTION CONTROL SYSTEM CONCEPT

Usually, all vehicle controllers are integrated based on the architecture of the vehicle control network. At the same time, the controllers are configured in different ways in order to perform their defined control functions. The goal of traction control is to increase the traction capability and move the wheels and the vehicle forward while maintaining stability of the vehicle. During the control action, the vehicle should not spin or drift away from a desired trajectory.

Figure 12.5 illustrates a diagram of a traction control system of a front-wheel-drive vehicle with an open differential in the drive axle where index "l" and "r" indicate left and right, respectively. Based on the speed difference of the two wheels, the speed differential controller, as it was considered in the previous section, equalizes the speed of both wheels by applying the brake of a spinning wheel. In this case, a bigger driving torque at the non-spinning wheel is implemented in a bigger tractive force that improves the vehicle traction. Thus, both drive wheels receive proper amounts of the torque to move the vehicle forward. The output of the speed differential controller is the braking torque difference between two wheels. The brake actuation can be hydraulic or electrical to achieve the braking torque target the controller needs. Thus, the traction improvement is undertaken by the braking torque vectoring.

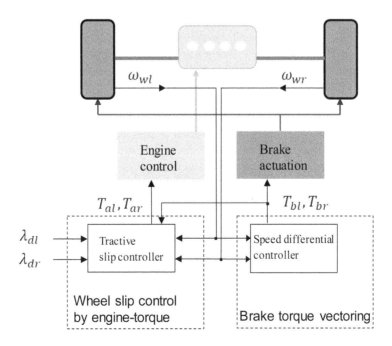

FIGURE 12.5   Configuration of a traction control system.

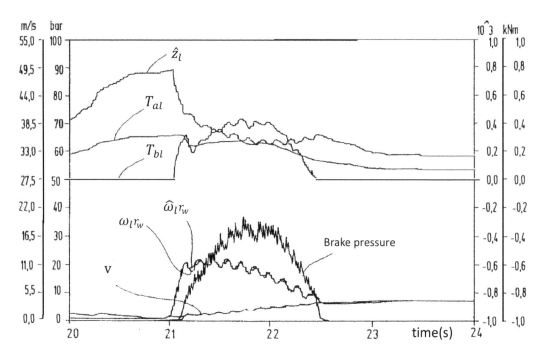

FIGURE 12.6   Experimental results of a traction controller in a straight-line motion.

In addition, the tractive slip control cooperates with the speed differential control. It controls the slip of the front drive axle to obtain the optimum/desired slip, and thus to provide a bigger traction force between the tires and the road surface. The slip of the front axle can be considered as the average of the two wheels. The tractive slip controller calculates the engine torque needed, incorporating the braking toque from the speed differential control. The calculated torque value is provided to the engine control system that adjusts the engine actuators to deliver the driving torque to the axle. If one of the wheels has too much tractive slip, the tractive slip controller will request the engine control to momentarily reduce the power; for example, the vehicle when it accelerates on an icy road slows down as the wheels lose traction. After the engine controller lowers the power, the tractive slip controller will maintain the given target slip and make the actual rotational speed of the wheels and the target rotational speed determined by the target slip close to each other.

The above-designed speed differential controller with the nonlinearities estimation was tested and verified in a traction control system on a passenger car with a front-wheel-drive system. Figure 12.6 presents the experimental results of the traction control tests that were published in the patent [4]. It particularly shows the effectiveness of the speed differential control with the estimation of the nonlinearities of the vehicle dynamics by the state observer. The left wheel starts to spin at $t = 21$ s, and the slip is rising rapidly due to the difference between the actual vehicle speed and the linear velocity computed by the rotational speed of the left wheel. The braking torque request determined by the controller, T_{bl}, comes soon after. The actual braking torque, the brake pressure in this case, comes up 100 ms later. Since the braking torque was realized by the hydraulic brake, the brake pressure was measured as illustrated in Figure 12.6. The torque, T_{al}, provided by the engine, decreases as the brake pressure is building up. As the build-up of the brake pressure continues, the wheel speed, $\omega_l r_w$, reaches the desired vehicle speed, v, at $t = 22.5$ s. The nonlinearities of the wheel dynamics are estimated by $\hat{z}_l$ during the control period. For the linear approximation of the model, the parameters are chosen as $g = 1$ and $h = 0$. It is shown further in reference [4] that the control is significantly slower without the compensation of the nonlinearities, i.e., the control takes one second longer with less accuracy in the same test.

REFERENCES

1. P. C. Müller, "Indirect Measurement of Nonlinear Effects by State Observers," in *Nonlinear Dynamics in Engineering Systems, IUTAM Symposium, Stuttgart*, Germany, 1989.
2. D. Söffker, J Bajkowski and P.C. Müller, "Detection of Cracks in Turborotors - A New Observer Based Method," *ASMS Journal of Dynamics Systems, Measurement and Control*, pp. 518–524, Sept. 1993.
3. T. D. Gillespie, Fundamentals of Vehicle Dynamics, Warrendale: SAE, 1992.
4. J. Yu, Verfahren und Vorrichtung zum Angaben der auf ein Rad wirkenden Stoermomente. Patent DE 10104599A1, Feburary 2001.

13 Vehicle Motion Control

In this chapter, the vehicle motion control refers to the control of vehicle motion along the longitudinal and lateral axes. Two control methods are studied, including vehicle speed and path-following control methods. Both controllers can be used in ADAS (Advance Driver Assist System) devices, for example the speed control can be used in an automatic cruise control system, and the path-following control can be utilized in a lane-keeping system.

The longitudinal control is based on the models described in Chapter 3 including the drivetrain, the brakes, the aerodynamic drag and tire-rolling resistance, as well as the influence of gravity when the vehicle is moving on a road with a nonzero inclination. The lateral control is based on the single-track model enriched with a vehicle position model, which considers the lateral forces influenced by the steering, the sideslip angle, the yaw rate and by the vehicle position relative to the trajectory path.

Particularly, autonomous vehicle navigation requires an intelligent control system for path following. Among other controls, such control system should include longitudinal and lateral controls. The controllers designed in this chapter utilize a modern control optimal technique that is advantageous to control of autonomous vehicles. The optimal control technique yields improved robustness in the presence of disturbances and unmodeled dynamics.

13.1 VEHICLE SPEED CONTROL

In Chapter 11 we reviewed the design of the linear quadratic optimal control. Now we apply the method to design a vehicle speed control that targets to achieve and maintain a given desired speed, v_d.

From Chapter 3, the vehicle motion is described by (3.35) as follows:

$$T_{af} + T_{ar} = T_{bf} + T_{br} + \left(F_{Ad} + Mg\sin\gamma_z\right)r_w + \left(F_{Rf} + F_{Rr}\right)r_w + Mar_w + \frac{J_{wf}a + J_{wr}a}{r_w} \tag{13.1}$$

In terms of the vehicle speed, we can rewrite the equation to

$$M_s\dot{v} = T_a - T_b - N_f \tag{13.2}$$

where T_a and T_b represent the driving torque and the braking torque at the wheels, and

$$M_s = Mr_w + \frac{J_{wf} + J_{wr}}{r_w}$$

$$N_f = (F_{Ad} + Mg\sin\gamma_z)r_w + \left(F_{Rf} + F_{Rr}\right)r_w$$

Assuming N_f is known, one can specify the control torque as

$$T_a - T_b = M_s(\dot{v}_d - u) + N_f \tag{13.3}$$

where u is the state-feedback control input. Substituting (13.3) into (13.2), we have the error system

$$\dot{e} = u \tag{13.4}$$

where $e = v_d - v$. In this case any linear controller can be designed. However, in vehicle dynamics, an uncertain disturbance often exists, so that a nonzero steady-state error may occur. An integrator

DOI: 10.1201/9781003134305-13

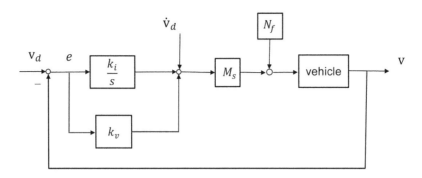

FIGURE 13.1   Vehicle speed control based on LQ design.

in the feedforward loop will effectively help reduce the error. In this way, the scalar error equation (13.4) becomes a vector system

$$\dot{e}_1 = e_2$$
$$\dot{e}_2 = u \tag{13.5}$$

where e_2 denotes the speed error. Equivalently (13.5) can be formulated in the standard state-space form

$$\begin{bmatrix} \dot{e}_1 \\ \dot{e}_2 \end{bmatrix} = \begin{bmatrix} 0 & 1 \\ 0 & 0 \end{bmatrix} \begin{bmatrix} e_1 \\ e_2 \end{bmatrix} + \begin{bmatrix} 0 \\ 1 \end{bmatrix} u \tag{13.6}$$

Now we can apply the linear-quadratic optimal controller as follows:

$$u = -k_i e_1 - k_v e_2 \tag{13.7}$$

which yields a closed-loop system shown in Figure 13.1.

In order to obtain the optimal vector of $(k_i\ k_v)$, one selects $\mathbf{Q}$ and $\mathbf{R}$ in the Riccati equation (11.65). The errors can be independently weighted if $\mathbf{Q}$ is chosen in a diagonal form

$$\mathbf{Q} = \begin{bmatrix} Q_i & 0 \\ 0 & Q_v \end{bmatrix} \tag{13.8}$$

Due to the simple form of the matrices in (11.62), the Riccati equation solution is readily found. Then, using this solution in (11.64) yields

$$\begin{bmatrix} k_i\ k_v \end{bmatrix} = \begin{bmatrix} \sqrt{\dfrac{Q_i}{R}} & \sqrt{2\sqrt{\dfrac{Q_i}{R}} + \dfrac{Q_v}{R}} \end{bmatrix} \tag{13.9}$$

From (13.3) and (13.7), the actual control torque is

$$T_a - T_b = M_s(\dot{v}_d + k_i e_1 + k_v e_2) + N_f \tag{13.10}$$

which shows that the torque $T_a - T_b$ is indirectly optimized by optimizing the control input u given by (13.7), i.e., the optimization is only with respect to the error system and state-feedback control input $u(t)$. Therefore, (13.10) represents basically a *suboptimal approach* with respect to the actual vehicle dynamics. The energy in $T_a - T_b$ may be expected to be smaller when $u(t)$ is larger.

In most driving situations, the vehicle parameters, such as those included in M_s and N_f, are not exactly known. Therefore, estimates are often in use. The torque in (13.3) is now specified as

$$T_a - T_b = \hat{M}_s(\dot{v}_d - u) + \hat{N}_f \tag{13.11}$$

This equation does exactly take the advantage of the linear quadratic optimal control, which guarantees the robustness mentioned in Section 11.6. The optimal control stays robust when a deviation in the controller gain happens. The estimates in (13.11) will introduce some coupling in the linear model (13.3) and lead to

$$\begin{bmatrix} \dot{e}_1 \\ \dot{e}_2 \end{bmatrix} = \begin{bmatrix} 0 & 1 \\ 0 & 0 \end{bmatrix} \begin{bmatrix} e_1 \\ e_2 \end{bmatrix} + \begin{bmatrix} 0 \\ 1 \end{bmatrix} \left[u + M_s^{-1} \left(-\Delta_M u + \Delta_M \dot{v}_d + \Delta_N \right) \right] \tag{13.12}$$

with

$$\Delta_M = \left(M_s - \hat{M}_s \right)$$

$$\Delta_N = \left(N_f - \hat{N}_f \right)$$

The term $(-\Delta_M u + \Delta_M \dot{v}_d + \Delta_N)$ is a nonlinear function of time and other parameters that represents an internal disturbance of the linearized error dynamics caused by modeling uncertainties, vehicle parameter variations, external disturbances, uncertainties in road-tire frictions, etc. Those variations can be interpreted as the gain deviation and it may be within the tolerance of the robustness. Since the controller has an integrator in the loop, it can effectively reduce the error to a certain level. A further study will be interesting, how a robust design can improve the control performance. In this regard, the next subsection presents an alternative for the robustness improvement using Lyapunov stability theory.

ROBUSTNESS ANALYSIS

From Chapter 11 we learned that Lyapunov's method is a useful technique and can be applied directly to nonlinear systems to analyze their stability. In Chapter 12, the tractive slip controller was designed using Lyapunov's method because of the nonlinear dynamic model of the wheel slip. Here, the overall closed-loop dynamics given by (13.12) remains nonlinear due to the inaccurate model-based compensation. We can use the same concept to study the robustness of the speed controller.

We begin with extending controller (13.7) to a different form

$$u = -k_i e_1 - k_v e_2 - \eta \tag{13.13}$$

where η is an auxiliary signal to be specified, and select a Lyapunov function in terms of the mass and the control error

$$V(e) = \frac{1}{2} M_s e_2^2(t) + \frac{1}{2} \hat{M}_s k_i e_1^2(t) \tag{13.14}$$

Differentiating (13.14) with respect to time, and considering (13.13), (13.11) and (13.3), leads to

$$\dot{V} = -k_v \hat{M}_s e_2^2 + e_2 \left\{ \Delta_M \dot{v}_d + \Delta_N - \hat{M}_s \eta \right\} \tag{13.15}$$

If the auxiliary signal can meet the following condition:

$$\eta = \eta_1 \text{sgn}(e_2)$$

$$\eta_1 \geq \hat{M}_s^{-1} \left(\max \left(|\Delta_M| \right) |\dot{v}_d| + \max \left(|\Delta_N| \right) \right) \tag{13.16}$$

where sgn(•) is the signum function, (13.15) becomes

$$\dot{V} \leq -k_v \hat{M}_s e_2^2 \tag{13.17}$$

This result shows that Lyapunov function V decreases continuously. If $\dot{V} = 0$, then $e_2 = 0$. Therefore, according to LaSalle's Theorem, the control error, e_2, is asymptotically stable. In fact, auxiliary signal η ensures the robustness of the closed-loop system. If η satisfies condition (13.16), the vehicle speed achieves the defined target speed.

One can see from the robustness analysis, that Lyapunov's method does ensure the stability, but not provide information about the performance of the control, such as transient response. Even with an integrator embedded in the controller, the analysis yields no indication on the transient behavior how quickly the disturbance can be suppressed.

To show the implementation of the vehicle speed control presented in this section, the following example demonstrates the control in a LabVIEW® simulation. Note that the robustness alternative is not implemented in this example, i.e., $\eta = 0$.

Example 13.1

Use LabVIEW to set up a simulation for the speed control. The vehicle model given before in (13.1) is based on the parameters of a pickup truck: $M = 2280$ kg, $r_w = 0.4$ m, $J_{wf} = 3.0$ kgm², $J_{wr} = 3.0$ kgm², $f_{Rf} = f_{Rr} = 0.014$. For the drag force, the parameters are $\rho = 1.2$ kg/m³, $v_{Ad} = 0.3$ and $A = 3.2$ m².

In this example, the optimal control is designed and simulated in the LabVIEW. Figure 13.2 shows the control implementation in the LabVIEW block diagram, which reflects the control structure illustrated in Figure 13.1. The input data is implemented on the left side of the block diagram. Equation (13.1) is used to simulate the vehicle model, which is implemented in the MathScript format on the left side of the block diagram. The controller is programmed in MathScript as well and placed on the right side of the block diagram. The controller parameters are inserted into the front panel, where one can set the values for the simulation. In the front panel, both the vehicle model and the controller parameters can be changed. More details can be seen in Appendix A.5.

The desired acceleration and velocity profiles are given in Figure 13.3, respectively. If only the acceleration is given, the velocity can be obtained by the integration, or vice versa, the acceleration can be calculated by the differentiation if the velocity is available first.

The vehicle speeds up with a constant acceleration of 3 m/s² in the first 4 seconds transitioning smoothly to a constant speed at 50 km/h, and keeps it for the next 5 seconds. Afterwards the velocity transitions smoothly when the deceleration changes from 0 to a constant deceleration of 3 m/s². The velocity then drops to zero value while the deceleration is maintained as 3 m/s². Figure 13.4 shows the total torque required at the wheels during the acceleration in the first second while Figure 13.5 illustrates the control errors. The results are referring to two different controller parameter sets. The solid line presents the control with the parameters chosen as $Q_i = 16$, $Q_V = 17$ and $R = 1$. In order to reduce the engine torque and save the energy consumption, control auxiliary input u needs to be increased, which means to decrease R according to the quadratic performance criterion given by (11.58). Note that here auxiliary control input u is the output of the optimization, not the torque itself. That result is presented by the dashed line showing the control for $Q_i = 16$, $Q_V = 17$ and $R = 0.01$. In this case, the torque during the acceleration has a faster response and reaches a lower level while the control error is also getting smaller. Selecting larger R will slow down the torque response and increase the amount of the torque. Changing the weighting matrix of $\mathbf{Q}$ will have similar effect in regard to the control errors. In this way the controller can be tuned with respect to the torque expenditure and the control error.

Another effect shown in this example is the robustness of the LQ design. In the vehicle control practice, vehicle parameters, such as the weight and inertia, are normally unknown. That will cause a deviation between the parameters used in the controller and vehicle model parameters as given by (13.12). For this reason, the simulation was run under the condition that in the controller the vehicle gross mass and the wheel inertia are assumed different than in the vehicle model, to demonstrate the control robustness aforementioned.

❏

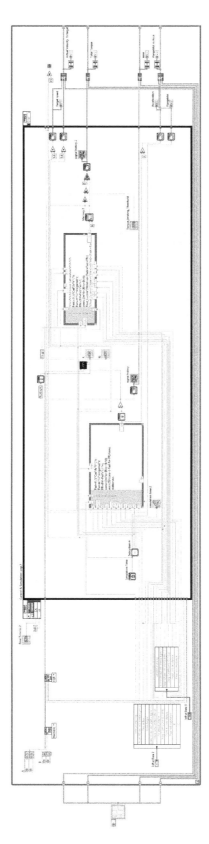

FIGURE 13.2 Block diagram of the speed control in LabVIEW.

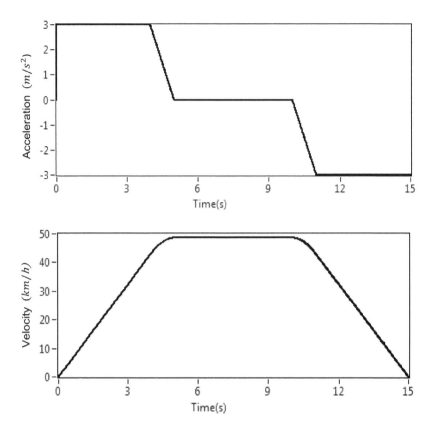

FIGURE 13.3 Acceleration and velocity profiles of the speed control.

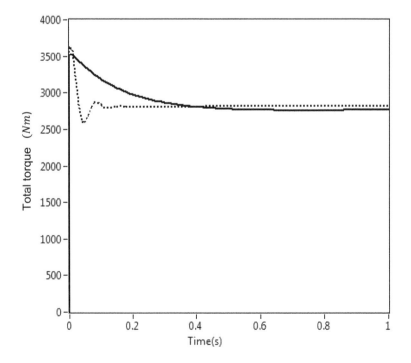

FIGURE 13.4 Total wheel torques during the acceleration.

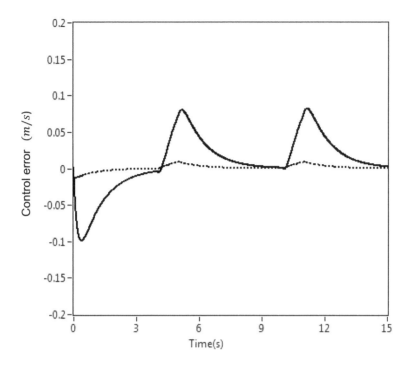

FIGURE 13.5   The speed control errors.

The LQ design results in the minimum energy in terms of e_1, e_2 and u. In this approach the energy of $T_a - T_b$ is actually not formally minimized, but influenced through variable control input u. This can be viewed as a suboptimal approach that provides a somewhat optimization with respect to the vehicle torque as well as to the control errors.

ENGINE TORQUE CALCULATION

The control input, $T_a - T_b$, is a combination of the driving and braking torques. During the vehicle speed control, a coordination between the both torques needs to be defined to handle the actuation exclusively. For example, the driving torque is active when the required control input is greater than zero. In this case the braking torque is assigned as inactive. The braking torque becomes active when the required control input is negative (accordingly, $T_a = 0$).

Considering the case when T_a is positive. As defined in (13.1), T_a is the driving torque that comes from vehicle dynamics equation, i.e., determined by the road conditions and vehicle inertia; Once driving torque T_a is determined by the speed controller, the required engine torque needs to be calculated. Ultimately the engine control will be active to deliver the engine torque to the drive wheels.

For a rear-wheel-drive vehicle, we know from Section 5.1.1 that the engine torque results in the driving torque at the axle, which was denoted as T_w in Chapter 5. Thus, we obviously obtain

$$T_w = T_a \tag{13.18}$$

Using (5.9), the engine torque can be calculated using values of T_a

$$T_e = \frac{1}{i_t i_d \eta_e} \left\{ T_a + \left(J_e i_t^2 i_d^2 + J_t i_d^2 + J_d \right) \frac{\dot{v}}{r_w} \right\} \tag{13.19}$$

One can see that keeping T_a and thus T_w minimized makes T_e lower. The notations used in (13.19) can be found in Chapter 5.

13.2 PATH-FOLLOWING CONTROL

The vehicle motion results from vehicle longitudinal and lateral dynamics. Motion control systems of autonomous vehicles usually control both longitudinal and lateral dynamics simultaneously. The longitudinal controller is responsible for regulating the vehicle speed while the lateral controller directs the vehicle for path tracking. Therefore, the vehicle speed controller developed in the previous section can be utilized as the longitudinal controller. For this reason, in this section, the path-following controller refers specifically to the lateral controller. It should be mentioned that the lateral controller also can be employed in a human driven vehicle, for example in lane keeping systems.

In autonomous vehicle navigation systems, the path planning is needed to achieve the moving target. The result of that can be described as a geometric path. The path following is the intermediate execution layer connecting such reference geometric path and the low-level vehicle control implemented through the steering, braking and the engine control. The path-following controller can utilize dynamic vehicle model as a base to provide a target input, for instance to the steering control. Subsequently, the steering control will ensure the steering target achieved, and hence the vehicle follows the desired path. In recent years, a decent number of research studies has focused on the path-following control [1, 2, 3]. We present alternative solutions based on a cascade-control structure utilizing torque-vectoring or rear steering. The method provides an analytical solution instead of computer programming method such as dynamic programming or model predictive control, This makes the proposed method more advantageous to the implementation in real time controls. The dynamic model defines the state variables relating to the position and orientation of the vehicle with respect to the roadway, i.e., to the driving path. Thereafter the path-following controllers are described.

DYNAMIC VEHICLE MODEL WITH POSITION AND ORIENTATION RELATIVE TO DRIVING PATH

The vehicle model is based on the single-track model which describes the vehicle dynamics in a horizontal two-dimensional space. In addition to vehicle sideslip angle and the yaw rate as the known state variables, the relative yaw angle and lateral offset of the vehicle from the centerline are introduced. The latter determines the orientation and the position of the vehicle. Figure 13.6 shows the driving path associated with the state variables. In this driving situation, the vehicle is steered away from the path centerline to the left. Here ψ_L is the relative yaw angle, i.e., the yaw angle between the vehicle centerline and the tangent to the path. y_L presents the lateral distance between the sensor lookahead position and the tangent to the path. l_s is distance between the sensor lookahead position and the vehicle CG.

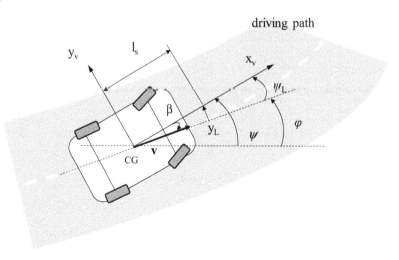

FIGURE 13.6   Position and orientation relative to the driving path.

Viewing the tangent to the path at the lookahead position provided by the sensor, it adds an angle of l_s/R_T subtended by the lookahead distance to the course angle. The relative yaw angle is the difference between the yaw angle and the course angle plus the subtended angle, that is

$$\psi_L = \psi - \left(\varphi_{st} + \frac{l_s}{R_T} \right) \tag{13.20}$$

In the steady state, the course angle, φ_{st}, is determined by the tangent to the path. Following (9.41), the rate of the relative yaw angle is given by

$$\dot{\psi}_L = \dot{\psi} - \dot{\psi}_{st} \tag{13.21}$$

The steady yaw rate can be expressed in terms of turning radius R_T, as explained by (9.39). We further consider velocity $\mathbf{v}$ originated from the CG of the vehicle to calculate the lateral velocity at the lookahead position, applying (2.44) to determine the velocity. Then the velocity along the y_L direction is expressed as

$$\dot{y}_L = \mathbf{v} \left(\beta + \psi_L \right) + l_s \, \dot{\psi} \tag{13.22}$$

Combining the two equations above, (13.21) and (13.22) and the single-track linear model (8.39) with rear steer, the dynamic vehicle model with position and orientation relative to driving path is presented in the following state-space form:

$$
\begin{bmatrix} \dot{\beta} \\ \ddot{\psi} \\ \dot{\psi}_L \\ \dot{y}_L \end{bmatrix} =
\begin{bmatrix}
-\dfrac{C_{\alpha f} + C_{\alpha r}}{Mv} & -1-\dfrac{C_{\alpha f}l_f - C_{\alpha r}l_r}{Mv^2} & 0 & 0 \\
-\dfrac{C_{\alpha f}l_f - C_{\alpha r}l_r}{I_z} & -\dfrac{C_{\alpha f}l_f^2 + C_{\alpha r}l_r^2}{I_z v} & 0 & 0 \\
0 & 1 & 0 & 0 \\
v & l_s & v & 0
\end{bmatrix}
\begin{bmatrix} \beta \\ \dot{\psi} \\ \psi_L \\ y_L \end{bmatrix}
$$
$$
+ \begin{bmatrix}
\dfrac{C_{\alpha f}}{Mv} & \dfrac{C_{\alpha r}l_r}{Mv} \\
\dfrac{C_{\alpha f}l_f}{I_z} & -\dfrac{C_{\alpha r}l_r}{I_z} \\
0 & 0 \\
0 & 0
\end{bmatrix}
\begin{bmatrix} \delta_f \\ \delta_r \end{bmatrix}
+ \begin{bmatrix} 0 \\ \dfrac{d_t}{I_z} \\ 0 \\ 0 \end{bmatrix} (F_{x\_RR} - F_{x\_LR})
+ \begin{bmatrix} 0 \\ 0 \\ -v \\ 0 \end{bmatrix} \dfrac{1}{R_T} \tag{13.23}
$$

While (8.39) is embedded into the above vehicle model, the following facts need to be considered and estimated in detail. First, the influence of the longitudinal tire forces of the steered front and rear wheels on lateral dynamics is less significant comparing to the influence of the lateral tire forces. Therefore, the longitudinal forces' lateral components, which are paired with the vehicle sideslip angle and steer angles in (8.39) are small enough to be neglected in (13.23). Another fact that should be taken into account is that the torque supplied to the front axle is equally distributed between the left and the right wheels by the axle open differential. This fact results in equal longitudinal forces of the front wheels, $F_{x\_LF} = F_{x\_RF}$, which are thus canceled from (8.39). Therefore, the longitudinal forces at the rear only remain to be controllable in (13.23). Finally, the front and the rear steering and the rear torque vectoring are used as the control actuators in the following controller design. An interesting point here is that both the rear steering and the rear torque vectoring have a similar impact on vehicle yaw dynamics because both actuations generate yaw moments about the normal axis of the vehicle. Therefore, each of them can be utilized as an alternative actuator. Also, in the following control design, it is assumed that the vehicle speed is constant during the control period since it can be regulated by the speed control that was considered before in Section 13.1.

13.2.1 CASCADE CONTROL DESIGN

First of all, the control target needs to be defined. In order to follow the driving path, the sensor lookahead position should be right on the path, i.e., lateral distance y_L should be zero. If the vehicle is in the steady state and y_L is zero, the tangent to the path is determined by the steady course angle $\left(\psi_{st} + \beta_{st}\right)$. As a result, the steady relative yaw angle from (13.20) can be calculated as

$$\psi_L = -\beta_{st} - \frac{l_s}{R_T} \tag{13.24}$$

This means that in the steady state the relative yaw angle always equals the vehicle sideslip angle offset by l_s/R_T. The relative yaw angle will not be zero as long as the sideslip angle does not equal to the offset. The geometric analysis shows that the zero condition, i.e., $\psi_L = 0$, can occur only if the steady-state sideslip angle is equal to the angle subtended by the lookahead distance at the turn center of the circular path. This condition depends on the lookahead distance, the turn radius and a particular velocity of the vehicle CG. For this reason, the ultimate goal of the path following is to meet the target

$$\begin{bmatrix} \psi_{Ld} \\ y_{Ld} \end{bmatrix} = \begin{bmatrix} -\beta_{st} - \dfrac{l_s}{R_T} \\ 0 \end{bmatrix} \tag{13.25}$$

where index d indicates the desired target value.

The sensor signals of the state variables in the vehicle model (13.23) are assumed to be available. The measurable signals can enable additional control mechanism, and often make a better control performance. The relative yaw angle ψ_L and lateral distance y_L are directly influenced by the vehicle dynamic states - the sideslip angle and the yaw rate, which are affected by front and rear steering inputs. From this perspective, the control can be implemented in a cascade control structure as shown in Figure 13.7.

The block diagram shows the cascade control with two controllers, two sets of sensors and one set of actuators acting on one of two processes in series. Such one set of actuators can be either the

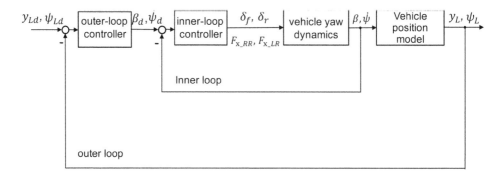

FIGURE 13.7 Cascade control of the path following.

front and rear steering or the front steering and the rear torque vectoring. The geometry of this cascade structure defines the inner control loop involving the state of $(\beta, \dot{\psi})$ and the outer control loop involving the states of (ψ_L, y_L). The outer-loop controller generates a control effort that serves as the target for the inner-loop controller. That controller in turn uses the steering or the torque differential to apply its control action directly to the dynamics of the inner-loop. The inner loop then generates the outputs that serve as the actual control inputs for the outer loop (see $\beta, \dot{\psi}$). The steering or torque differential that the inner-loop controller uses to maintain the sideslip angle and the yaw rate serve as the actuators, which act directly on the inner control loop and indirectly on the outer control loop.

The path-following control based on model (13.23) is particularly amenable to cascade control because such control of the sideslip angle and yaw rate can maintain the vehicle position and yaw angle without any additional actuators. The defined cascade control can meet two objectives. The first objective is to control the relative yaw angle and the lateral distance via the action of the inner-control loop around a secondary state measurement. The second objective is to reduce the system order and the sensitivity of the state variables of the outer loop. In this way, a standard feedback controller can be applied. In the following, the inner-loop controller and the outer-loop controller are designed, respectively. The cascade control method presented is based upon the description in reference [4].

OUTER-LOOP CONTROLLER

The outer-loop controller defines the target that the inner-loop controller is required to achieve (ψ_L, y_L). Based on the vehicle position model, the outer-loop controller is now designed. In previous Section 13.1 we applied the linear quadratic optimal control to the vehicle speed control. Now, we apply the same method to the path-following control, which leads to the linear quadratic optimal control with output target (Section 11.7).

The vehicle position model can be understood as the lower part of (13.23), that is

$$
\begin{bmatrix} \dot{\psi}_L \\ \dot{y}_L \end{bmatrix} = \begin{bmatrix} 0 & 0 \\ v & 0 \end{bmatrix} \begin{bmatrix} \psi_L \\ y_L \end{bmatrix} + \begin{bmatrix} 0 & 1 \\ v & l_s \end{bmatrix} \begin{bmatrix} \beta \\ \dot{\psi} \end{bmatrix} + \begin{bmatrix} -v \\ 0 \end{bmatrix} \frac{1}{R_T} \tag{13.26}
$$

In this equation, the sideslip angle and the yaw rate are considered as the control inputs, which can be denoted by $\mathbf{u} = \begin{bmatrix} \beta & \dot{\psi} \end{bmatrix}^T$. Further, we define the state variables

$$
\mathbf{x} = (\psi_L \quad y_L)^T, \quad z_0 = \frac{1}{R_T}
$$

and introduce the matrices accordingly

$$
\mathbf{A} = \begin{bmatrix} 0 & 0 \\ v & 0 \end{bmatrix}, \quad \mathbf{B} = \begin{bmatrix} 0 & 1 \\ v & l_s \end{bmatrix}, \quad \mathbf{B_z} = \begin{bmatrix} -v & 0 \end{bmatrix}^T.
$$

According to the criteria given by (11.66), one can verify that $(\mathbf{A}, \mathbf{B})$ is controllable. The state weighting matrix Q can be chosen so that $\left(A, \sqrt{Q} \right)$ is observable.

Since the state variables are measurable, the output target can be defined to be equal to the steady state of the state vector, i.e.,

$$
\mathbf{x}(\infty) = \mathbf{y}_d = \begin{bmatrix} -\beta_{st} - \dfrac{l_s}{R_T} \\ 0 \end{bmatrix} \tag{13.27}
$$

the linear quadratic optimal controller is defined by

$$\mathbf{u}(t) = -\mathbf{K}\ \mathbf{x}(t) + \mathbf{u}_0 \tag{13.28}$$

with the performance index

$$J = \int_0^\infty \left\{ (\mathbf{x} - \mathbf{x}(\infty))^T \mathbf{Q}(\mathbf{x} - \mathbf{x}(\infty)) + (\mathbf{u} - \mathbf{u}(\infty))^T \mathbf{R}(\mathbf{u} - \mathbf{u}(\infty)) \right\} dt \tag{13.29}$$

The objective of the optimal control is to track the desired relative yaw angle and lateral position of the vehicle by minimizing the deviation to its steady states as well as the control efforts to the steady control states.

Let the feedback gain matrix be

$$\mathbf{K} = \begin{bmatrix} k_{out1} & k_{out2} \\ k_{out3} & k_{out4} \end{bmatrix} \tag{13.30}$$

The matrix above is to be calculated through (11.64) and Riccati equation (11.65) introduced in Chapter 11:

$$\mathbf{K} = \mathbf{R}^{-1}\ \mathbf{B}^T\ \mathbf{P} \tag{13.31}$$

$$\mathbf{A}^T \mathbf{P} + \mathbf{P}\mathbf{A} - \mathbf{P}\mathbf{B}\mathbf{R}^{-1}\mathbf{B}^T\mathbf{P} + \mathbf{Q} = 0 \tag{13.32}$$

Because of the output target given by (13.27), $\mathbf{u}_0$ is specified by (11.78) as follows:

$$\mathbf{u}_0(t) = \begin{bmatrix} \beta_{st} - \left(\beta_{st} + \dfrac{l_s}{R_T}\right)k_{out1} \\[2mm] -\left(\beta_{st} + \dfrac{l_s}{R_T}\right)k_{out3} + \dfrac{v}{R_T} \end{bmatrix} \tag{13.33}$$

where $\mathbf{C}$ is equal to a unit matrix. Thus, the complete outer-loop controller (13.28) can be detailed to

$$\mathbf{u}(t) = \begin{bmatrix} -k_{out1}\ \psi_L\ -k_{out2}\ y_L + \beta_{st} - (\beta_{st} + \dfrac{l_s}{R_T})k_{out1} \\[2mm] -k_{out3}\ \psi_L\ -k_{out4}\ y_L - (\beta_{st} + \dfrac{l_s}{R_T})k_{out3} + \dfrac{v}{R_T} \end{bmatrix} \tag{13.34}$$

The outer-loop control is optimized with respect to the error states and the control measure in performance index (13.29). The selection of the state weighting matrix $\mathbf{Q}$ affects the transient response of the states, i.e., the relative yaw rate and the lateral distance. The control weighting matrix, $\mathbf{R}$, alters the magnitude of the control inputs that are the vehicle sideslip angle and the yaw rate. Selecting larger $\mathbf{R}$ matrix can ensure the control inputs smaller and smooth. The sideslip angle and the yaw rate will be required to stay small and smooth for the inner-loop control, which is described in the next section.

13.2.2 Inner-Loop Control via Front Steering and Rear Torque Vectoring

In this approach, the front steering and rear torque vectoring are utilized as the control actuators while the rear steer angle $\delta_r = 0$. The torque on the rear axle can be either driving torque or braking torque.

The inner-loop controller receives the output of the outer-loop controller, and takes it as a target to achieve, i.e., the desired sideslip angle and the desired yaw rate are

$$
\begin{bmatrix} \beta_d(t) \\ \dot{\psi}_d(t) \end{bmatrix} = \begin{bmatrix} -k_{out1}\ \psi_L\ -k_{out2}\ y_L + \beta_{st} - (\beta_{st} + \dfrac{l_s}{R_T})k_{out1} \\[4mm] -k_{out3}\ \psi_L\ -k_{out4}\ y_L - (\beta_{st} + \dfrac{l_s}{R_T})k_{out3} + \dfrac{v}{R_T} \end{bmatrix} \tag{13.35}
$$

Note that these desired parameters are not in a steady state since they are just the control targets for the inner-loop control.

The inner-loop control is based on the yaw dynamics model:

$$
\begin{bmatrix} \dot{\beta} \\ \dot{\psi} \end{bmatrix} = \begin{bmatrix} -\dfrac{C_{\alpha f}+C_{\alpha r}}{Mv} & -1-\dfrac{C_{\alpha f}l_f - C_{\alpha r}l_r}{Mv^2} \\[4mm] -\dfrac{C_{\alpha f}l_f - C_{\alpha r}l_r}{I_z} & -\dfrac{C_{\alpha f}l_f^2 + C_{\alpha r}l_r^2}{I_z v} \end{bmatrix} \begin{bmatrix} \beta \\ \dot{\psi} \end{bmatrix} + \begin{bmatrix} \dfrac{C_{\alpha f}}{Mv} & 0 \\[4mm] \dfrac{C_{\alpha f}l_f}{I_z} & \dfrac{d_t}{I_z r_w} \end{bmatrix} \begin{bmatrix} \delta_f \\ \Delta T_r \end{bmatrix} \tag{13.36}
$$

where the torque differential between right and left wheels is indicated by ΔT_r, that is

$$
\Delta T_r = (F_{x\_RR} - F_{x\_LR})r_w \tag{13.37}
$$

Here, we apply the linear quadratic optimal control again. For the state-space form, we define the states and matrices:

$$
\mathbf{x}(t) = \begin{bmatrix} \beta & \dot{\psi} \end{bmatrix}^T, \mathbf{u} = \begin{bmatrix} \delta_f & \Delta T_r \end{bmatrix}^T
$$

and

$$
\mathbf{A} = \begin{bmatrix} -\dfrac{C_{\alpha f}+C_{\alpha r}}{Mv} & -1-\dfrac{C_{\alpha f}l_f - C_{\alpha r}l_r}{Mv^2} \\[4mm] -\dfrac{C_{\alpha f}l_f - C_{\alpha r}l_r}{I_z} & -\dfrac{C_{\alpha f}l_f^2 + C_{\alpha r}l_r^2}{I_z v} \end{bmatrix}, \mathbf{B} = \begin{bmatrix} \dfrac{C_{\alpha f}}{Mv} & 0 \\[4mm] \dfrac{C_{\alpha f}l_f}{I_z} & \dfrac{d_t}{I_z r_w} \end{bmatrix}
$$

With the control target $\mathbf{x}_d(t) = \begin{bmatrix} \beta_d & \dot{\psi}_d \end{bmatrix}^T$ and the control error as

$$
\mathbf{e}(t) = \mathbf{x}(t) - \mathbf{x_d}(t) \tag{13.38}
$$

the model equation (13.36) is translated to

$$
\dot{\mathbf{e}}(t) = \mathbf{A}\mathbf{e}(t) - \dot{\mathbf{x}}_d(t) + \mathbf{A}\mathbf{x}_d(t) + \mathbf{B}\mathbf{u}(t) \tag{13.39}
$$

The total control signal, $\mathbf{u}$, can be expressed as the sum of an error state control, $\mathbf{u}_e$, and an auxiliary control signal, $\mathbf{u}_0$, which counteracts the second and third terms on the right side of the above equation:

$$
\mathbf{u}(t) = \mathbf{u}_e(t) + \mathbf{u}_0(t) \tag{13.40}
$$

where

$$
\mathbf{u}_0(t) = \mathbf{B}^{-1}\left(\dot{\mathbf{x}}_d(t) - \mathbf{A}\mathbf{x}_d(t)\right) \tag{13.41}
$$

Differentiating target value $\mathbf{x_d}$ given by (13.35), the first term of the above equation is calculated as follows:

$$\dot{\mathbf{x}}_{\mathrm{d}} = \begin{bmatrix} -k_{out1}\ \dot{\psi}_L\ -k_{out2}\ \dot{y}_L \\ -k_{out3}\ \dot{\psi}_L\ -k_{out4}\ \dot{y}_L \end{bmatrix} \tag{13.42}$$

where the derivative signals of $\dot{\psi}_L$ and $\dot{y}_L$ can be obtained by a differentiating filter or by using (13.21) and (13.22), i.e.,

$$\dot{\mathbf{x}}_{\mathrm{d}} = \begin{bmatrix} -k_{out1}\left(\dot{\psi}-\dot{\psi}_{st}\right)-k_{out2}\ (\mathrm{v}\ \left(\beta+\psi_L\right)+l_s\ \dot{\psi}) \\ -k_{out3}\left(\dot{\psi}-\dot{\psi}_{st}\right)-k_{out4}\ (\mathrm{v}\ \left(\beta+\psi_L\right)+l_s\ \dot{\psi}) \end{bmatrix} \tag{13.43}$$

According to the criteria given by (11.66), one can verify that $(\mathbf{A},\ \mathbf{B})$ is controllable. The state weighting matrix Q can be chosen so that $\left(A,\ \sqrt{Q}\right)$ is observable.

Now, we define the linear quadratic optimal controller for the error state

$$\mathbf{u}_{\mathrm{e}}\left(t\right)=-\mathbf{K}\ \mathbf{e}\left(t\right) \tag{13.44}$$

with the performance index

$$J = \int_0^\infty \left\{\mathbf{e}^T\left(t\right)\mathbf{Q}\mathbf{e}\left(t\right)+\mathbf{u}_{\mathrm{e}}^T\left(t\right)\mathbf{R}\mathbf{u}_{\mathrm{e}}\left(t\right)\right\}dt \tag{13.45}$$

The feedback gain matrix, denoted as

$$\mathbf{K} = \begin{bmatrix} k_{in1} & k_{in2} \\ k_{in3} & k_{in4} \end{bmatrix} \tag{13.46}$$

will be calculated through Riccati equation (11.64) and (11.65). Using (13.40), we obtain the inner-loop controller

$$\begin{bmatrix} \delta_f \\ \Delta T_r \end{bmatrix} = \begin{bmatrix} -k_{in1} & -k_{in2} \\ -k_{in3} & -k_{in4} \end{bmatrix}\begin{bmatrix} \beta-\beta_d \\ \dot{\psi}-\dot{\psi}_d \end{bmatrix}+\mathbf{B}^{-1}\left(\begin{bmatrix} \dot{\beta}_d \\ \ddot{\psi}_d \end{bmatrix}-\mathbf{A}\begin{bmatrix} \beta_d \\ \dot{\psi}_d \end{bmatrix}\right) \tag{13.47}$$

The torque on the rear axle can be driving torque during acceleration or braking torque during deceleration. The torque vectoring of torque differential ΔT_r is to be realized accordingly. For the driving torque vectoring, the amount of the torque distributed on the rear axle is available from the engine control system (see Section 13.1) while the braking torque vectoring uses the measurement through the vehicle braking system. Once the torque differential is determined, the individual wheel torque can be calculated.

The inner-loop control is optimized with respect to the error states and the error- feedback control measure in the performance index (13.45). The selection of state weighting matrix $\mathbf{Q}$ affects the transient response of the error states that are the sideslip angle error and the yaw rate error. Control weighting matrix $\mathbf{R}$ alters the magnitude of the control inputs that serves as an error control. Note that the ultimate control inputs, which are the front steering and the rear torque differential, are not directly optimized since they are not formally minimized in performance index. Therefore, the proposed approach is basically a *suboptimal approach* with respect to the actual vehicle front and rear torque differential. However, a minimized and slower response of the state errors in the side-slip angle and the yaw rate control can be achieved by enlarging the state weighting matrix while a smaller magnitude of error control $\mathbf{u}_e$ can be reached by selecting larger control weighting matrix.

13.2.3 INNER-LOOP CONTROL VIA FRONT AND REAR STEERING

Here, the front and rear steering are used as the control actuators while $F_{x\_RR} = F_{x\_LR}$ is on the rear. The inner-loop control is based on the yaw dynamics model:

$$\begin{bmatrix} \dot{\beta} \\ \ddot{\psi} \end{bmatrix} = \begin{bmatrix} -\dfrac{C_{\alpha f}+C_{\alpha r}}{Mv} & -1-\dfrac{C_{\alpha f}l_f-C_{\alpha r}l_r}{Mv^2} \\ -\dfrac{C_{\alpha f}l_f-C_{\alpha r}l_r}{I_z} & -\dfrac{C_{\alpha f}l_f^2+C_{\alpha r}l_r^2}{I_z v} \end{bmatrix}\begin{bmatrix} \beta \\ \dot{\psi} \end{bmatrix} + \begin{bmatrix} \dfrac{C_{\alpha f}}{Mv} & \dfrac{C_{\alpha r}}{Mv} \\ \dfrac{C_{\alpha f}l_f}{I_z} & -\dfrac{C_{\alpha r}l_r}{I_z} \end{bmatrix}\begin{bmatrix} \delta_f \\ \delta_r \end{bmatrix} \qquad (13.48)$$

Similar to the inner-loop controller of the rear torque vectoring presented in the previous section, we define the states and matrices of the state-space form as follows:

$$\mathbf{x}(t)=\begin{bmatrix} \beta & \dot{\psi} \end{bmatrix}^T, \quad \mathbf{u}=\begin{bmatrix} \delta_f & \delta_r \end{bmatrix}^T$$

and

$$\mathbf{A}=\begin{bmatrix} -\dfrac{C_{\alpha f}+C_{\alpha r}}{Mv} & -1-\dfrac{C_{\alpha f}l_f-C_{\alpha r}l_r}{Mv^2} \\ -\dfrac{C_{\alpha f}l_f-C_{\alpha r}l_r}{I_z} & -\dfrac{C_{\alpha f}l_f^2+C_{\alpha r}l_r^2}{I_z v} \end{bmatrix}, \quad \mathbf{B}=\begin{bmatrix} \dfrac{C_{\alpha f}}{Mv} & \dfrac{C_{\alpha r}}{Mv} \\ \dfrac{C_{\alpha f}l_f}{I_z} & -\dfrac{C_{\alpha r}l_r}{I_z} \end{bmatrix}$$

Because one of the actuators is the rear steering, $\mathbf{u}$ and $\mathbf{B}$ are changed comparing to the rear torque vectoring given by (13.36). Applying the same approach that was used before for the inner-loop control with the front steering and the rear torque vectoring actuations, we obtain the following controller for the inner-loop control via the front steering and the rear steering:

$$\begin{bmatrix} \delta_f \\ \delta_r \end{bmatrix} = \begin{bmatrix} -k_{in1} & -k_{in2} \\ -k_{in3} & -k_{in4} \end{bmatrix}\begin{bmatrix} \beta-\beta_d \\ \dot{\psi}-\dot{\psi}_d \end{bmatrix} + \mathbf{B}^{-1}\left(\begin{bmatrix} \dot{\beta}_d \\ \ddot{\psi}_d \end{bmatrix} - \mathbf{A}\begin{bmatrix} \beta_d \\ \dot{\psi}_d \end{bmatrix}\right) \qquad (13.49)$$

As learned from the previous section, the controller given by (13.49) can be also considered as a *suboptimal approach* with respect to the actual vehicle front and rear steering. The smaller and slower response of the state errors in the sideslip angle and the yaw rate can be achieved by enlarging the state weighting matrix while the smaller magnitude of the error control $\mathbf{u}_e$ can be reached by selection of larger control weighting matrix. On the other hand, $\dot{\mathbf{x}}_d$ and $\mathbf{x}_d$ signals are ensured to be minimized and smooth by the weighting matrix in the outer-loop control. Since the rear steer angle may be constrained by its maximum, it is desirable to minimize the rear steer angle so that it will not saturate at the maximum value. For this reason, the measured and calculated signals in the control law above should be minimized, smoothed and even made slowly transient.

ESTIMATION OF SIDESLIP ANGLE

In the inner-loop control, the design is based on the yaw dynamics model (13.36) or (13.48), which has the state vector $\mathbf{x} = (\beta \quad \dot{\psi})^T$. In the case that the sideslip angle sensor is not available, this angle needs to be estimated. As described in Chapter 11, the Kalman filter can be applied to the state estimation.

Based on the availability of the measurement, output matrix $\mathbf{C}(t)$ in (13.36) needs to be specified. It appears that the yaw rate signal must be available. In today's vehicle technology, the yaw rate sensor is a common sensor equipped on the vehicle. Therefore, output matrix is set to $\mathbf{C}(t) = (0 \quad 1)$. For the purpose of designing a Kalman filter, the unknown state variable is considered as a random process and disturbed by white noise v(t) with intensity matrix $\mathbf{V}(t)$. The noise processes are usually assumed to be independent and stationary. The only measurement used is the yaw rate, which is subjected to the measurement noise $w(t)$ assumed to be white. In this case, intensity matrix $\mathbf{W}(t)$ becomes a scalar W(t). Hence, the state-space form of the yaw dynamics model is formulated as

$$\dot{\mathbf{x}}\ (t) = \mathbf{A}\ \mathbf{x}(t) + \mathbf{B}\mathbf{u}(t) + \mathbf{v}(t)$$
$$\mathbf{y}(t) = \mathbf{C}\ \mathbf{x}(t) + w(t) \tag{13.50}$$

where, index t is added to emphasize that the system is time-varying due to the noise. System matrices $\mathbf{A}$, $\mathbf{B}$ and $\mathbf{C}$ are constant due to the vehicle constant speed during the control period. The output is no longer a vector because of the single measurement of the yaw rate. The state excitation noise vector has components v_β and v_ψ for the sideslip angle and the yaw rate, respectively, and they can be introduced as follows:

$$\mathbf{v}(t) = \begin{bmatrix} v_\beta(t) \\ v_\psi(t) \end{bmatrix}$$

Since the random processes $\mathbf{v}(t)$ and $w(t)$ are assumed stationary, intensity matrices $\mathbf{V}(t)$ and W(t) are constant, i.e.,

$$V(t) = \begin{bmatrix} V_\beta & 0 \\ 0 & V_\psi \end{bmatrix}, W(t) = W$$

Now we can design a Kalman filter, which provides an estimate of the sideslip angle. In this case, the steady-state Kalman filter is applicable because of constant parameters in Riccati equation. Using (11.47), (11.48) and (11.49) from Chapter 11, the Kalman filter is defined as

$$\dot{\hat{\mathbf{x}}}(t) = \mathbf{A}\ \hat{\mathbf{x}}(t) + \mathbf{B}\mathbf{u}(t) + \mathbf{K}(\dot{\psi}(t) - \hat{\dot{\psi}}(t)) \tag{13.51}$$

$$\mathbf{K} = \mathbf{P}\mathbf{C}^T W^{-1} \tag{13.52}$$

$$\mathbf{0} = \mathbf{A}\mathbf{P} + \mathbf{P}\mathbf{A}^T - \mathbf{P}\mathbf{C}^T W^{-1}\mathbf{C}\mathbf{P} + \mathbf{V} \tag{13.53}$$

The gain matrix, $\mathbf{K}$, results in a 2×1 vector. The covariance matrix, $\mathbf{P}$, reaches the steady state calculated by Riccati equation (13.53). In this equation, intensity $\mathbf{V}$ indicates the randomness of the state that the yaw rate sensor is trying to measure while intensity W is an indicator of the random noise in the measurement.

An example of LabVIEW simulation is provided in Section 14.3 to illustrate the estimation of the sideslip angle. In that simulation, the state vector is estimated based on the designed Kalman filter. The sideslip angle is one of the state variables in the vehicle-trailer dynamics system that is also considered in Chapter 14.

* * *

In conclusion to Section 13.2, the path-following control is designed based on a cascade control structure, which features an outer-loop controller and an inner-loop controller. Both controllers are based on the same method of the LQ design.

Based on the type of the preferable control actuator, two control methods were developed for the inner-loop controller. The path-following control via front steering and rear torque vectoring was presented by (13.34) and (13.47) while the path-following control via front and rear steering was given by (13.34) and (13.49). Note that (13.47) and (13.49) are identical in the equation formulation, but distinguish in **K** and **B** matrices. By selecting the state as well as the control weighting matrices, the tracking errors and the magnitude of the control signals can be optimized.

It should be also emphasized that the linear quadratic optimal control design ensures the convergence of state variables to a desired target, i.e.,

$$
\lim_{t \to \infty} \begin{bmatrix} \psi_L \\ y_L \end{bmatrix} = \begin{bmatrix} \psi_{Ld} \\ y_{Ld} \end{bmatrix}
$$

$$
\lim_{t \to \infty} \begin{bmatrix} \beta \\ \dot{\psi} \end{bmatrix} = \begin{bmatrix} \beta_d(\infty) \\ \dot{\psi}_d(\infty) \end{bmatrix} = \begin{bmatrix} \beta_{st} \\ \dot{\psi}_{st} \end{bmatrix}
$$

(13.54)

The second equation above comes out from (13.35) when the signals β and $\dot{\psi}$ are in the steady state. Thus, the ultimate target of the yaw dynamics control is to reach the steady states of the yaw rate and the sideslip angle. This indicates an important property of the control target for the yaw stability.

At the end of Section 13.2, the estimation of the sideslip angle by using Kalman filter was described. The proposed method is not only applicable in the path-following control, but also can be applied to any other control design where the vehicle yaw dynamics model is in use.

REFERENCES

1. Chr. Chatzikomis, A. Sorniotti, P. Gruber, M. Zanchetta, D. Willans and B. Balcombe, "Comparison of Path Tracking and Torque-Vectoring Controllers for Autonomous Electric Vehicles," IEEE Transactions on Intelligent Vehicles, vol. 3, No. 4, December 2018
2. R. Attia, R. Orjuela, and M. Basset, "Combined Longitudinal and Lateral Control for Automated Vehicle Guidance," Veh. Syst. Dyn., vol. 52, No. 2, 2014
3. C. M. Filho, D. F. Wolf, V. Grassi, and F. S. Osorio, "Longitudinal and Lateral Control for Autonomous Ground Vehicles," IEEE Intelligent Vehicles Symposium, Dearborn, MI, USA, 2014
4. J. Yu, Lateral Control in Path-tracking of Autonomous Vehicles, US Patent Application 17330536 , 2021.

14 Vehicle Stability Control

14.1 YAW STABILITY CONTROL

An important aspect of lateral dynamics and thus path-following is yaw stability control. It refers to control of yaw dynamics, which is characterized mainly by the lateral acceleration, the sideslip angle and the yaw rate. The ultimate goal is to control vehicle states defined by the above-listed three parameters to their desired targets, for example, to the steady states. The yaw stability control has been a big focus since the introduction of the very first on-board active systems that regulate vehicle stability, such as the electronic stability control (ESC, also known as ESP, electronic stability program) designed in the early 1990s. In the following, we examine some control schemes based on the single-track model first, and then present a practical implementation of the control actuators accompanied by a case study simulation. At the end, the yaw stability control for autonomous vehicles is discussed.

14.1.1 YAW RATE TARGET

The objective of the yaw stability control is to improve both vehicle stability and handling during a turn by preventing vehicle from spinning about the normal axis and drifting out of the path curve. In order to accomplish so, the actual yaw rate needs to be controlled to a defined target, and thus the yaw rate target should be determined mathematically first. There are different ways to solve such problem. In this section, we discuss two possible ways, one approach is based on the steady states of the yaw rate and the sideslip angle, while another one is using a model reference.

Steady-State-Based Approach

As described in Chapter 9, the vehicle transient response to driver's steering input exhibits vehicle dynamic characteristics, which are essential factors for the driver's perception of how good or bad the vehicle behaves. Therefore, the driver's perception should be included in vehicle performance evaluation. Based on vehicle testing results and subjective assessment [1], the vehicle dynamic behavior is well received by drivers in the presence of the following two conditions:

- $\left(\dot{\psi} / \delta_f\right)_{st}$, which is the steady value of the yaw gain factor, is sufficient large
- $T_{\dot{\psi}_{max}}$, which is the peak response time (Figure 9.7), is sufficient small.

Unfortunately, there is no single objective criterion to evaluate yaw dynamics. Criterion $\left(\dot{\psi} / \delta_f\right)_{st}$ as the steady value of the yaw rate gain factor cannot be too large since it may lead to a strong vehicle reaction even when the driver unintentionally makes a slight steering input. If $\left(\dot{\psi} / \delta_f\right)_{st}$ is too small, the driver has to perform a large steering input to achieve an intended vehicle reaction. For too small $T_{\dot{\psi}_{max}}$, vehicle will react too fast while the vehicle reacts slow if $T_{\dot{\psi}_{max}}$ is too large. These two conditions also imply that the vehicle response behaves less oscillatory and intends to be smoother. For these reasons, the steady-state yaw rate could be a proper choice as the yaw rate target.

In Chapter 9, the steady-state yaw rate was described, and the steering angle, the vehicle speed and the turn radius defined their interrelationship given by (9.38). For constant steering input, the vehicle travels on a circle path at a constant speed, and the vehicle remains stable and stays on the course. Recalling Figure 9.10, the course angle keeps the vehicle on track, but the yaw angle cannot follow the course angle because of the vehicle sideslip angle. So, it is readily recognized that the course angle should be the yaw target, denoted as ψ_d,

$$\psi_d = \psi + \beta \tag{14.1}$$

This means that the yaw angle always has a steady error equals to β. Since $\dot{\beta} = 0$ in the steady state, the *yaw rate target* directly follows the steady-state yaw rate

$$\dot{\psi}_d = \dot{\psi}_{st} \tag{14.2}$$

Referring back to (9.39), we have

$$
\dot{\psi}_d = \frac{v\,\delta_H}{i_s\left(v^2 U_g + l_f + l_r\right)}
$$

$$
= \frac{v\,\delta_H}{i_s(l_f + l_r)\left(1 + \dfrac{v^2}{v_{ch}^2}\right)} \tag{14.3}
$$

At the same time, the sideslip angle results in the steady state since both the yaw angle and the sideslip angle associate to each other. Accordingly, the desired vehicle sideslip angle is defined by (9.42)

$$
\beta_d = \frac{-l_f M v^2 + C_{\alpha r}\left(l_f + l_r\right)l_r}{C_{\alpha r}\left(l_f + l_r\right)\left(v^2 U_g + l_f + l_r\right)}\,\delta_f \tag{14.4}
$$

Due to the nature of the steady state, a corrective measure needs to be considered when using the steady states of the yaw rate and the sideslip angle as the control targets. In general, the response of vehicle dynamics in time domain is a transient process. The steady states of the yaw rate as well as the sideslip angle are influenced by their response time, for example, by using $T_{\dot{\psi}}$ for the yaw rate in Figure 9.7. To achieve the steady state of $T_{\dot{\psi}}$ requires some time. For this reason, a time delay in achieving the steady state as the control target needs to be included in control design.

The steady-state of the yaw rate and the sideslip angle leads to a steady lateral acceleration of the vehicle. Per normal dynamics presented in Chapter 10, the lateral acceleration is the input for the roll movement, as demonstrated by (10.54). Hence, the quicker the lateral acceleration is reaching the steady-state the less vehicle roll vibration occurs.

Model-Reference-Based Approach

Since vehicle dynamics is highly nonlinear, a designed control system will behave satisfactorily only over a limited operation range of vehicle parameters and road conditions. As often happens, the need in yaw stability control occurs outside of the limited operation range. Therefore, using a mathematical model as a reference may be useful. In such case, the target yaw rate and the target of the sideslip angle are specified by means of their model that will produce the desired output for the vehicle. Such mathematical model can be simulated on a vehicle on-board computer.

The bicycle model can represent vehicle dynamics in a wide range of operation, and it is simple to implement. Therefore, the bicycle model is preferred to be used as reference model. However, the parameters need to be chosen in order to match the target yaw rate and the sideslip angle:

$$
\begin{bmatrix} \dot{\beta}_d \\ \ddot{\psi}_d \end{bmatrix} =
\begin{bmatrix} a_{11} & a_{12} \\ a_{21} & a_{22} \end{bmatrix}
\begin{bmatrix} \beta_d \\ \dot{\psi}_d \end{bmatrix} +
\begin{bmatrix} b_1 \\ b_2 \end{bmatrix} \delta_f \tag{14.5}
$$

where the steering angle δ_f is used as the input to the reference model. The parameters can be calculated as in model (8.41) and then adjusted by the experimental data, for example:

$$a_{11} = -\frac{C_{\alpha f} + C_{\alpha r}}{Mv}, \quad a_{12} = -1 - \frac{C_{\alpha f}l_f - C_{\alpha r}l_r}{Mv^2}$$

$$a_{21} = -\frac{C_{\alpha f}l_f - C_{\alpha r}l_r}{I_z}, \quad a_{22} = -\frac{C_{\alpha f}l_f^2 + C_{\alpha r}l_r^2}{I_z v}$$

$$b_1 = \frac{C_{\alpha f}}{Mv}, \quad b_2 = \frac{C_{\alpha f}l_f}{I_z}$$

where stiffnesses $C_{\alpha f}, C_{\alpha r}$ can be chosen in their linear range, and nominal values for M, I_z can be used. In this way, the stability controller will force an unstable vehicle to follow a linear behavior specified by the model as if the vehicle instability is not present. The vehicle speed utilized in the model is the actual speed of the vehicle, however, it should be emphasized that too high vehicle speeds will create vibrating behavior of the vehicle yaw rate control.

Limitation of Control Target

Neither the steady-state-based approach nor the reference model approach considers whether the vehicle lateral force to support the yaw movement is available or not. This lateral force depends on the type of the road surface and its condition, i.e., the force is limited by the peak friction coefficient of the tire-road surface. On dry roads, the lateral force that vehicle can withstand is much higher than the force on slippery roads. Thus, the tires on roads with low peak friction coefficient are unable to provide bigger lateral forces to support high yaw rates. Therefore, for the same steering input, the yaw rate that a vehicle demonstrates on dry roads cannot be achieved on slippery roads. Hence, the target yaw rate must be bounded by the tire-road friction. The lateral acceleration reflects the tire-road friction directly and is usually measurable, so it can be taken for the boundary consideration. Following the vehicle lateral acceleration calculated in (2.52), and remembering that $\dot{\beta} = 0$ in the steady state, we have

$$a_y = \dot{v} \sin\beta + v \cos\beta \, \dot{\psi} \tag{14.6}$$

In the case that the sideslip angle is small, $\beta \approx 0$, the acceleration can be approximated as follows:

$$a_y = \dot{v}\beta + v\dot{\psi} \tag{14.7}$$

From the relation between the lateral acceleration and the tire-road friction, we know

$$a_y \leq \mu_H g, \tag{14.8}$$

where μ_H is the peak friction coefficient. Thus, both the sideslip angle and the yaw rate are bounded by the limit of the tire-road peak friction coefficient. Due to small values of the sideslip angle, the second term of (14.7) contributes the most.

Using (14.7) and (14.8), we choose the upper bound of the target yaw rate as

$$\left| \dot{\psi}_{d\_max} \right| = \frac{\mu_H g}{v} \tag{14.9}$$

The choice of the upper bound for the sideslip angle can be based on practical experience. Usually, the sideslip angle is difficult to measure, and it is more or less considered as a perception of a driver.

From common driving experience, a smaller sideslip angle would make driving more comfortable. We can choose

$$\left|\beta_{d\_max}\right| = 6° \ldots 9° \tag{14.10}$$

The limitations should be assigned taking in consideration the type of the vehicle. For the passenger cars and the pick-up trucks, the limits are different.

Summarizing the above results, the boundaries for the target sideslip angle and the target yaw rate are given by

$$\left|\dot{\psi}_d\right| \le \left|\dot{\psi}_{d\_max}\right|$$
$$\left|\beta_d\right| \le \left|\beta_{d\_max}\right| \tag{14.11}$$

respectively.

14.1.2 STATE FEEDBACK CONTROL

In this subsection, a state feedback control is presented, in which a feedback controller with a feed-forward compensation is designed. The technique used here transforms the complicated control design problem of the yaw stability into a much simple problem, in which the equivalent linear model is utilized, so that the well-established linear control theory can be applied to develop stabilizing controllers. We start first with the model formulation.

The Model Formulation

In Chapter 8, several vehicle models were introduced, including linear and nonlinear models. For the purpose of the yaw stability control, the model defined by (8.39) will be used in the following. The model has control inputs of the wheel brakes as well as the rear-wheel steering angle.

First, we will reformulate the model to make it suitable for the state-feedback control design. From (8.39), a constant 2×2 matrix $\mathbf{M}$ is introduced, which contains the mass and the moment of inertia of the vehicle

$$\mathbf{M} = \begin{bmatrix} M & 0 \\ 0 & I_z \end{bmatrix} \tag{14.12}$$

By defining $\mathbf{x} = \begin{pmatrix} \beta & \dot{\psi} \end{pmatrix}^T$ as the state vector, (8.39) becomes

$$\mathbf{M}\dot{\mathbf{x}}(t) = \mathbf{A}(t)\mathbf{x}(t) + \mathbf{b}(t)\delta_f + \mathbf{u}(t), \tag{14.13}$$

where

$$\mathbf{A}(t) = \begin{bmatrix} -\dfrac{C_{\alpha f} + C_{\alpha r}}{v} & -M - \dfrac{C_{\alpha f}l_f - C_{\alpha r}l_r}{v^2} \\ -C_{\alpha f}l_f + C_{\alpha r}l_r & -\dfrac{C_{\alpha f}l_f^2 + C_{\alpha r}l_r^2}{v} \end{bmatrix}, \quad \mathbf{b}(t) = \begin{bmatrix} \dfrac{C_{\alpha f}}{v} \\ C_{\alpha f}l_f \end{bmatrix}$$

and the total control input is summarized to

$$\mathbf{u}(t) = \begin{bmatrix} \dfrac{C_{\alpha r}}{v} \\ -C_{\alpha r}l_r \end{bmatrix} \delta_r + \begin{bmatrix} \dfrac{\delta_f}{v} & 0 \\ l_f\delta_f - d_t & -d_t \end{bmatrix} \begin{bmatrix} F_{x\_LF} \\ F_{x\_LR} \end{bmatrix} + \begin{bmatrix} \dfrac{\delta_f}{v} & 0 \\ l_f\delta_f + d_t & d_t \end{bmatrix} \begin{bmatrix} F_{x\_RF} \\ F_{x\_RR} \end{bmatrix} \tag{14.14}$$

where the influence of the sideslip angle and the steered rear wheels on longitudinal tire forces is considered to be less significant comparing to the influence of the steered front wheels. Therefore, the longitudinal forces' lateral components, which are paired with the vehicle sideslip angle and rear steer angle are small enough to be neglected in (14.14). The steering input, δ_f, is given by the driver of the vehicle and remains unchanged by the controller. The rear steering represented by δ_r, front braking forces F_{x_LF}, F_{x_RF} and rear braking forces F_{x_LR}, F_{x_RR} are basically the controllable inputs. They can be provided through predefined actuators, such as a rear steer and a hydraulic brake unit. Once control input $\mathbf{u}(t)$ is determined, the rear steering angle, δ_r, and the braking forces, F_{xf} and F_{xR}, can be determined. Thus, the first step of the controller design is to find an appropriate control input, $\mathbf{u}(t)$, to meet the control target. Afterwards it can be specified, which actuator(s) is to apply in order to achieve the target. A practical implementation will be explained later in Subsection 14.1.4.

One notable property of model (14.13) is the linearity in its parameters. Due to this fact, the model can be formulated as

$$\mathbf{M}\dot{\mathbf{x}}(t) = \mathbf{Y}(t)\mathbf{p} + \mathbf{u}(t) \tag{14.15}$$

where $\mathbf{Y}(t)$ is a 2×3 regression matrix

$$\mathbf{Y}(t) = \begin{bmatrix} Y_{11} & Y_{12} & Y_{13} \\ Y_{21} & Y_{22} & Y_{23} \end{bmatrix}$$

$$= \begin{bmatrix} -\dfrac{\beta}{v} - \dfrac{l_f}{v^2}\dot{\psi} + \dfrac{\delta_f}{v} & -\dfrac{\beta}{v} + \dfrac{l_r}{v^2}\dot{\psi} & -\dot{\psi} \\ -l_f\beta - \dfrac{l_f^2}{v}\dot{\psi} + l_f\delta_f & -\dfrac{l_r^2}{v}\dot{\psi} + l_r\beta & 0 \end{bmatrix} \tag{14.16a}$$

and $\mathbf{p}$ is a 3×1 parameter vector

$$\mathbf{p} = (C_{\alpha f} \quad C_{\alpha r} \quad M)^T \tag{14.16b}$$

Those parameters in $\mathbf{p}$, the tire cornering stiffness and the gross vehicle mass, are considered unknown, and vehicle dynamics is considered linear in the unknown terms. This kind of linearity will be conveniently used in the controller design throughout the following sections.

State Feedback Control

The feedback linearization is a commonly used controller design technique for nonlinear systems. The idea of this technique is to transform nonlinear system dynamics into linear dynamics, so that linear design techniques can be applied. Some applications and examples of such control approach can be found in control of robot manipulators [2]. Here, we use a similar principle to convert a complicated control design to an equivalent linear model by using a feedforward compensation.

In regard to (14.13), designing a controller is to find $\mathbf{u}(t)$ so that state vector $\mathbf{x}(t)$ moves toward a desired target. As discussed in the previous section, the steady states should be selected as the desired targets. By denoting the desired steady states as

$$\mathbf{x}_d(t) = [\beta_d \quad \dot{\psi}_d]^T \tag{14.17}$$

the control error can be defined

$$\mathbf{e}(t) = \mathbf{x}_d(t) - \mathbf{x}(t) \tag{14.18}$$

Following the above-given definition, the derivative of the desired target is equal to zero, i.e., $\dot{\mathbf{x}}_d(t) = 0$. Hence, the differentiation of the error in (14.18) leads to

$$\dot{\mathbf{e}}(t) = -\dot{\mathbf{x}}(t) \tag{14.19}$$

Now, substituting model equation (14.13) into (14.19) yields

$$\mathbf{M}\dot{\mathbf{e}}(t) = -\mathbf{A}(t)\mathbf{x}(t) - \mathbf{b}(t)\delta_f - \mathbf{u}(t) \tag{14.20}$$

The first two terms on the right side of the above equation present the time-varying dynamics of the system, and can be cancelled by a feedforward compensation in the control law to be designed.

The control input function is defined as follows:

$$\mathbf{u}(t) = -\mathbf{A}(t)\mathbf{x}(t) - \mathbf{b}(t)\delta_f - \mathbf{u}_0 \tag{14.21}$$

where the parameters in $\mathbf{A}$ and $\mathbf{b}$ are assumed to be known and $\mathbf{u}_0$ is an auxiliary control signal to be specified. With the control definition above, (14.20) becomes a linear system

$$\mathbf{M}\dot{\mathbf{e}}(t) = \mathbf{u}_0 \tag{14.22}$$

for which $\mathbf{u}_0$ can be chosen as any linear controller. One option is to select $\mathbf{u}_0$ as the proportional feedback,

$$\mathbf{u}_0 = -\mathbf{K}\mathbf{e}(t) \tag{14.23}$$

where $\mathbf{K}$ is a 2×2 control gain matrix. Then, the closed-loop error equation becomes

$$\mathbf{M}\dot{\mathbf{e}}(t) + \mathbf{K}\mathbf{e}(t) = 0 \tag{14.24}$$

Since $\mathbf{M}$ is a constant diagonal matrix, the simplest way is to choose $\mathbf{K}$ in a diagonal form so that the error equation is completely decoupled, i.e.,

$$\begin{aligned} M\dot{e}_1(t) + k_1 e_1(t) &= 0 \\ I_2 \dot{e}_2(t) + k_2 e_2(t) &= 0 \end{aligned} \tag{14.25}$$

By choosing positive k_1 and k_2, the error vector, $\mathbf{e}(t)$, will converge to zero, regardless of values of $\mathbf{e}(0)$. Figure 14.1 presents a block diagram of the state control.

In fact, feedback control input (14.21) does convert the control problem for a time-varying system into a simple control design of a linear system. The algorithm computes a model-based control law

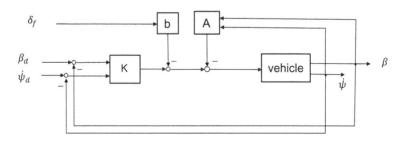

FIGURE 14.1   State feedback control diagram.

to compensate the time-varying dynamics of the vehicle that is controlled. In the end, the result is reduced to a simple linear closed-loop system, for which the control error converges to zero by selecting the linear controller parameters.

On the other hand, to do such compensation, the parameters of the vehicle model should be known. That is often a limitation in practical applications of the control. Indeed, in many cases, some parameters in system (14.13) are not known accurately because they depend on driving and road conditions, or some parameters vary due to vehicle loading. Those parameters are given by vector $\mathbf{p}$ in (14.16b). The tire cornering stiffness is influenced by the tire slip angle as well as the tire normal reaction, etc., and will not be a constant when vehicle turns at higher accelerations. The gross vehicle mass changes due to the loading of the vehicle. One way to deal with the variation is to determine the best estimates for those parameters, denoted as $\hat{\mathbf{p}}$. Now instead of (14.21), we apply a control law

$$\mathbf{u}(t) = -\hat{\mathbf{A}}(t)\mathbf{x}(t) - \hat{\mathbf{b}}(t)\delta_f + \mathbf{K}\mathbf{e}(t) \tag{14.26}$$

to system (14.20). After defining the deviation between the estimate and the true value of the parameters as

$$\tilde{\mathbf{A}}(t) = \hat{\mathbf{A}}(t) - \mathbf{A}(t); \ \tilde{\mathbf{b}}(t) = \hat{\mathbf{b}}(t) - \mathbf{b}(t) \tag{14.27}$$

the closed-loop equation for the system results in

$$\mathbf{M}\dot{\mathbf{e}}(t) + \mathbf{K}\mathbf{e}(t) = \tilde{\mathbf{A}}(t)\mathbf{x}(t) + \tilde{\mathbf{b}}(t)\delta_f \tag{14.28}$$

On the right side of this equation, the errors are driven by the mismatch between the estimate and the model so that the errors will never go exactly to zero. The behavior of such system is undetermined, it could cause instability of the system even if $\mathbf{K}$ is selected for good stability of the left-hand side. To solve this problem, in the next section we study robustness issues and extend the state feedback controller to a robust controller.

14.1.3 ROBUST YAW STABILITY CONTROLLER

The vehicle model used for the controller design in the previous section was derived under certain assumptions, which were introduced before in Section 3.2.2 and Chapter 9. Those are: the lateral acceleration is lower than 0.4g, the tire cornering stiffness is linear to the wheel slip angle, the vehicle sideslip angle is small, no roll dynamic impact considered. In many real driving maneuvers, these assumptions often cannot be held, so that the system features uncertainties, which may be unknown or time varying. An exact assessment of the uncertainties may technically become too complex and lead to an unusable result for the controller design. For this reason, the robust control arises as a suitable approach and to provide and demonstrate advantageous performance of the vehicle under such control.

The robust controller designed in this section is obtained based on the state-feedback controller described in the previous section. We use Lyapunov stability analysis to ensure the robustness of the design.

Consider the model of the yaw dynamics given by (14.13) in the previous section:

$$\mathbf{M}\dot{\mathbf{x}}(t) = \mathbf{A}(t)\mathbf{x}(t) + \mathbf{b}(t)\delta_f + \mathbf{u}(t) \tag{14.29}$$

in which the control feedback can be defined as

$$\mathbf{u}(t) = -\hat{\mathbf{A}}(t)\mathbf{x}(t) - \hat{\mathbf{b}}(t)\delta_f - \mathbf{u}_0 \tag{14.30}$$

We will see later that auxiliary control signal $\mathbf{u}_0$ needs to be specified differently in order to guarantee the robustness of the controller. Following definition (14.18) and substituting (14.30) into (14.13), then the closed-loop error equation is given by

$$\mathbf{M}\dot{\mathbf{e}}(t) = \tilde{\mathbf{A}}(t)\mathbf{x}(t) + \tilde{\mathbf{b}}(t)\delta_f + \mathbf{u}_0 \qquad (14.31)$$

The next step is to select auxiliary control signal $\mathbf{u}_0$ so that the control error can converge to zero.
Recalling (14.15) and (14.16a), (14.31) can be reformulated as

$$\mathbf{M}\dot{\mathbf{e}}(t) = \mathbf{Y}(t)\tilde{\mathbf{p}} + \mathbf{u}_0 \qquad (14.32)$$

with the parameter vector error

$$\tilde{\mathbf{p}} = \hat{\mathbf{p}} - \mathbf{p} \qquad (14.33)$$

So far, the control error equation has been formed, we now use Lyapunov stability analysis to show the error vector $\mathbf{e}$ will be asymptotically stable with the right choice of auxiliary control signal $\mathbf{u}_0$. First, we select the mass and inertia-related Lyapunov function

$$V = \frac{1}{2}\mathbf{e}^T(t)\,\mathbf{M}\,\mathbf{e}(t) \qquad (14.34)$$

Differentiating (14.34) with respect to time leads to

$$\dot{V} = \mathbf{e}^T(t)\,\mathbf{M}\,\dot{\mathbf{e}}(t) \qquad (14.35)$$

Substituting (14.32) into the above equation yields

$$\dot{V} = \mathbf{e}^T(t)\{\mathbf{Y}(t)\tilde{\mathbf{p}} + \mathbf{u}_0\} \qquad (14.36)$$

In order to be able to make $\mathbf{e}(t)$ asymptotically stable, $\dot{V}$ needs to be at least negative semidefinite. We chose

$$\mathbf{u}_0 = -\mathbf{K}\mathbf{e}(t) - \mathbf{d} \qquad (14.37)$$

where $\mathbf{K}$ is a diagonal, positive-definite 2×2 matrix

$$\mathbf{K} = \begin{bmatrix} k_1 & 0 \\ 0 & k_2 \end{bmatrix} \qquad (14.38)$$

with positive constants k_1 and k_2. $\mathbf{d} = (d_1 \ d_2)^T$ is a 2×1 vector whose elements satisfy

$$d_i = \left\{ \sum_{j=1}^{3} |Y_{ij}| p_{mj} + \eta_i \right\} \mathrm{sgn}(e_i), \ i = 1,\ 2 \qquad (14.39)$$

where $\mathrm{sgn}(\bullet)$ is the signum function and η_i are positive constants. Y_{ij} and p_{mj} are defined following (14.16a) and (14.33), where $|Y_{ij}|$ is the absolute value of each component in $\mathbf{Y}$ matrix and

$$p_{mj} = \max\left(|\tilde{p}_j|\right). \qquad (14.40)$$

Considering the definitions from (14.37) to (14.40), the time derivative of the Lyapunov function in (14.36) becomes

$$\dot{V} \leq -\mathbf{e}^T(t)\,\mathbf{K}\,\mathbf{e}(t) - \sum_{i=1}^{2}\eta_i |e_i| \qquad (14.41)$$

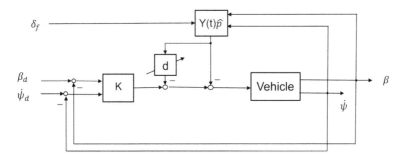

FIGURE 14.2 Robust yaw stability control.

This result shows Lyapunov function V is decreasing for all time. The only chance for $\dot{V} = 0$, would be $\mathbf{e} = 0$. The open-loop system considered here is autonomous, therefore, from LaSalle's Theorem, control error $\mathbf{e}$ is asymptotically stable. Hence, the complete robust controller is given by (14.30) and (14.37), where the robustness is ensured by upper boundary $\mathbf{d}$ of uncertainties. Figure 14.2 illustrates the block diagram of the robust yaw stability control.

From the above design of the robust controller, one can see that it has a fixed structure that yields stable performance for a class of systems featuring uncertainties. In addition, robust controller is simpler to implement, and does not require many tuning parameters. The following will show that the robust controller can be extended using an adaptive parameter estimation scheme, in which an uncertain constant parameter is being adapted during a control cycle.

Parameter Adaptation with Robustness

Here, we study parameter identification during the robust control. Normally a parameter identification and adaptation are performed in adaptive controls. Adaptive control algorithms with complete parameters adaptation may not be useful here since not all unknown parameters are constant. Also, adaptive control requires a rigorous proof for the stability, which results in the global convergence of the control error and parameter estimation error. In vehicle dynamics applications that are under consideration in this section, the tire cornering stiffnesses can vary depending on tire-road conditions, and the gross mass of the vehicle is a constant parameter during a driving maneuver. Therefore, the focus is on adaptation of the constant parameter.

Recalling the model formulation given by (14.15), the first term on the right side of the equation, $\mathbf{Y}(t)\mathbf{p}$, separates the model into the known signal terms and the unknown parameters. As indicated before, the tire cornering stiffness may not be constant during any single maneuver, especially when the vehicle is at high lateral acceleration. The gross mass, M, of the vehicle is however constant during a motion. Therefore, it can be estimated. From (14.15) we have

$$\mathbf{M}\dot{\mathbf{x}}(t) = \mathbf{Y}_r(t)\mathbf{p}_r + \mathbf{y}_a M + \mathbf{u}(t) \tag{14.42}$$

where $\mathbf{Y}_r(t)$ and $\mathbf{p}_r$ are the subsets of $\mathbf{Y}(t)$ and $\mathbf{p}$ respectively,

$$\mathbf{Y}_r(t) = \begin{bmatrix} -\dfrac{\beta}{v} - \dfrac{l_f}{v^2}\dot{\psi} + \dfrac{\delta_f}{v} & -\dfrac{\beta}{v} + \dfrac{l_r}{v^2}\dot{\psi} \\[2ex] -l_f\beta - \dfrac{l_f^2}{v}\dot{\psi} + l_f\delta_f & -\dfrac{l_r^2}{v}\dot{\psi} + l_r\beta \end{bmatrix} \tag{14.43}$$

$$\mathbf{P}_r = \begin{bmatrix} C_{\alpha f} & C_{\alpha r} \end{bmatrix}^T$$

and $\mathbf{y}_a(t) = \begin{bmatrix} -\dot{\psi} & 0 \end{bmatrix}^T$.

Now we apply the same controller given by (14.30) to (14.42); this leads to the following error equation:

$$\mathbf{M}\dot{\mathbf{e}}(t) = \mathbf{Y}_r(t)\tilde{\mathbf{p}}_r + y_a\tilde{M} + \mathbf{u}_0 \tag{14.44}$$

with the parameter errors between the estimate and true values

$$\tilde{\mathbf{p}}_r = \hat{\mathbf{p}}_r - \mathbf{p}_r$$
$$\tilde{M} = \hat{M} - M \tag{14.45}$$

Following the same way through Lyapunov stability analysis as was done in the previous section, we select Lyapunov function

$$V = \frac{1}{2}\mathbf{e}^T(t)\,\mathbf{M}\,\mathbf{e}(t) + \frac{1}{2}k_m^{-1}\tilde{M}^2 \tag{14.46}$$

where k_m is a positive parameter. Differentiating (14.46) with respect to time leads to

$$\dot{V} = \mathbf{e}^T(t)\,\mathbf{M}\,\dot{\mathbf{e}}(t) + k_m^{-1}\tilde{M}\dot{\tilde{M}} \tag{14.47}$$

Substituting (14.44) into (14.47) yields

$$\dot{V} = \mathbf{e}^T\{\mathbf{Y}_r(t)\tilde{\mathbf{p}}_r + \mathbf{u}_0\} + \mathbf{e}^T y_a\tilde{M} + k_m^{-1}\,\dot{\tilde{M}}\tilde{M} \tag{14.48}$$

As same as the robust controller, we chose

$$\mathbf{u}_0 = -\mathbf{K}\mathbf{e}(t) - \mathbf{d}, \tag{14.49}$$

where $\mathbf{K}$ is a diagonal, positive-definite 2×2 matrix

$$\mathbf{K} = \begin{bmatrix} k_1 & 0 \\ 0 & k_2 \end{bmatrix} \tag{14.50}$$

with positive constants k_1 and k_2. $\mathbf{d} = (d_1 \; d_2)^T$ is a 2×1 vector whose elements satisfy

$$d_i = \left\{ \sum_{j=1}^{2}|Y_{ij}|\,p_{mj} + \eta_i \right\}\mathrm{sgn}(e_i), \; i = 1, 2 \tag{14.51}$$

with the same parameters defined by (14.39).

Considering the definitions from (14.49) to (14.51), the time derivative of the Lyapunov function in (14.48) becomes

$$\dot{V} \le -\mathbf{e}^T(t)\,\mathbf{K}\,\mathbf{e}(t) - \sum_{i=1}^{2}\eta_i|e_i| + \left(k_m^{-1}\dot{\tilde{M}} + \mathbf{e}^T y_a\right)\tilde{M} \tag{14.52}$$

In order to be able to make $\mathbf{e}(t)$ asymptotically stable, $\dot{V}$ needs to be at least negative semidefinite. This becomes true if the last term of above equation equals to zero, i.e.,

$$k_m^{-1}\dot{\tilde{M}} + \mathbf{e}^T y_a = 0$$

Since the unknown mass is a constant, that results in $\dot{\tilde{M}} = \dot{\hat{M}}$. It follows

$$\dot{\hat{M}} = -k_m \mathbf{e}^T \mathbf{y}_a \qquad (14.53)$$

which is the parameter adaptation law for the vehicle gross mass. Now (14.52) is the same as (14.41) showing that the control error $\mathbf{e}(t)$ is asymptotically stable. Since $\dot{V}$ is negative semidefinite, so V is then upper bounded. From (14.46), the parameter error $\tilde{M}$ is hence bounded. At this point, there is no guarantee that the estimate will converge to the true value.

Comparing to the robust controller to the robust controller with parameter adaptation, it is seen that they are basically same, but with a minor difference in (14.51) and with the parameter adaptation given by (14.53). As mentioned before, the parameter identification and adaptation occur in adaptive controls. In reference [2], adaptive controllers with the parameter identification and adaptation were studied for robotic manipulators, for which it was showed that the parameter error convergence can be achieved if the regression matrix is sufficiently excited.

14.1.4 PRACTICAL IMPLEMENTATION OF CONTROL INPUTS

So far for the yaw stability control, two different controllers were designed – the state-feedback and the robust controller. Control input $\mathbf{u}(t)$ was specified based on the state-feedback method. In practical engineering, the control actuation can be implemented by either steering or braking. In this section we study how the calculated control input can be provided by the actuators. We present two options: a brake-only control, and a hybrid approach that includes both the brake control and the rear-steering.

Brake-Only Actuation

The relation between the control law and the actuation is determined by (14.14). If the control actuation is only done by engaging brakes only, i.e. $\delta_r = 0$, (14.14) results in

$$\mathbf{u}(t) = \begin{bmatrix} \dfrac{\delta_f}{v} & 0 \\ l_f\delta_f - d_t & -d_t \end{bmatrix} \begin{bmatrix} F_{x\_LF} \\ F_{x\_LR} \end{bmatrix} + \begin{bmatrix} \dfrac{\delta_f}{v} & 0 \\ l_f\delta_f + d_t & d_t \end{bmatrix} \begin{bmatrix} F_{x\_RF} \\ F_{x\_RR} \end{bmatrix} \qquad (14.54)$$

From a mathematical point of view, one needs to find a solution of (14.54) for F_{x_LF}, F_{x_RF}, F_{x_LR} and F_{x_RR} with two linear equations. In terms of finding the solution, (14.54) is an underdetermined equation system, i.e., the number of variables is greater than the number of equations, and in this case the equation will have an infinite number of solutions. Therefore, a mathematical calculation with a physical understanding of solutions will be necessary.

From the physical point of view, $\mathbf{u}(t)$ is calculated by the controller so that the vehicle can reach the steady state of the yaw rate and the sideslip angle during a dynamic maneuver. During the control, the vehicle is either understeered or oversteered, and $\mathbf{u}(t)$ must be able to create a resultant braking torque and a yaw movement to force the vehicle to move toward the yaw rate and sideslip angle targets. In order to achieve this goal, the actual yaw rate of the vehicle needs to be increased or decreased. Therefore, the braking forces are applied only on one side of the vehicle, either left side or right side. Hence, there are only two solutions coming into consideration:

$$\mathbf{u}(t) = \begin{bmatrix} \dfrac{\delta_f}{v} & 0 \\ l_f\delta_f - d_t & -d_t \end{bmatrix} \begin{bmatrix} F_{x\_LF} \\ F_{x\_LR} \end{bmatrix}, \quad F_{x\_RF} = F_{x\_RR} = 0 \qquad (14.55)$$

or

$$\mathbf{u}(t) = \begin{bmatrix} \dfrac{\delta_f}{v} & 0 \\ l_f\delta_f + d_t & d_t \end{bmatrix} \begin{bmatrix} F_{x\_RF} \\ F_{x\_RR} \end{bmatrix}, \quad F_{x\_LF} = F_{x\_LR} = 0 \tag{14.56}$$

Here, $\left(F_{x\_LF} \ F_{x\_LR}\right)^T$ and $\left(F_{x\_RF} \ F_{x\_RR}\right)^T$ can be directly obtained by using the inverse matrix method. However, when using the inverse matrix method, the solution will have a singularity at $\delta_f = 0$, where the calculated braking forces become undefined. As described later on, the singularity will be avoided in the proposed method.

For the brake-only actuation, the first question that should be addressed is on which vehicle side the wheels should be braked? In order to achieve control target, control input $\mathbf{u}(t) = (u_1 \ u_2)^T$ calculated by the controller is required to actuate vehicle brake. From (14.54), one can interpret the physical meaning of the two components of the control input vector. Component u_1 indirectly reveals the tire forces required in the lateral direction, in which both the right and the left braking contribute the same. The second component, u_2, actually specifies the required yaw moment. The right and the left braking contribute to the actual yaw moment in opposite direction corresponding directly to whether an increase or decrease of the yaw moment is needed. We can take the advantage of component u_2, and use it to determine on which side the brake should be applied. There are three possible cases for u_2:

$u_2 < 0$: Right braking.
Considering the coordinate system given in Figure 8.5, $u_2 < 0$ indicates a decrease of the yaw moment, hence the yaw rate. In this case, the right braking contributes to the clockwise rotation of the vehicle, and the yaw rate will decrease when the vehicle is making the left turn but will increase when the vehicle is steered to the right. Figure 14.3 shows these two scenarios. In the left turn the yaw rate is getting smaller while it is getting larger in the right turn of the vehicle.

$u_2 > 0$: Left braking.
Considering the coordinate system given in Figure 8.5, $u_2 > 0$ indicates increase of the yaw moment, hence the yaw rate. In this case, the left braking contributes to the counterclockwise rotation of the vehicle, and the yaw rate will increase during the left turn of the vehicle but will decrease during the right turn. Figure 14.4 shows these two scenarios. In the left turn, the yaw rate is getting larger while it is getting smaller in the right turn.

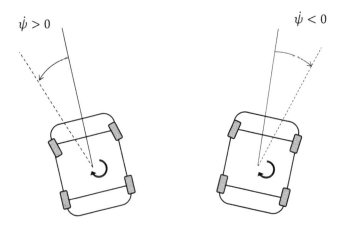

FIGURE 14.3 Right side braking.

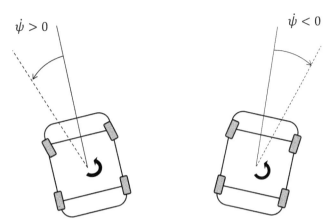

FIGURE 14.4   Left side braking.

$u_2 = 0$: No braking
For $u_2 = 0$, there is no braking torque needed for the yaw movement. Therefore, the braking forces can be set to zero.

Note that the decision on engaging a brake is not based on the vehicle dynamic behavior at the moment of the control action, such as vehicle oversteer or understeer. The decision making requires only a calculated value of u_2. The above-discussed yaw rate change depends only on a coordinate system chosen for the vehicle modeling.

After the side braking is selected, the braking force applied to each wheel needs to be quantified. Ideally, the solution given by (14.55) and (14.56) can be used directly for that purpose. Since the brake-only actuation is considered, the longitudinal tire forces must be negative. However, the individual forces calculated from (14.55) and (14.56) might occur to be positive and also can become very large when the front steering angle approaches to zero. Therefore, it is impractical to implement solution of (14.55) and (14.56) to quantify the braking forces. As an alternative, we can redistribute the required torque between the front and the rear wheels to meet the purpose of the yaw stability control, and, at the same time, to keep the control stable by avoiding positive values of the individual longitudinal tire forces. Here, we show how to distribute the braking force between the front and rear wheels and to provide the stability analysis.

Knowing control input $\mathbf{u}(t) = (u_1 \ u_2)^T$ calculated by the controller, we solve (14.55) and (14.56) without explicit calculation of the force at each wheel, that is

$$F_{x\_LF} + F_{x\_LR} = \frac{u_2 - l_f u_1 \mathrm{v}}{-d_t} \tag{14.57}$$

and

$$F_{x\_RF} + F_{x\_RR} = \frac{u_2 - l_f u_1 \mathrm{v}}{d_t} \tag{14.58}$$

respectively. We can distribute the braking forces between the front and rear wheels so that (14.57) for the left braking and (14.58) for the right braking is still valid. This total amount of the front and rear forces generates the yaw moment needed to stabilize the vehicle. The question of how much of the yaw rate should be supplied by the front wheel braking and how much of the yaw rate should be provided by the rear wheel braking can be addressed through a vehicle dynamic behavior analysis. If the vehicle is oversteered, the front wheel may have more braking force, while the rear wheel may have a bigger braking force if the vehicle is understeered. Let $F^*_{x\_LF}$, $F^*_{x\_RF}$ and $F^*_{x\_LR}$, $F^*_{x\_RR}$ be the actual distributed braking force between the front and rear wheels, respectively. Hence, the force distribution is summarized as follows:

$$
F_{x\_LF}^{*} + F_{x\_LR}^{*} =
\begin{cases}
F_{x\_LF} + F_{x\_LR} & for \quad F_{x\_LF} + F_{x\_LR} < 0 \\
F_{x\_LR}^{*} = \dfrac{u_2}{-d_t} & for \quad F_{x\_LF} + F_{x\_LR} \geq 0
\end{cases}
\tag{14.59a}
$$

for the left wheel braking, and

$$
F_{x\_RF}^{*} + F_{x\_RR}^{*} =
\begin{cases}
F_{x\_RF} + F_{x\_RR} & for \quad F_{x\_RF} + F_{x\_RR} < 0 \\
F_{x\_RR}^{*} = \dfrac{u_2}{d_t} & for \quad F_{x\_RF} + F_{x\_RR} \geq 0
\end{cases}
\tag{14.59b}
$$

for the right wheel braking. The front braking force is zero by setting $F_{x\_LF}^{*} = 0$ or $F_{x\_RF}^{*} = 0$ if the calculated total force is not negative. In this case, only the rear wheel is braking during the control. Based on this force distribution approach, the actual control input will no longer satisfy (14.55) and (14.56). The deviation can be evaluated using (14.54). Since either $(F_{x\_LF} + F_{x\_LR})$ or $(F_{x\_RF} + F_{x\_RR})$ remain unchanged or the front braking force is zero, the deviation will occurs only in the term $\delta_f F_{x\_LF}$ and $\delta_f F_{x\_RF}$ respectively. Defining Δu_1 and Δu_2 as the deviation of u_1 and u_2 defined by the controller, then we have

$$
\Delta u_1 =
\begin{cases}
\dfrac{\delta_f}{v}\left(F_{x\_LF}^{*} - F_{x\_LF} \right) & \text{left braking} \\
\dfrac{\delta_f}{v}\left(F_{x\_RF}^{*} - F_{x\_RF} \right) & \text{right braking}
\end{cases}
\tag{14.60a}
$$

and

$$
\Delta u_2 =
\begin{cases}
l_f \delta_f \left(F_{x\_LF}^{*} - F_{x\_LF} \right) & \text{left braking} \\
l_f \delta_f \left(F_{x\_RF}^{*} - F_{x\_RF} \right) & \text{right braking}
\end{cases}
\tag{14.60b}
$$

From this approach, the braking forces actually applied are different than the forces calculated from (14.55) and (14.56). Now we consider Lyapunov function to analyze stability and to determine conditions under which the control errors remain stable.

Let us review the same Lyapunov function given by (14.34) and its derivative again. The derivative of the Lyapunov function results in

$$
\dot{V} \leq -\mathbf{e}^{T}(t)\,\mathbf{K}\,\mathbf{e}(t) - \mathbf{e}^{T}(t)\Delta \mathbf{u} - \mathbf{e}^{T}(t)\mathbf{d}
\tag{14.61}
$$

where the same controller definition as (14.37) with different specification for $\mathbf{d} = (d_1 \ d_2)^{T}$ component is used. The control input deviation, $\Delta \boldsymbol{u}(t) = (\Delta u_1 \ \Delta u_2)^{T}$, emerges due to the actual brake distribution through $F_{x\_LF}^{*}$ and $F_{x\_RF}^{*}$. The calculation of the deviation is given in (14.60a) and (14.60b). In order to compensate the deviation caused by $\Delta \boldsymbol{u}(t)$, vector $\mathbf{d}$ can be specified as

$$
d_i = \left\{ \sum_{j=1}^{3} |Y_{ij}| \, p_{mj} + |\Delta u_i| + \eta_i \right\} \mathrm{sgn}(e_i), \ i = 1,\, 2
\tag{14.62}
$$

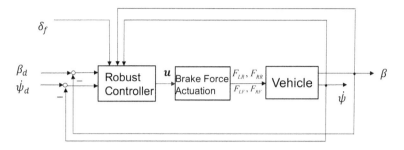

FIGURE 14.5   The brake-only control.

by adding term $|\Delta u_i|$ to (14.39). According to (14.60), Δu_i is a combination of known variables. Thus, the derivative given by (14.61) becomes

$$\dot{V} \le -\mathbf{e}^T(t)\ \mathbf{K}\ \mathbf{e}(t) - \sum_{i=1}^{2} \eta_i |e_i| \tag{14.63}$$

which is the same result as (14.41) showing that the control error $\mathbf{e}(t)$ is asymptotically stable. So far we have seen that the braking forces distribution does have an impact on the stability, but can keep the control stable by adapting upper boundary d for the deviations.

For vehicle dynamics presented by (14.13) with the robust controller given by (14.30) and (14.37) with the definition of (14.62), the error $\mathbf{e}(t)$ of the closed-loop control can be asymptotically stable by applying the brake according to (14.59). Controller gains k_i and η_i guarantee the sufficient conditions for the boundedness of the control errors. As aforementioned, the goal of the yaw stability control is to stabilize the rotational movement of the vehicle about the normal axis, and the controlled vehicle can achieve stable condition without reaching the exact yaw rate and sideslip angle targets. Figure 14.5 shows the diagram of the closed-loop of the brake-only control.

Brake and Rear-Steer Actuation

The brake-only control is actually the standard from today's technology, most vehicles are controlled only by the brakes. In passenger cars, the brake actuation is usually performed by a hydraulic system. During the control, the controller commands the hydraulic system to modulate the pressure. The response of the hydraulic pressure can be slower than the controller intends. Hence, the influence of the brake-only control may be too slow to avoid significant control deviations. The rear-steer control is based on the electronic steering of the rear axle wheels, and can act quicker than the hydraulic modulation. As an example, the rear-steering is used for model following control in reference [3]. Some technical references consider the front-steering for the vehicle stabilization [4].

We consider the scenario where the rear-steer actuator is activated in addition to the brake actuator. Looking again at the control input (14.14), the calculated control request of $\mathbf{u}(t) = (u_1\ u_2)^T$ should be distributed among the rear-steer and the brakes. The question now is how should $\mathbf{u}(t)$ be distributed? The technical approach here is to split the control input into two portions. The first portion of the input is delivered by the rear steering while the second portion is provided by the brakes, as shown in the following equation:

$$\mathbf{u}(t) = \mathbf{u}_r(t) + \mathbf{u}_b(t) \tag{14.64}$$

where

$$\mathbf{u}_r(t) = \begin{bmatrix} \dfrac{C_{\alpha r}}{v} \\ -C_{\alpha r} l_r \end{bmatrix} \delta_r$$

$$\mathbf{u}_b(t) = \mathbf{u}(t) - \mathbf{u}_r(t) \qquad (14.65)$$

$$= \begin{bmatrix} \dfrac{\delta_f}{v} & 0 \\ l_f \delta_f - d_t & -d_t \end{bmatrix} \begin{bmatrix} F_{x\_LF} \\ F_{x\_LR} \end{bmatrix} + \begin{bmatrix} \dfrac{\delta_f}{v} & 0 \\ l_f \delta_f + d_t & d_t \end{bmatrix} \begin{bmatrix} F_{x\_RF} \\ F_{x\_RR} \end{bmatrix}$$

In the above equation, $\mathbf{u}_r = (u_{r1} \quad u_{r2})^T$ is the first term on the right side of (14.14) and refers to the first portion of the control input produced by the rear steering. $\mathbf{u}_b = (u_{b1} \quad u_{b2})^T$ is the rest of the control input to be produced by the brakes, either the right-side or left-side braking. From the components of the calculated control input, $\mathbf{u}(t)$, we know that u_1 indicates the lateral force needed and u_2 is the total moment required to correct the yaw movement. Since u_2 is physically more dominant than u_1, it is used to decide on which side of the vehicle should be braked. In this context, we define

$$\delta_r = \begin{cases} -\delta_{rc}\,\mathrm{sgn}(u_2) & \text{for} \quad 0 \le \delta_{rc} < 4° \\ 0 & \text{if} \quad u_2(u_2 - u_{r2}) < 0 \\ 0 & \text{if} \quad \delta_f \delta_r < 0 \end{cases} \qquad (14.66)$$

where δ_{rc} is a parameter for the control gain of the rear steering. In this way, we basically pre-define the rear steer angle, which shall produce a yaw moment to increase or decrease the yaw rate of the vehicle; for example in the case of $u_2 < 0$, the rear steer angle results in $\delta_r > 0$ and the contributed moment will be given by $u_{r2} < 0$. A negative value of u_{r2} shall decrease the yaw moment. The second condition in (14.66) is to ensure that the rear steering should not be performed if the torque amount provided by u_{r2} is higher than u_2. Due to safety reasons, the rear steer must be limited to a small range, typically $\delta_r < 4°$. Another safety reason that must be considered is the direction of the rear steering. In Chapter 9, we studied two configurations – in-phase and out-of-phase steering. Especially at high speeds, the phase of the rear steering impacts significantly vehicle dynamics. If the rear wheels are steered out-of-phase with the front wheels, it will increase the yaw rate so that the vehicle tends to move outward and possibly constitutes an oversteer maneuver. However, the in-phase steering can improve vehicle performance in turning. This fact is expressed by the third condition in (14.66). The rear steering is set to zero if the rear is steered out-of-phase with the front steering, i.e., $\delta_f \delta_r < 0$.

The closed-loop control with the rear steer controller presented by (14.66) can be analyzed in the same way as the brake-only control. Unlike in the brake-only control, component $\mathbf{u}_b = (u_{b1} \quad u_{b2})^T$ in (14.64) needs to be used for the braking torque distribution, i.e.,

$$F_{x\_LF} + F_{x\_LR} = \frac{u_{b2} - l_f u_{b1} v}{-d_t} \qquad (14.67)$$

and

$$F_{x\_RF} + F_{x\_RR} = \frac{u_{b2} - l_f u_{b1} v}{d_t} \qquad (14.68)$$

for the left-side or the right-side braking, respectively. The positive or negative sign of u_{b2} determines which vehicle side is to be braked. The braking torque distribution between the front and rear wheels follows the same approach as in the brake-only control given by (14.59). Letting $F^*_{x\_LF}$, $F^*_{x\_RF}$ be the actual distributed braking force on the front wheels and $F^*_{x\_LR}$, $F^*_{x\_RR}$ on the rear wheels, we can view stability of the control. Defining Δu_1 and Δu_2 to be the deviation of u_1 and u_2, then we have

$$\Delta u_1 = \begin{cases} \dfrac{\delta_f}{v}\left(F^*_{x\_LF} - F_{x\_LF}\right) & \text{left braking} \\[3mm] \dfrac{\delta_f}{v}\left(F^*_{x\_RF} - F_{x\_RF}\right) & \text{right braking} \end{cases} \tag{14.69a}$$

and

$$\Delta u_2 = \begin{cases} l_f \delta_f \left(F^*_{x\_LF} - F_{x\_LF}\right) & \text{left braking} \\[3mm] l_f \delta_f \left(F^*_{x\_RF} - F_{x\_RF}\right) & \text{right braking} \end{cases} \tag{14.69b}$$

The control input deviation, $\Delta \mathbf{u}(t) = (\Delta u_1 \ \Delta u_2)^T$, emerges due to the actual brake distribution through forces $F^*_{x\_LF}$ and $F^*_{x\_RF}$, is treated in the same way as in the brake-only actuation.

Similar to the brake-only control, the conclusion can be made in regard to the stability. For vehicle dynamics presented by (14.13) with the robust controller given by (14.30) and (14.37), the error, $\mathbf{e}(t)$, of the closed-loop control can be asymptotically stable by applying the rear steering as shown in (14.65) through (14.66) and employing the brake distribution following (14.67) through (14.68). The control loop of the hybrid brake and rear-steer controller is shown in Figure 14.6. The advantage of the rear steering here is to be able to provide a quick and limited yaw moment, and to allow the brake control to generate the rest of yaw moment needed to stabilize the vehicle. A case study will be described in the following section.

14.1.5   A Case Study of Lane-Change Maneuver

In the following, we study a driving maneuver with the robust yaw stability control, which begins with the brake-only intervention and later considers the hybrid brake and rear-steer intervention. The study is based on the simulation environment of LabVIEW® and CarSim®. The driver's input of the steering wheel angle is modeled as a sine wave to simulate the driver's steering behavior during an emergency lane change. The steering wheel angle is the only input for the vehicle model running in CarSim.

The simulation is running in LabVIEW with a vehicle model inputted from the CarSim simulation package. Additionally, the CarSim model was defined to export its data to LabVIEW. A pickup truck model in CarSim is selected, and its model parameters are unknown to the controller. The controller uses the following key parameters: $M = 2280$ kg, $I_z = 5050$ kgm², $l_f = 1.39$ m,

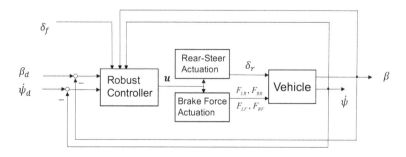

FIGURE 14.6   Hybrid brake and rear-steer control.

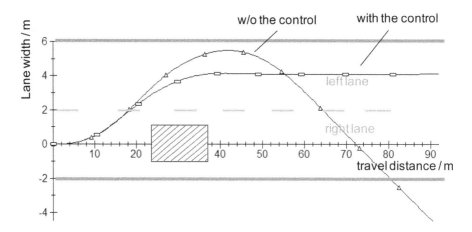

FIGURE 14.7   Top view on the simulated vehicle path during the lane change maneuver.

$l_r = 1.81$ m, $C_{\alpha f} = 100000$ N / rad, $C_{\alpha r} = 160000$ N / rad. The vehicle maneuver starts at a speed of $v = 80$ km / h.

Example of the Brake-Only Actuation

The lane change is a typical vehicle driving maneuver on roads. In some urgent situations, the driver must react and change the lane quickly to avoid an obstruction. Figure 14.7 illustrates the lane change by a vehicle from the right lane to the left lane. In the right lane, an unexpected obstacle is sighted, the obstacle could be any object or a stopped car on the road. At the travel distance zero in Figure 14.7, the driver abruptly reacts to steer the vehicle to the left. Most likely in this situation the vehicle would skid and the driver would panic. Figure 14.8 shows the steering input animating a sudden reaction the driver may apply in such critical condition. The driver steers quickly to the left lane and then steers back in order to keep straight in the left lane. Under this steering input the vehicle is oversteered, and may skid sideways. If the driver loses the control (see w/o the control in Figure 14.7), the vehicle will move to an unwanted direction, as the simulation result shows. If the yaw stability control is applied, the vehicle will remain in control and move to the left lane maintaining its stable condition.

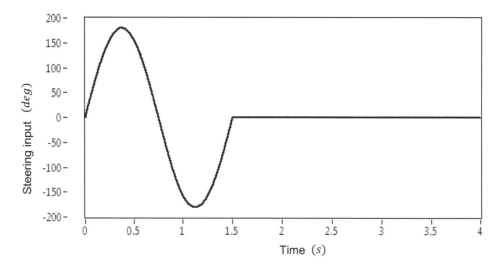

FIGURE 14.8   The input of the steering wheel.

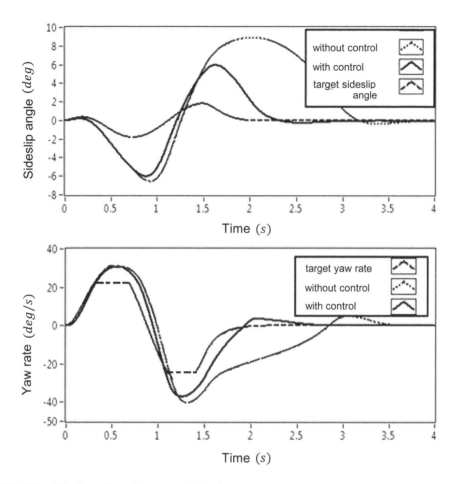

FIGURE 14.9   Sideslip angle and yaw rate during the maneuvers.

Figure 14.9 illustrates the vehicle sideslip angles and yaw rates in both maneuvers, with and without the control. The dashed line indicates the target yaw rate and sideslip angle, respectively. The yaw rate target hits the limit bounded by the tire-road friction when reaching above 22° / s. The sideslip angle target is below its limit during the maneuver. The dotted line in Figure 14.9 corresponds to the unstable trajectory path shown before in Figure 14.7, in which the vehicle is driven without the yaw stability control. In this case, the sideslip angle and the yaw rate enlarge while turning and cannot come back on track by the steering input from the driver. The solid line in Figure 14.9 shows the controlled outputs when the stability control of the vehicle is in action, where the actual yaw rate and slip angle are controlled to follow the targets. Even though the control targets are not exactly met, the vehicle remains stable. This emphasizes the goal of the control – to stabilize the vehicle. The accuracy of the control to the target becomes secondary. However, such conclusion should be assessed in each road conditions and driving situations.

The lateral acceleration and speed of the vehicle during the maneuvers are shown in Figure 14.10. The maneuver with the control engaged shows that the lateral acceleration remains in the range because the sideslip angle and the yaw rate are controlled.

In order to detail the brake-only control intervention and actuation, the total amount requested for the yaw moment is plotted in Figure 14.11, where the requested torque is expressed through the control input u_2. The calculated control input u_2 indicates the yaw moment needed for achieving the yaw target. During the time period from $t = 0.55$ s to $t = 1.05$ s, u_2 is negative and requests the

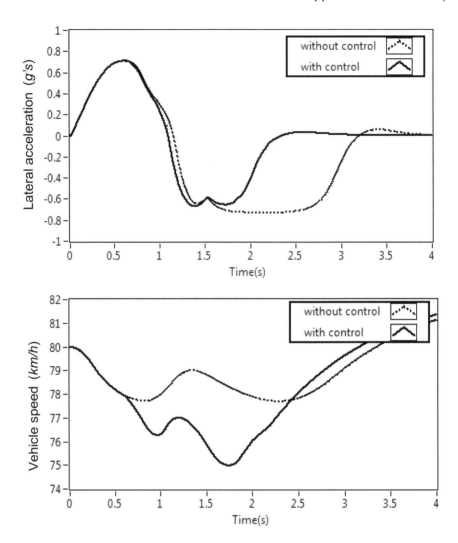

FIGURE 14.10 Acceleration and speed during the maneuvers.

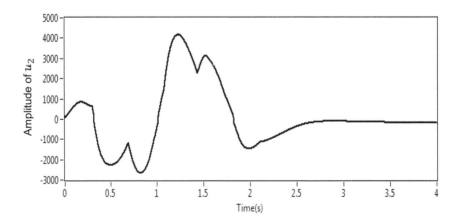

FIGURE 14.11 Total yaw moment requested.

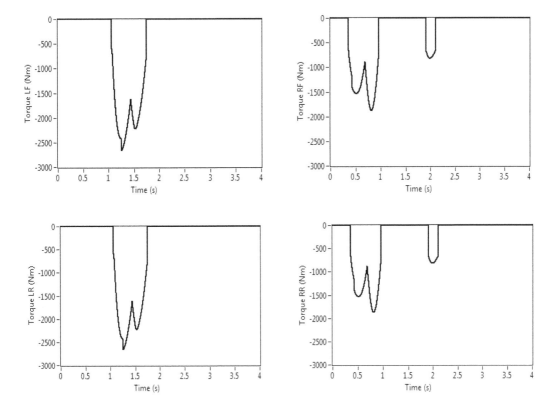

FIGURE 14.12 Braking torque at each wheel.

braking on the right side of the vehicle. The applied braking torques are shown in Figure 14.12. In this example, the braking torque on both right-front (RF) and right-rear (RR) wheels are equally distributed, the maximum magnitude of the torque is below 2000 Nm.

On the other hand, from $t = 0.55$ s up to $t = 0.8$ s the vehicle is oversteered because the actual yaw rate is higher than the target while it is understeered from $t = 0.8$ s to $t = 1.05$ s in the negative direction of the yaw rate. The right-side braking lowers the yaw rate and the sideslip angle in oversteer, then increases the yaw rate and the sideslip angle when vehicle is in understeer. In similar way, the left-side braking is needed when u_2 is positive. From $t = 1.05$ s to $t = 1.8$ s the vehicle is oversteered all the time, and the left-side braking lowers the yaw rate and sideslip angle. The torque on both left-front (LF) and left-rear (LR) wheels are equally distributed below 2600 Nm. One can see that the decision on which side to brake is not based upon oversteer or understeer of the vehicle. The decision is based on the direction of the calculated yaw moment.

In this example, the control starts once the control error is above a very low threshold in order to demonstrate the whole control behavior. In practice, the pre-defined threshold needs to be experimentally verified and programmed for the control intervention so that the control is not activated frequently.

The brake actuation plays an essential role in the control implementation. From today's technology, the brake actuation is mostly a hydraulic system in passenger cars. A hydraulic unit, called modulator, is installed between the master cylinder and the wheel brakes. Based on the control command from the above-designed controller, the hydraulic modulator builds up the pressure to create the braking torque. The response time of the pressure buildup can be large, especially when high pressure is required. This will impact on the control performance, i.e., the controller takes a longer time to stabilize the vehicle.

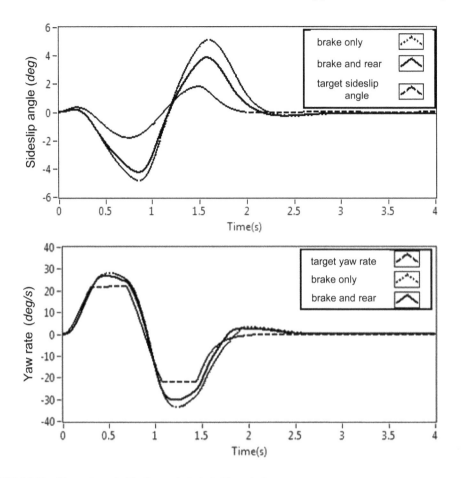

FIGURE 14.13   Yaw rate and sideslip angle in hybrid control.

Example of the Brake and Rear-Steer Actuation

Now we consider the hybrid intervention in which the rear steering is engaged simultaneously with the brake actuation. The exact same maneuver as in the previous section is executed, the yaw moment requested by the controller will be partially provided by the rear-steer and the rest is done by the brakes. As described in the previous section, the requested yaw moment executed by the rear-steer system is aligned with the requested control input of u_2 with taking in consideration the direction of rear-steer angle δ_r. In this example, the maximum angle of $|\delta_r| = 1°$ is chosen.

Figure 14.13 illustrates the sideslip angles and the yaw rates for the same maneuver with the brake-only control and the brake with the rear-steer control. The dashed line indicates the target yaw rate and the target sideslip angle as before. The dotted line indicates the brake-only control while the solid line shows the brake and the rear-steer control. Both the actual yaw rate and the actual sideslip angle follow the targets better with the hybrid brake and rear-steer control. The yaw rate reacts smaller in the amplitude so that the vehicle is not skidding away as much as with the use of the brake-only control. The vehicle sideslip angle is also better controlled. This controlled behavior of the vehicle can be actually expected according to the analysis of the rear steering presented in Chapter 9.

Since the rear-steer covers a part of the total yaw moment requested, the braking torque at each wheel is also reduced as seen by comparing Figure 14.14 with Figure 14.12.

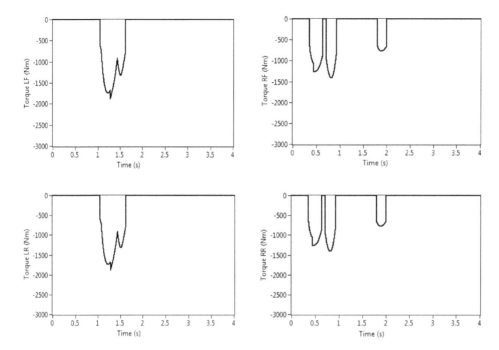

FIGURE 14.14 Braking torque at each wheel in the hybrid brake and rear-steer control.

14.1.6 Yaw Stability Control in Autonomous Vehicle

So far, the yaw stability control in human driven vehicles was presented in the book. In this case the front steering input and the path-following are actually controlled by a driver. The yaw stability controller is an additional system to enhance vehicle stability when the vehicle becomes unstable in extreme driving situations. The same method can be employed for autonomous vehicles, i.e., the yaw stability control needs to be implemented in addition to a path-following control in which the vehicle is controlled to follow the path defined.

In autonomous vehicles, the steering and the powertrain as well as the braking controls are generated as automated processes based on available sensors and advanced control algorithms. In extreme driving situations, a completely autonomous vehicle (no driver involved) not only needs to be controlled stable, but also to follow the road path at the same time, such as collision avoidance or emergent lane change on the road. In such situations, the path is still generated in the first place. Ideally, therefore, the controller should have the capability to meet both requirements on the stability control and path following. In the following, we propose a yaw stability control by extending the path-following control designed in Section 13.2. to the torque vectoring between the front wheels and between the rear wheels in an autonomous vehicle. In order to have a complete explanation, we reiterate first the key design features of the method of *path-following control via rear torque vectoring* with an emphasis on the wheel torque distribution by the extended control actuators at the end.

As described in Section 13.2, the vehicle model is based on the assumption that the influence of the longitudinal tire forces of the steered front and rear wheels on vehicle lateral dynamics is less significant comparing to the influence of the lateral tire forces, so that the controllable inputs were reduced to the torque differential at the rear wheels. That assumption can still be acceptable for autonomous vehicles since the steer angle change is performed in a controlled way in those vehicles. The model approximation is sufficient for the control accuracy, as long as the stabilization of the vehicle can be achieved, which is the goal of the yaw stability control.

Using the Method of Path-Following Control

Consider the model given (13.23) with adding the front brake as one of the control actuators:

$$
\begin{bmatrix} \dot{\beta} \\ \ddot{\psi} \\ \dot{\psi}_L \\ \dot{y}_L \end{bmatrix} = \begin{bmatrix} -\dfrac{C_{\alpha f}+C_{\alpha r}}{Mv} & -1-\dfrac{C_{\alpha f}l_f - C_{\alpha r}l_r}{Mv^2} & 0\ 0 \\ -\dfrac{C_{\alpha f}l_f - C_{\alpha r}l_r}{I_z} & -\dfrac{C_{\alpha f}l_f^2 + C_{\alpha r}l_r^2}{I_z v} & 0\ 0 \\ & 0\ 1\ 0\ 0 & \\ & v\ l_s\ v\ 0 & \end{bmatrix} \begin{bmatrix} \beta \\ \dot{\psi} \\ \psi_L \\ y_L \end{bmatrix}
$$

$$
+ \begin{bmatrix} \dfrac{C_{\alpha f}}{Mv} \\ \dfrac{C_{\alpha f}l_f}{I_z} \\ 0 \\ 0 \end{bmatrix} \delta_f + \begin{bmatrix} 0 \\ \dfrac{d_t}{I_z} \\ 0 \\ 0 \end{bmatrix} \left[(F_{x\_RF} + F_{x\_RR}) - (F_{x\_LF} + F_{x\_LR}) \right] + \begin{bmatrix} 0 \\ 0 \\ -v \\ 0 \end{bmatrix} \dfrac{1}{R_T}
$$

$$(14.70)$$

Here we keep the same assumptions as in Section 13.2 except that the forces now are not equally distributed between the left and right wheels of the front axle. In extreme driving condition, in which the yaw stability control is supposed to intervene, the engine torque should be turned off, i.e., $T_a = 0$, in order to avoid any increase of the vehicle speed. Therefore, the front and rear forces in the model are merely braking forces. The control actuators are the front steering and braking forces on all wheels. Define the brake actuator as

$$
\Delta T = \left[(F_{x\_RF} + F_{x\_RR}) - (F_{x\_LF} + F_{x\_LR}) \right] r_w
$$

$$(14.71)$$

here, ΔT actually is the torque differential between the total torques on the right side and left side wheels. As shown later in this subsection, the required torque vectoring, represented by torque differential ΔT, will result in either the right side braking or the left side braking of the vehicle. The braking torque distribution will end up between the front and the rear wheels on one side of the vehicle. The cascade control structure, which was presented before in Section 13.2, is still the same as the original path-following control with rear torque vectoring, adding the front braking as an additional control actuator. Figure 14.15 shows the cascade control diagram with the additional actuator.

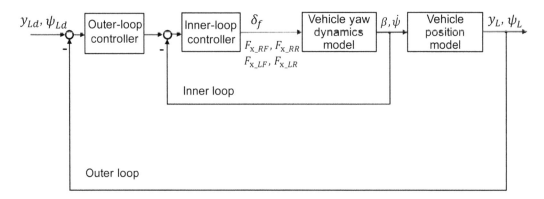

FIGURE 14.15 Diagram of yaw stability cascade control.

The desired sideslip angle and the yaw rate for the inner-loop control are the controller outputs of the outer-loop, as given by (13.35)

$$
\begin{bmatrix} \beta_d(t) \\ \dot{\psi}_d(t) \end{bmatrix} = \begin{bmatrix} -k_{out1}\ \psi_L - k_{out2}\ y_L + \beta_{st} - (\beta_{st} + \frac{l_s}{R_T})k_{out1} \\ -k_{out3}\ \psi_L - k_{out4}\ y_L - (\beta_{st} + \frac{l_s}{R_T})k_{out3} + \frac{v}{R_T} \end{bmatrix} \tag{14.72}
$$

Going through the same derivation process as shown in Section 13.2 by using (13.35) through (13.47) to design the inner-loop control, the yaw dynamics controller becomes very close in its action to the controller given by (13.47)

$$
\begin{bmatrix} \delta_f \\ \Delta T \end{bmatrix} = \begin{bmatrix} -k_{in1} & -k_{in2} \\ -k_{in3} & -k_{in4} \end{bmatrix} \begin{bmatrix} \beta - \beta_d \\ \dot{\psi} - \dot{\psi}_d \end{bmatrix} + \mathbf{B}^{-1}\left(\begin{bmatrix} \dot{\beta}_d \\ \ddot{\psi}_d \end{bmatrix} - \mathbf{A} \begin{bmatrix} \beta_d \\ \dot{\psi}_d \end{bmatrix} \right) \tag{14.73}
$$

with the only difference in actuator ΔT. If the braking torques can be distributed to all wheels so that controller equation (14.73) is satisfied during the control time period, then yaw stability is guaranteed meaning that the sideslip angle and the yaw rate will reach the steady state $\begin{bmatrix} \beta_d & \dot{\psi}_d \end{bmatrix}^T$ while the vehicle keeps the path followed.

Braking Torque Distribution

For the distribution of torque differential ΔT between the wheels, the same control method described previously in Section 14.1.4 is used here. The method was introduced for the yaw stability control of human driven vehicle, and offers a flexible braking torque distribution between all wheels. Since the vehicle dynamics model used for human driven vehicle was different, the result of the torque distribution is also different. The braking torque distribution here for autonomous vehicle relies on the same physical meaning of ΔT as u_2 in human driven vehicle, but mathematically it results in different forms due to the model difference.

In order to achieve control target (14.72), the torque differential, ΔT, is calculated by the controller (14.73) to specify the required yaw moment. The right and left braking contribute to the yaw moment in opposite direction corresponding directly to whether an increase or decrease in the yaw moment is needed. Since only the braking forces are to apply, (14.71) can be directly used to determine, on which vehicle side the brake should be applied. There are three cases for ΔT can be introduced:

$\Delta T < 0$: Right braking.
The condition $\Delta T < 0$ indicates a decrease of the yaw moment, hence, the yaw rate. In this case, the vehicle is supposed to turn clockwise, and the yaw rate decreases during the left turn but increases in the negative direction during the right turn of the vehicle. The right side braking will effectively be used as (14.71) results in

$$
F_{x\_RF} + F_{x\_RR} = \frac{\Delta T}{r_w} < 0 \tag{14.74}
$$

$\Delta T > 0$: Left braking.
The condition of $\Delta T > 0$ indicates an increase of the yaw moment, hence the yaw rate. In this case, the vehicle is supposed to turn counterclockwise, and the yaw rate increases during the left turn but decreases in the negative direction during the right turn of the vehicle. The left side braking will effectively contribute to that. Equation (14.71) leads to

$$
F_{x\_LF} + F_{x\_LR} = -\frac{\Delta T}{r_w} < 0 \tag{14.75}
$$

$\Delta T = 0$: No braking.
For $\Delta T = 0$, there is no torque needed for the yaw movement thus no brake apply.

Once the total braking torque on one side is calculated, it can be distributed between the front and rear brakes. The individual torque amount is determined during the control action by analyzing the vehicle dynamic behavior characterized by the vehicle speed and the understeer gradient.

In conclusion, the method previously designed for the path-following control via rear torque vectoring was extended to yaw stability control of autonomous vehicles. The difference between these two controls exists in the application of the actuators. The autonomous vehicle stability control performs the torque vectoring through all brakes while the engine torque is zero. It still guarantees the path-following, but slows down the speed of the vehicle due to the braking. However, during the control, autonomous vehicles can reach the steady-state of the yaw rate and sideslip angle. The method proposed in this section has a significant advantage for autonomous vehicle applications because the method can achieve the same target as the yaw stability control designed for human driven vehicles. At the same time, the employment of the method reduces the complexity of autonomous vehicle control architecture by eliminating another stability controller and a coordination between vehicle controllers. The controller output allows for the sidewise braking and the flexible torque distribution between the front and rear brakes.

14.2 ROLLOVER CONTROL

In vehicle safety technology there was always a greater focus on vehicle rollover, and it has become an important research and engineering area in vehicle dynamics and control.

In the following sections the vehicle rollover behavior is first analyzed, and a controller is then designed to prevent rollover. The rollover analysis is concerned with the question: what are the vehicle and operational parameters to characterize the rollover behavior and what is the vehicle response to changes of those parameters? Control design mostly concerns how to improve the vehicle response. The vehicle response includes a transient period with an oscillatory and overshoot behavior as well as the steady state period. We begin with characteristics of the roll model so that one can visualize how a vehicle can be controlled to modify its roll response in a desirable direction.

14.2.1 ROLLOVER ANALYSIS

The rollover involves a complex interaction of forces acting on the vehicle as seen from material discussed in Chapter 10. It can be only modeled until a certain point. Once a rollover happens, the roll movement cannot be predicted and modeled. We continue the study of the normal dynamics, and further view the rollover in an analytical way.

As we studied the normal dynamics in Chapter 10, the deflections of the suspension and tires are also considered there. In Figure 10.6 the roll plane of a vehicle in cornering maneuver is illustrated, where the defections of the suspension and tires modeled by the dampers and springs. While the total lateral force acts in the ground plane to counterbalance the lateral accelerations at the sprung and unsprung masses of the vehicle, a roll moment is created. If the roll continues, the inside wheels lift up and the weight load of the vehicle shifts toward the outside wheels. The rollover begins when the inside wheels leave the ground.

Using the three-dimensional dynamic equations (10.55), (10.56) and (10.54) and notations of technical parameters from Chapter 10, we have

$$\dot{\beta} - \frac{m_v(h_s - h_r)}{Mv}\dot{\phi} = -\left(1 + \frac{C_{\alpha f}l_f - C_{\alpha r}l_r}{Mv^2}\right)\dot{\psi} - \frac{C_{\alpha f} + C_{\alpha r}}{Mv}\beta - \frac{m_v g}{Mv}\phi + \frac{C_{\alpha f}}{Mv}\delta_f \quad (14.76)$$

$$I_{z\_v}\ddot{\psi} = -\frac{C_{\alpha f}l_f^2 + C_{\alpha r}l_r^2}{v}\dot{\psi} - \left(C_{\alpha f}l_f - C_{\alpha r}l_r\right)\beta + C_{\alpha f}l_f\delta_f \quad (14.77)$$

$$I_{x\_v}\ddot{\phi} + m_v(h_s - h_r)^2\ddot{\phi} = m_v a_y(h_s - h_r) + m_v g(h_s - h_r)\phi - d_\phi\dot{\phi} - c_\phi\phi \quad (14.78)$$

The first two equations, (14.76) and (14.77), present the sideslip angle and the yaw response, respectively, that is similar to response of the bicycle model presented in the study of lateral dynamics in Chapter 8. However, the roll angle gets involved into the sideslip angle response. This is because the yaw movement produces the lateral acceleration, which in turn causes the roll motion. The roll motion alters the sideslip angle, and hence the yaw response through the variation of the lateral tire forces arising from the lateral weight load transfer and suspension deflection. The third equation, (14.78), states the roll response to the lateral acceleration as the input. In today's vehicle technology, a lateral acceleration sensor is already a standard equipment; therefore, this sensor signal can be very useful for modeling vehicle dynamics and its control applications. Because of the availability of lateral acceleration measurement, (14.78) is decoupled from its dependency on the yaw and the sideslip angle, so that the equation is in the form of a second-order linear system.

To analyze the dynamic response of the roll motion, we start with the *steady-state* motion and then view the *transient response* behavior. The steady-state motion is easily understood and provides a direct insight of the vehicle dynamic characteristics and their relationship in the roll plane.

Assume that a vehicle is in a left turn. Due to the flexibility, the outside suspension and tire are compressed from their original equilibrium position, because the weight load on the out wheels is increasing. At the same time, the inside wheel suspension and the tire are released and subjected to a less load than at the equilibrium state. Considering the single-axle model, the steady-state roll angle can be obtained as shown in (10.40). The load transfer is given then by (10.45)

$$\Delta F_N = \frac{\left(c_\phi + m_v g h_r\right) m_v a_v \left(h_s - h_r\right)}{2 d_t \left(c_\phi - m_v g \left(h_s - h_r\right)\right)} + \frac{m_v a_v h_r}{2 d_t} \tag{14.79}$$

When the vehicle is driven at a straight steady condition, each side of the vehicle takes half of the vehicle weight. When rollover begins and the inside wheels lifted up, a half of the vehicle weight from the inside wheels transfers to the outside wheels, i.e.,

$$\Delta F_N = \frac{Mg}{2} \tag{14.80}$$

Considering the sprung mass to be close to the gross mass, $m_v \cong M$, (14.79) becomes

$$\frac{m_v g}{2} = \frac{\left(c_\phi + m_v a_v h_r\right) m_v a_v \left(h_s - h_r\right)}{2 d_t \left(c_\phi - m_v g \left(h_s - h_r\right)\right)} + \frac{m_v a_v h_r}{2 d_t} \tag{14.81}$$

Since the lateral acceleration is the input in the roll motion, (14.81) can be expressed as follows:

$$\frac{a_y}{g} = \frac{d_t}{h_s} \frac{1}{1 + \left(1 - \frac{h_r}{h_s}\right)\left[\dfrac{1 + \dfrac{m_v g h_r}{c_\phi}}{1 - \dfrac{m_v g \left(h_s - h_r\right)}{c_\phi}} - 1\right]} \tag{14.82}$$

where $a_y = a_v$ in steady condition (see Section 10.3). The first term on the right-hand side of the above equation (the fraction of $\frac{d_t}{h_s}$) is determined by two basic dimension parameters of the vehicle – the wheel track and the CG height. The second term is primarily influenced by roll stiffness c_ϕ because of its high value. When the roll stiffness is getting larger, the second term decreases considerably. If the vehicle is considered to be rigid, i.e., $c_\phi \to \infty$, the second term simplifies to 1, and (14.82) has only the first term remained

$$\frac{a_y}{g} = \frac{d_t}{h_s} \tag{14.83}$$

This ratio, the half of the wheel track to the height of CG, indicates the roll stability condition and is named as the *static rollover threshold*. For a rigid vehicle model, the rollover will occur if the lateral acceleration normalized to the gravity acceleration becomes larger than d_t/h_s. The higher the CG and the smaller the wheel track is, the easier the rollover will occur. Taking into account the roll stiffness, c_ϕ, the static rollover threshold is reduced by the second term on the right-hand side of (14.82). Lower stiffness values will make the rollover easier.

Consider both (14.82) and (14.83), the rollover condition is formulated as

$$\frac{a_r}{g} \geq \frac{d_t}{h_s} \frac{1}{1 + \left(1 - \dfrac{h_r}{h_s}\right)\left[\dfrac{1 + \dfrac{m_v g h_r}{c_\phi}}{1 - \dfrac{m_v g(h_s - h_r)}{c_\phi}} - 1\right]} \tag{14.84}$$

where a_r is denoted as the *rollover threshold* of the lateral acceleration.

The steady-state analysis is useful for basic understanding purposes, when only the lateral acceleration is considered as a cause of the rollover. However, the roll velocity and the vehicle lateral acceleration also contribute to rollover, thus *transient response* is also an important aspect of the roll response. Dynamic effects through a sudden or cyclic application of steering can considerably increase the propensity to roll over. In case of a sudden change of the lateral acceleration, a certain period of time will be required for transient response to decay and for the roll angle to level off at the steady value. This transient period can be short or long and excessively oscillatory depending on the vehicle parameters in (14.78).

As described earlier, (14.78) is a standard second-order linear system and behaves as the quadratic lag. Based on its constant coefficients, *damping ratio* ζ can be calculated as

$$\zeta = \frac{d_\phi}{2\sqrt{(I_{x\_v} + m_v(h_s - h_r)^2)(c_\phi - m_v g(h_s - h_r))}} \tag{14.85}$$

Depending on ζ, the roll angle responding to a suddenly applied lateral acceleration can exhibit an oscillatory or nonoscillatory behavior as shown in Figure 11.3 and Figure 11.4, respectively. Thus, parameters d_ϕ and c_ϕ will ultimately determine the behavior of the roll response. Since the lateral acceleration is assumed as a step input, the *rollover threshold* of the roll angle can be calculated by substituting (14.84) into (10.40)

$$\phi_r \geq \frac{d_t}{h_s} \frac{m_v g(h_s - h_r)}{c_\phi} \tag{14.86}$$

If the vehicle is considered as a rigid body, i.e., roll stiffness $c_\phi \to \infty$, the roll angle in (14.86) will be zero. If the *transient response* is oscillatory, the roll angle has an overshoot after it reaches the steady value. The damping ratio affects the amplitude of the oscillation. The higher the damping ratio is, the lower amplitude of the vehicle mass oscillation will get, which implies that the rollover threshold will increase.

The rollover conditions given by (14.84) and (14.86) provide actually only a reference for the rollover thresholds. In control practice, those parameters need to be analyzed and tested through experiments on vehicles or virtual vehicle simulations. The simulation has the advantage that the risk caused by rollover can be avoided and the development time reduces as well. In today's automotive industry, sensor technology has been advanced for recent decades. As mentioned before, the lateral acceleration sensor has become a standard equipment in the most vehicles, and the roll rate sensor is also available on the market. However, the roll angle sensor has not been in use so far. The next section will describe on how the roll angle signal can be estimated based on other available sensor signals.

14.2.2 ROLL ANGLE ESTIMATION

For the estimation, it is to assume that the lateral acceleration and the roll rate signals are available. Now we define a system state as $\mathbf{x} = (\phi \quad \dot{\phi})^T$, the input of $u = a_y$, and the matrices

$$\mathbf{A} = \begin{bmatrix} 0 & 1 \\ \dfrac{-c_\phi + m_v g(h_s - h_r)}{I_{x\_v} + m_v(h_s - h_r)^2} & \dfrac{-d_\phi}{I_{x\_v} + m_v(h_s - h_r)^2} \end{bmatrix}$$

$$\mathbf{B} = \begin{bmatrix} 0 \\ \dfrac{m_v(h_s - h_r)}{I_{x\_v} + m_v(h_s - h_r)^2} \end{bmatrix}, \quad \mathbf{C} = \begin{bmatrix} 0 & 1 \end{bmatrix}$$

then (14.78) can be formulated in the state-space form

$$\dot{\mathbf{x}} = \mathbf{A}\mathbf{x} + \mathbf{B}u$$
$$y = \mathbf{C}\mathbf{x} \tag{14.87}$$

In (14.78), all parameters are constant. Hence, (14.87) is a time-invariant system with an unknown state of the roll angle. In the previous chapter, it is shown that a state observer can be used to estimate unknown states. The observability of system (14.87) can be verified by applying (11.26). As the result, the state observer is designed as

$$\dot{\hat{\mathbf{x}}}(t) = \mathbf{A}\hat{\mathbf{x}}(t) + \mathbf{B}u(t) + \mathbf{K}_o\big(y(t) - \mathbf{C}\hat{\mathbf{x}}(t)\big) \tag{14.88}$$

where $\mathbf{K}_o$ is the gain matrix to be chosen. By using the same lateral acceleration as the input and the output measurement of the roll rate, the estimated state of $\hat{\mathbf{x}}$ converges to $\mathbf{x}$. Figure 14.16 shows the design block diagram that graphically represents (14.88) of the observer. As seen, the difference between the actual roll rate as the model output and the estimate of the model output is feedbacked to the observer. The observer estimates both the roll angle and roll rate signals.

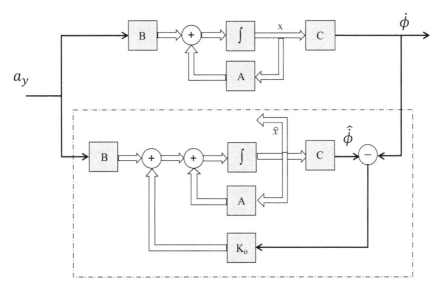

FIGURE 14.16   State observer for the roll angle estimation.

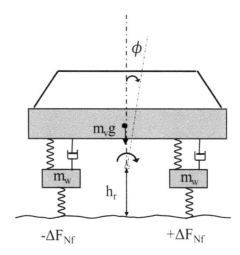

FIGURE 14.17   Rollover plane of a single-axle model.

14.2.3 ROLLOVER CONTROL

The purpose of the rollover control is to minimize the trend of the rollover and to prevent the vehicle rolling over as early as possible. Viewing the roll plane shown in Figure 14.17, the rollover can be understood in terms of the vehicle roll angle. When the wheels on one side of the vehicle start to lift up and the roll angle increases, activating the rollover control keeps the increase of the roll angle as low as possible. The control actuation can be implemented via vehicle steering or braking; we focus here on the brake intervention.

From the control point of view, the control target here cannot be a pre-defined steady state or desired value the controller has to follow since the control action is to stop the roll tendency and prevent the rollover of the vehicle.

From rollover dynamics analysis, it is known that the minimizing of the roll angle can reduce the tendency to the rollover. According to the linear relation given by (14.78), the roll angle will get smaller if the lateral acceleration decreases. In order to achieve this goal, the vehicle speed needs to be reduced. For example, increasing the actual longitudinal braking force will reduce the vehicle speed, and hence the lateral acceleration. By using such force influence, a controller can be defined as follows:

$$
\begin{aligned}
F_{x\_LF} &= p_{LF}F_{LF\_ys} \\
F_{x\_LR} &= p_{LR}F_{LR\_ys} \qquad \text{for left braking}
\end{aligned}
\tag{14.89}
$$

or

$$
\begin{aligned}
F_{x\_RF} &= p_{RF}F_{RF\_ys} \\
F_{x\_RR} &= p_{RR}F_{RR\_ys} \qquad \text{for right braking}
\end{aligned}
\tag{14.90}
$$

where F_{RR_ys}, F_{RF_ys}, F_{LF_ys} and F_{LF_ys} are the braking force of each wheel calculated from yaw stability control, respectively. p_{RR}, p_{RF}, p_{LF} and p_{LF} are positive parameters to be alternatively specified greater than 1.0. In this way, the longitudinal tire force can be enlarged while the potential lateral force is reduced. The decision whether to perform braking on the left or right, can be decided by the yaw rate direction.

Figure 14.18 shows the control implementation in a block diagram. The rollover controller is integrated with the yaw stability controller and processes the data and sensor signals as given by (14.86). The block "Torque transfer" receive all torque requests from the rollover and yaw stability

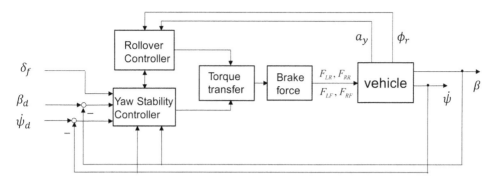

FIGURE 14.18   Rollover control implementation in a block diagram.

controllers, and addresses the priority of the actuation. In most cases, the yaw stability control is already in action before the rollover control activated. While cornering, the vehicle usually exceeds the yaw target before a need in the rollover controller emerges. However, there are still situations where the rollover control is needed without the yaw stability control activation. For example, on a surface with a low friction, the vehicle could drift to the side while the yaw rate and sideslip angle remain acceptable. In this case, the rollover control can select the maximum available braking forces on the surface to reduce the lateral acceleration. At the time, where both controllers are required to be active, the rollover control will have the priority to transmit the torque to the brake actuation.

Rollover Control Simulation

For the case study in Section 14.1.5, a simulation environment in LabVIEW and CarSim was set up to simulate the yaw stability control. Here, the rollover control is demonstrated on the same simulation setup. The simulation is performed for a so-named *sine and dwell* test maneuver. The *sine and dwell* steering input has a shape shown in Figure 14.19. The steering wheel is steered quickly to the first peak and then steered back, where the steering wheel is paused at the second peak for 0.5s. The vehicle data used for the simulation is the same as in Section 14.1.5. An additional weight of 400 kg is loaded up to the vehicle so that the height of the center of the gravity reaches $h_s = 1.3$ m. After the loading, the longitudinal coordinate of CG did not change.

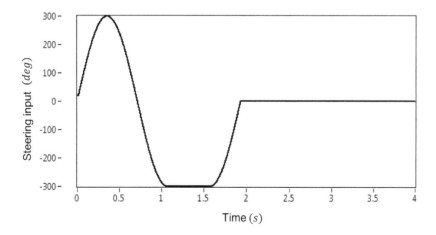

FIGURE 14.19   Sine and dwell steering input.

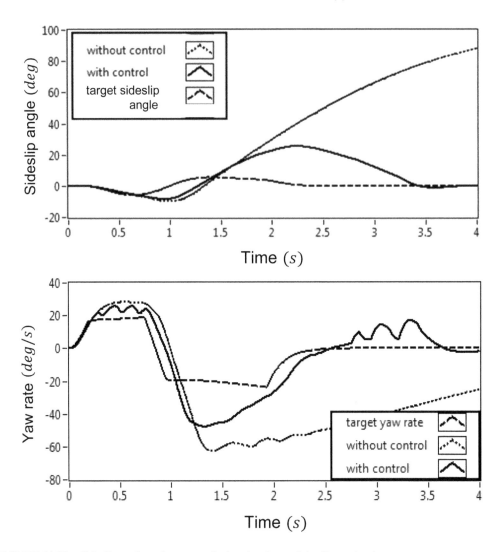

FIGURE 14.20 Sideslip angle and yaw rate during the sine and dwell steering input.

Figure 14.20 illustrates the vehicle sideslip angles and yaw rates in maneuvers with and without the rollover control. The dashed line indicates the targets for the yaw rate and the sideslip angle, respectively. In addition, the yaw rate target hits the limit bounded by the tire-road friction when reaching above 22°/ s. The sideslip angle target is below the limit during the maneuver. The dotted line corresponds to an unstable situation, in which the vehicle is driven without the stability control. The sideslip angle and yaw rate enlarge while the vehicle turning, and the vehicle cannot come back on the track by applying a steering input. The solid line in Figure 14.20 shows the application of the yaw stability control and the rollover control of the vehicle, where the actual yaw rate and sideslip angle are controlled to follow the targets. In the simulation, the rollover threshold is set for $\phi_r = 3^{\circ}$. Accordingly, at $t = 0.25$ s, the rollover begins to develop so that the left wheels lift up, the rollover control becomes activated. During the rollover control, the braking forces coming from the yaw stability controller are overwritten by the ones from the rollover controller, which apply the maximum of the braking forces on the right side wheels. The rollover control is active until $t = 0.8$ s and starts again at $t = 1.4$ s. The activation time moment of the rollover control can be seen in the roll angle analysis presented in Figure 14.22.

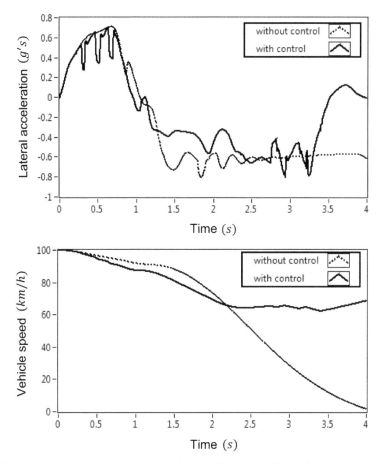

FIGURE 14.21   lateral acceleration and speed during the sine and dwell steering input.

The lateral acceleration and the speed of the vehicle during the maneuvers with the activated controls and without the control are shown in Figure 14.21. The maneuver analysis with the control shows that the acceleration remains in the desired range while the sideslip angle and the yaw rate are controlled. During the rollover control, the lateral acceleration also goes down as shown in Figure 14.21.

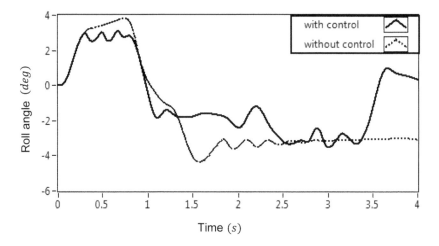

FIGURE 14.22   Roll angle during the sine and dwell steering input.

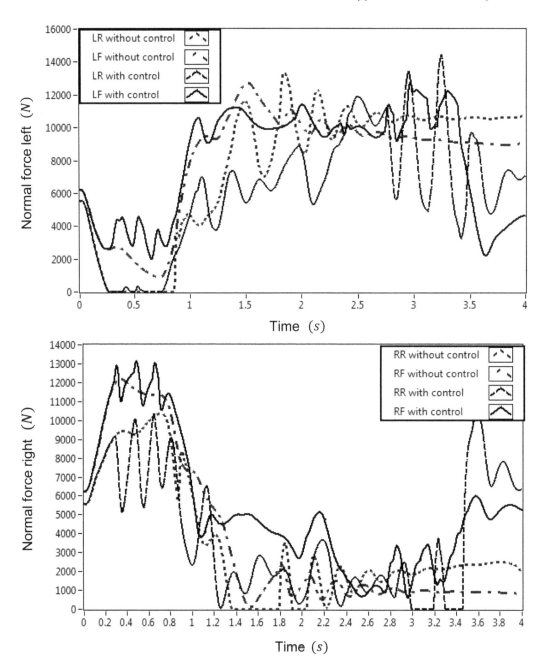

FIGURE 14.23 Normal forces during the rollover control.

The roll angle displayed in Figure 14.22 behaves linear to the lateral acceleration. Observing the maneuver without the rollover control, one can see that the roll angle reaches to the role angle threshold, which activates the rollover control. Figure 14.23 presents the normal reactions on the wheels. Without the control, the wheels on the left side of the vehicle begins to lift up at $t = 0.25$ s. The rollover controller engages and holds the wheels on the ground as the solid line and the dashed line show the normal reactions at the front and rear wheels, respectively. Afterwards the vehicle is steered to the opposite direction. At $t = 1.4$ s, the right side wheels lift up again if the control is not in action, and the vehicle tends to roll over. If the rollover control is active, the right wheels remain on the ground.

14.3   STABILIZATION OF VEHICLE-TRAILER SYSTEM

Stability control of vehicle-trailer systems has been widely studied in technical literature. The design methods involved braking as well as a steering control. Typically, a state feedback controller is employed with the use of a reference model or steady-states as the control targets. This kind of control may have two disadvantages. First, the control is based on all states which have no correlation between the towing vehicle and the trailer swing that can be controlled in a decoupled way. Second, the vehicle and the trailer in the modeling are usually viewed as two rigid bodies coupled through the hitch. However, the stiffness of mechanical coupling between these two is not taken into account, and may impose significant influence to the dynamics of the movement. Ultimately, these factors will impact the effectiveness of the control.

While vehicle is towing a trailer, the trailer could start to oscillate around the hitch. It can be caused by road conditions, a lane change maneuver, a quick steering input, etc. The vehicle tends to become unstable as the oscillation is getting stronger. Damping the oscillation will prevent the instability of both the towing vehicle and the trailer. Therefore, the technical idea here is to focus on the trailer itself by observing trailer dynamics and controlling its motion. An approach presented in the following section considers the towing of a semi-trailer with the purpose to mitigate the trailer's oscillation maintaining the stability of the vehicle-trailer system.

14.3.1   TRAILER STABILIZATION THROUGH REAR STEERING

Figure 14.24 illustrates the single-track model of a vehicle and semi-trailer system, which was introduced in Chapter 8. Once a yaw movement of the towing vehicle occurs, the trailer will swing to a side. Therefore, the oscillation is resulted in the variation of the hitch angle. The purpose of the

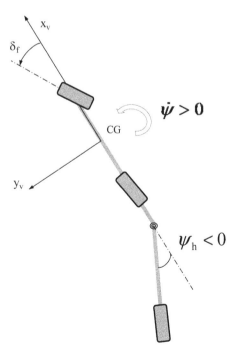

FIGURE 14.24   Top view of the vehicle and trailer model.

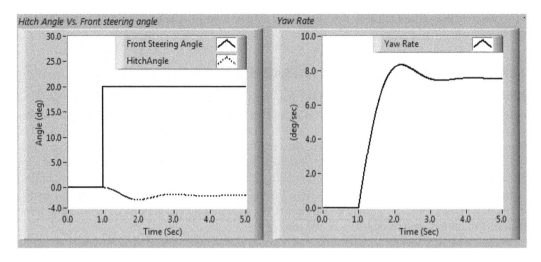

FIGURE 14.25   Hitch angle and the yaw rate responses to the front steering input.

trailer stabilization is to minimize the trailer's oscillation, and thus to prevent the instability of both the towing vehicle and the trailer as early as possible.

From the geometry point of the view, any yaw movement of the towing vehicle can prompt a change of the hitch angle. If the vehicle turns to the left from a straight line, $\dot{\psi} > 0$, and the hitch angle will decrease, i.e., $\psi_h < 0$. While the vehicle turns to the right, $\dot{\psi} < 0$, the hitch angle will increase, hence $\psi_h > 0$. This pattern is also reflected in the dynamic behaviour of the hitch angle responding to the yaw rate change of the vehicle. The yaw rate change can be viewed as a reaction to the steering input. Figure 14.25 shows the dynamic response of the hitch angle to the front steering input in a LabVIEW simulation. While the vehicle is moving in a straight line, the front steering angle is turned through 20° to the positive direction beginning at t = 1 s. The yaw rate immediately goes in a positive jump while the hitch angle goes to negative.

The model parameters of a light truck (towing vehicle) and a semi-trailer are used in the simulation: $M = 2817$ kg, $I_z = 10712$ kgm^2, $M_t = 638$ kg, $I_t = 2150$ kgm^2, $l_f = 1.7$ m, $l_r = 1.2$ m, $l_h = 0.9$ m, $l_t = 3.6$ m, $l_k = 0.1$ m $C_{\alpha f} = 66000$ N / rad, $C_{\alpha r} = 144000$ N / rad, $C_{\alpha t} = 48000$ N / rad. The vehicle is moving at a speed of v = 80 km / h.

The simulation is run in LabVIEW as shown in Figure 14.26. Here, the vehicle and the semi-trailer model given by (8.56) is implemented in MathScript in the block diagram. The While-loop and Event-Structure are used to manage the parameter change and to repeat the simulation. The model parameters are to insert in the front panel. The notations of the technical parameters used in Figure 14.26 and further in this section can be found in Chapter 8. The block diagram associated with Figure 14.26 is detailed by Figure A6.1 in Appendix A.6.

Steady State of Hitch Angle

In vehicle lateral dynamics as well as vehicle roll dynamics, the steady-state behavior has been an important component in control designs. The steady-states of the sideslip angle and the yaw rate are defined as the desired targets of the yaw stability control. It is readily to see in vehicle-trailer model (8.56) that the trailer is reaching steady state when the sideslip angle and the yaw rate are steady, i.e., the hitch angle becomes steady. Hence, the steady hitch angle can be viewed as a desired target for stabilizing the trailer. Note that the front steering is a known input from the driver while the rear

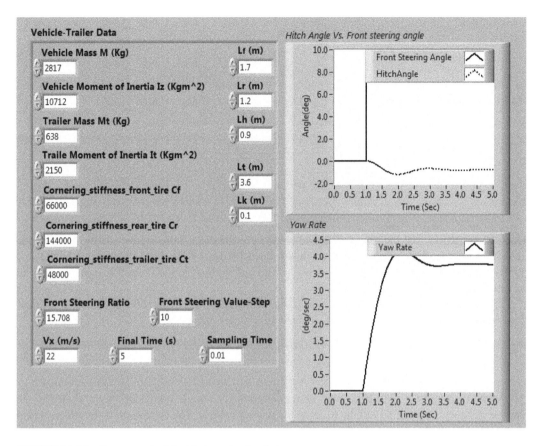

FIGURE 14.26   Front-panel of the hitch angle simulation.

steering is taken as the control input. In the steady state, the state variables are constant, i.e., $\dot{x} = 0$ on the left side of (8.56). From this condition in the open-loop control, we have

$$\left(C_{\alpha f}+C_{\alpha r}+C_{\alpha t}\right)\beta+\left((M+M_t)\mathrm{v}+\frac{C_{\alpha f}l_f-C_{\alpha r}l_r-(l_k+l_r+l_h+l_t)C_{\alpha t}}{\mathrm{v}}\right)\dot{\psi}-C_{\alpha t}\psi_h=C_{\alpha f}\delta_f\ (14.91)$$

$$\left(C_{\alpha f}\left(l_r+l_h+l_f\right)+C_{\alpha r}l_h\right)\beta+\left(\frac{C_{\alpha f}l_f\left(l_f+l_r+l_h\right)-C_{\alpha r}l_rl_h}{\mathrm{v}}+M\mathrm{v}(l_r+l_h)\right)\dot{\psi}=C_{\alpha f}\left(l_f+l_r+l_h\right)\delta_f\ (14.92)$$

$$C_{\alpha t}\left(l_k+l_t\right)\beta+\left(-\frac{C_{\alpha t}\left(l_t+l_r+l_h+l_k\right)\left(l_t+l_k\right)}{\mathrm{v}}+M_t\mathrm{v}l_t\right)\dot{\psi}-\left(l_k+l_t\right)C_{\alpha t}\psi_h=0\qquad(14.93)$$

Combining (14.91) and (14.93) through the hitch angle, it results in

$$\left(C_{\alpha r}+C_{\alpha f}\right)\beta+\left((M+M_t)\mathrm{v}-M_t\mathrm{v}\frac{l_t}{(l_k+l_t)}+\frac{C_{\alpha f}l_f-C_{\alpha r}l_r}{\mathrm{v}}\right)\dot{\psi}=C_{\alpha f}\delta_f\qquad(14.94)$$

Now, (14.92) and (14.94) can be solved for the steady yaw rate and sideslip angle. However, it is more interesting to know how those steady states relate to the steady states of the towing vehicle when moving without the trailer. For this purpose, we investigate here a special case to gain the insight into the dynamics of the towing vehicle and the trailer.

Assuming

$$l_t \approx (l_k + l_t) \tag{14.95}$$

i.e., the CG of the trailer is on the normal axis of the trailer wheel, and then solving (14.92) and (14.94), the steady states of the yaw rate and sideslip angle have the solution yield in

$$\dot{\psi}_{st} = \frac{\text{v}\,\delta_f}{(\text{v}^2 U_g + l_f + l_r)} \tag{14.96}$$

$$\beta_{st} = \frac{-l_f M \text{v}^2 + C_{\alpha r}\left(l_f + l_r\right)l_r}{C_{\alpha r}\left(l_f + l_r\right)\left(\text{v}^2 U_g + l_f + l_r\right)}\delta_f \tag{14.97}$$

One can recognize that the above-obtained results are the exact same as the steady states of the vehicle bicycle model without trailer, as shown in (9.39) and (9.42). Physically, it means that the steady states of the towing vehicle are identical with and without trailer if the trailer can be viewed as a lumped mass on the wheels.

Following (14.93) with condition (14.95), the steady hitch angle, denoted as ψ_{h_st}, is a linear combination of the steady sideslip angle and the steady yaw rate

$$\psi_{h\_st} = \beta_{st} - \left(\frac{\left(l_t + l_r + l_h\right)}{\text{v}} - \frac{M_t \text{v}}{C_{\alpha t}}\right)\dot{\psi}_{st} \tag{14.98}$$

The second term on the right side of the above equation is an angle subtended by the trailer length and the distance from the CG to the hitch of the towing vehicle. The third term is the trailer wheel slip angle, α_t, which is caused by the lateral force of the trailer tire. The above result verifies (8.50), which presents the relation of the hitch angle to the vehicle sideslip angle. Figure 14.27 displays a geometrical explanation of the steady state of the vehicle-trailer system. Vehicle velocity $\mathbf{v}$ and trailer velocity $\mathbf{v}_t$ are aligned with tangent line at their position on the trajectory path curve, respectively. Considering each angle defined in the vehicle-fixed coordinate system, the sum of the trailer wheel slip angle and the hitch angle is equal to the sum of the sideslip angle and the subtended angle $(l_t + l_r + l_h)/R_T$:

$$\alpha_t - \psi_{h\_st} = -\beta_{st} + \frac{\left(l_t + l_r + l_h\right)}{R_T} \tag{14.99}$$

which confirms (14.98).

Rear Steer Control Design

From the study above, we know that the steady state of the towing vehicle is not influenced by the trailer under the assumption that trailer is a lumped mass at the wheel center. Therefore, stability of the towing vehicle can be controlled independently from the trailer. That will add a significant advantage to the overall control system on the towing vehicle because in the yaw stability control of the towing vehicle can remain the same as it was designed in the previous sections of the book, i.e., independent of the presence of the trailer.

Based on the discussion in the beginning of this Section 14.3.1, the yaw rate of the towing vehicle has a significant impact on the hitch angle, where the rear steering can affect the yaw rate directly. Those dynamic characteristics can be utilized for the control strategy of trailer stability. When trailer sway occurs, the trailer moves side to side, i.e., the hitch angle is oscillating. If the control is designed to follow a defined state, for example an equilibrium point, the control needs to actuate

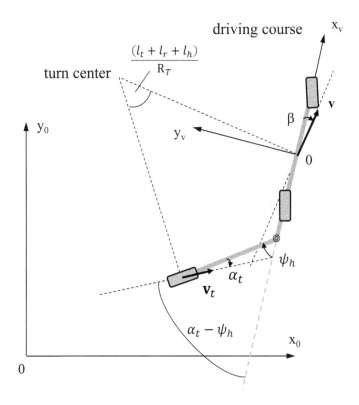

FIGURE 14.27 Geometrical motion display of the steady vehicle-trailer system.

over the time as long as the trailer is oscillating. Therefore, the control target here is not to control the hitch angle exactly to the steady hitch angle, but to bound it in a desired range thus mitigating the oscillation in order to prevent the trailer becoming unstable.

We define the controller as

$$
\delta_r = \begin{cases} -\delta_{rc}\ \mathrm{sgn}\left(\psi_h - \psi_{h\_st}\right) & \text{if } \left|\psi_h - \psi_{h\_st}\right| \geq \psi_{h\_max} \\ 0 & \text{if } \left|\psi_h - \psi_{h\_st}\right| < \psi_{h\_max} \end{cases}
\tag{14.100}
$$

where δ_{rc} is a parameter for the control gain of the rear steering. In this way, we basically pre-define the rear steer angle, which shall produce a yaw moment to increase or decrease the yaw rate of the trailer; for example in case of $\psi_h < 0$, the rear steer angle results in $\delta_r > 0$. A positive rear steer angle shall create a negative yaw moment. The second condition in (14.100) is to ensure that the rear steering should not be performed if the hitch angle is within the pre-defined range by ψ_{h_max}, which is a threshold parameter. Due to safety reasons, the rear steer must be limited to a small range, typically $\delta_r < 4°$. In the controller, it is not considered whether the rear steer is in-phase or out-of-phase with the front steering. In both cases a yaw moment will be created, which is needed to dampen the oscillation of the trailer. The defined controller is actually a *P-Controller* with a constant error correction because it only reacts to the direction of the oscillation without involving the amplitude of oscillation. The advantage is that the rear steering control for the trailer is decoupled from the towing vehicle, i.e., it can take place without the towing vehicle being controlled. However, in case of the towing vehicle being controlled at the same time, the required rear steer input needs to be considered and coordinated with the towing vehicle control.

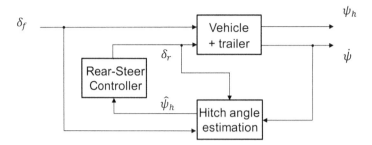

FIGURE 14.28   Trailer stabilization.

One of the key elements of the above-defined controller is the measurement of the hitch angle, which triggers the control. However, in today's automotive technology, the hitch angle sensor has not been used commonly so far. Therefore, the approach here is to estimate the hitch angle based on other available sensor signals. As learned in the previous chapters, the Kalman-filter would be an appropriate method for this kind of application. Should be the estimated hitch angle used, the closed-loop control can be used as it is shown in Figure 14.28. Considering the front and rear steering as the inputs, and the yaw rate signal as the output, the hitch angle can be estimated. The next section details the estimation block shown in Figure 14.28.

14.3.2   HITCH ANGLE ESTIMATION

The method is based on the single-axle trailer model used in the previous sections. For the estimation, it is assumed that the towing vehicle yaw rate is available. The vehicle-trailer model is given by (8.56), in which $\mathbf{x} = (\beta \quad \dot{\psi} \quad \dot{\psi}_h \quad \psi_h)^T$ is the state vector and $\mathbf{u} = (\delta_f \quad \delta_r)^T$ is the control input, respectively. The hitch angle is not measurable since no such kind of sensor exists. The state vector, hence the hitch angle, will be estimated by using the Kalman filter. First, in order to apply the Kalman filter, (8.56) is re-formulated into a standard state-space form.

Keeping the same system matrix as

$$
\mathbf{M}_T = \begin{bmatrix}
(M + M_t)\mathrm{v} & -M_t(l_r + l_h + l_t) & -M_t l_t & 0 \\
M\mathrm{v}(l_r + l_h) & I_z & 0 & 0 \\
M_t \mathrm{v}\, l_t & -M_t l_t(l_r + l_h + l_t) - I_t & -(M_t l_t^2 + I_t) & 0 \\
0 & 0 & 0 & 1
\end{bmatrix}
$$

we define

$$
\mathbf{A} = \mathbf{M}_T^{-1} \mathbf{A}_T
$$

$$
\mathbf{B} = \mathbf{M}_T^{-1} \mathbf{B}_T
$$

Based on the availability of the measurement as discussed before, the output matrix $\mathbf{C}(t)$ needs to be specified. If the only sensor to be used is the yaw rate sensor, the output matrix is set to $\mathbf{C}(t) = (0 \quad 1 \quad 0 \quad 0)$. In this case, not only the hitch angle, but also the hitch angle rate as well as the vehicle sideslip angle are estimated.

For the purpose of designing the Kalman filter, it is necessary to model the unknown hitch angle as a random process. It is assumed that the hitch angle is modeled by white noise v(t) with intensity matrix $\mathbf{V}(t)$ as a random process that is independent and stationary. The only measurement in use is the yaw rate, which is subjected to the measurement noise $w(t)$ assumed to be white. In this case, intensity matrix $\mathbf{W}(t)$ becomes a scalar $W(t)$. Hence, the state-space form for the vehicle-trailer model is formulated as

$$\dot{\mathbf{x}}(t) = \mathbf{A}(t)\mathbf{x}(t) + \mathbf{B}(t)\mathbf{u}(t) + \mathbf{v}(t)$$
$$y(t) = \mathbf{C}(t)\mathbf{x}(t) + w(t),$$

(14.101)

where index t is added to emphasis that the system is time-varying due to the noise and the vehicle speed that is varying. The output is no longer a vector because of the single measurement. The state excitation noise vector has only one component for the hitch angle:

$$\mathbf{v}(t) = \begin{bmatrix} v_1(t) \\ v_2(t) \\ v_3(t) \\ v_4(t) \end{bmatrix}.$$

Since the random processes $\mathbf{v}(t)$ and $w(t)$ are stationary so that intensity matrices $\mathbf{V}(t)$ and $W(t)$ are constant, i.e.,

$$V(t) = \begin{bmatrix} V_{11} & 0 & 0 & 0 \\ 0 & V_{22} & 0 & 0 \\ 0 & 0 & V_{33} & 0 \\ 0 & 0 & 0 & V_{44} \end{bmatrix}, \quad W(t) = W$$

Now we can design the Kalman filter which provides an estimate of the hitch angle. Using (11.35), (11.41) and (11.42), the Kalman filter is defined as

$$\dot{\hat{\mathbf{x}}}(t) = \mathbf{A}(t)\,\hat{\mathbf{x}}(t) + \mathbf{B}(t)\mathbf{u}(t) + \mathbf{K}(t)(\dot{\psi}(t) - \hat{\dot{\psi}}(t))$$

(14.102)

$$\mathbf{K}(t) = \mathbf{P}(t)\mathbf{C}^T(t)W^{-1}$$

(14.103)

$$\dot{\mathbf{P}}(t) = \mathbf{A}(t)\mathbf{P}(t) + \mathbf{P}(t)\mathbf{A}^T(t) - \mathbf{P}(t)\mathbf{C}^T(t)W^{-1}\mathbf{C}(t)\mathbf{P}(t) + \mathbf{V}$$

(14.104)

The gain matrix, $\mathbf{K}(t)$, results in a 4×1 vector. The time-varying covariance matrix, $\mathbf{P}(t)$, is calculated by Riccati equation (14.104). In this equation, intensity V indicates the randomness of the hitch angle that the Kalman filter is trying to estimate. Intensity W is an indicator of the random noise in making the measurement. Figure 14.29 shows the block diagram of the hitch angle estimation that is built using (14.102), (14.103) and (14.104).

14.3.3   SIMULATION AND ANALYSIS OF TRAILER STABILIZATION

Figure 14.30 displays the front panel of the simulation program in LabVIEW. The controller parameters are inserted into the front panel, where one can set the values for the simulation. In the front

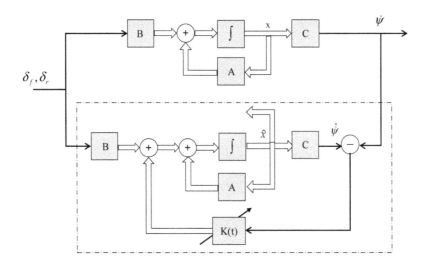

FIGURE 14.29   Kalman-filter designed for the hitch angle estimation.

panel, both the vehicle model and the controller parameters can be changed. Figure 14.31 shows the implementation in the LabVIEW block diagram, which reflects the control structure illustrated before in Figure 14.28. The input data is implemented on the left side of the block diagram. Equation (14.101) is used to simulate the combined vehicle and semi-trailer model. The model matrices **A**, **B** and **C** are entered into the "state-space model" in the middle of the block diagram. The "CD Construct Stochastic Model" on the left side of the block diagram specifies the stochastic model while the "CD Construct Noise Model" defines the statistical behavior of the process noise vector **v**(t) and the measurement noise vector w(t). The Continuous Kalman Filter from LabVIEW's library is used for the states estimation. The controller is programmed in MathScript as well and is placed on the bottom of the block diagram. On the right side of the block diagram are the signal formats

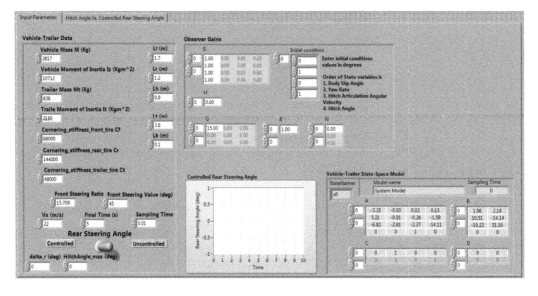

FIGURE 14.30   Front panel of the trailer stabilization.

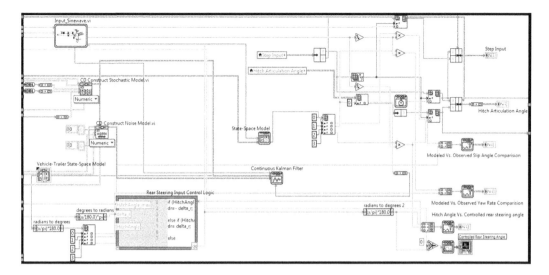

FIGURE 14.31   LabVIEW block diagram of the trailer stabilization.

and the outputs. The VI (LabVIEW term meaning Virtual Implementation) on the top of the left side inputs the driver's steering signal.

The driver's steering input is shaped as shown in Figure 14.32. By rotating the steering wheel, the front steer angle reaches quickly to its first peak of 45°, from which the steer angle immediately goes to another peak of −45°, and then gets back to zero. Since the vehicle was originally driven straight, the steady-state hitch angle is zero, hence $\psi_{h\_st} = 0$.

Figure 14.33 illustrates the hitch angles in the maneuvers with and without the rear steer control. In the left graph, the solid line presents the hitch angle without the control while the dashed line indicates the one with the control. As seen, the rear steer control is able to reduce the oscillation of the hitch angle. In this simulation, $\delta_{rc} = 2°$ and $\psi_{h\_max} = 3°$ are chosen, which can be seen in the right graph. The right graph also shows the comparison between the modeled and estimated hitch angles. Both the estimated and the modeled angles remain acceptably close. Figure 14.34 presents the estimation of the yaw rate and vehicle sideslip angle during the same simulation.

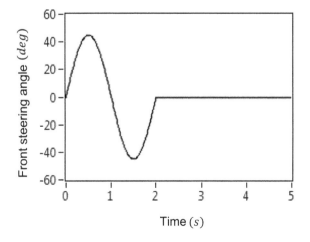

FIGURE 14.32   Sinewave as the steering input.

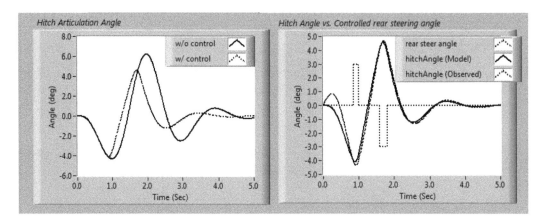

FIGURE 14.33 Hitch angle dynamic behavior.

The above-presented simulation demonstrated the control behavior of the proposed controller to dampen trailer oscillations around the equilibrium point. As seen, the hitch angle amplitude keeps decreasing while the rear steer angle is controlled. The controller is not making the oscillation amplitude to zero, which would require a permanent control around the equilibrium point since the trailer always creates a hitch angle while it is moving.

In this section, trailer dynamics and its steady state were analyzed. The study shows that the steady state of the towing vehicle is not influenced by the trailer under the assumption that trailer is a lumped mass at the wheel center. Therefore, stability of the towing vehicle can be controlled independently from the trailer. A LabVIEW simulation program demonstrated the effectiveness of a simple controller to mitigate the oscillation of the trailer, where the required signal of the hitch angle, is estimated by the designed Kalman filter.

While studying the path-following control in Section 13.2, two controllers of the rear steering and the rear torque vectoring were designed. Both methods have the same effect balancing the

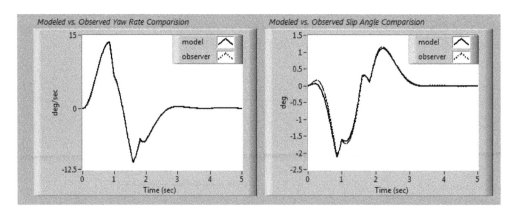

FIGURE 14.34 Estimation of the yaw rate and the vehicle sideslip angle when towing the trailer.

yaw moment through the control of the rear wheels. The analogy can be carried over to the above-presented section, i.e., the rear torque vectoring on the towing vehicle can be applied to stabilize the trailer. That will be an interesting and promising topic for future studies.

REFERENCES

1. M. Mitschke, Dynamik der Kraftfahrzeuge, Band C: Fahrverhalten, Berlin: Springer-Verlag, 1990.
2. F. L. Lewis, C. T. Abdallah and D. M. Dawson, Control of Robot Manipulators, New York: Macmillan Publishing Company, 1993.
3. Masato Abe, Vehicle Handling Dynamics: Theory and Application, Boston, New York, London: Elsevier, 2009.
4. Dean C. Karnopp, Donald L. Margolis, Engineering Applications of Dynamics, Hoboken, New Jersey: John Wiley & Sons, Inc. 2008.

Appendix A
LabVIEW Implementations for Simulation

In this appendix, the LabVIEW implementations are given for some examples with simulation program in previous chapters.

A.1 LabVIEW PROGRAM OF EXAMPLE 4.1

The implementation for $\mu_{x,B}$-curve simulation in LabVIEW is set up as follows. Figure A1.1 displays the front panel of the program while Figure A1.2 shows the block diagram. In the front panel, the road surface can be selected, and each surface number corresponds to a different sets of parameters C1, C2 and C3. Those parameters are the inputs to the block diagram so that the associated $\mu_{x,B}$-curve simulated. Figure A1.3 lists the MathScript codes of the equations, which the calculation of $\mu_{x,B}$-curve is based on.

❏

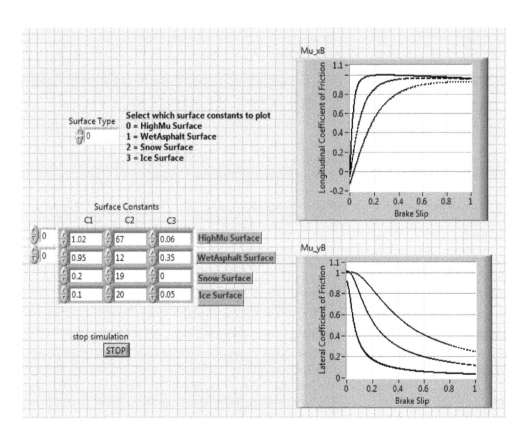

FIGURE A1.1 Program front-panel for $\mu_{x,B}$-curve simulation.

DOI: 10.1201/9781003134305-A

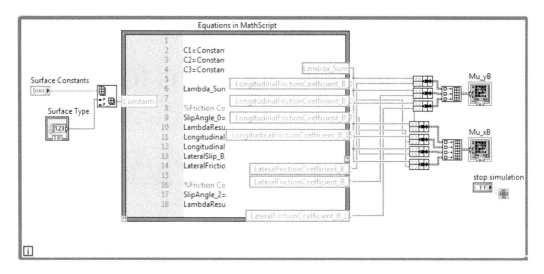

FIGURE A1.2 Program block-diagram for $\mu_{x,B}$-curve simulation.

```
                                    Equations in MathScript
          1
          2   C1=Constants(1);
          3   C2=Constants(2);
          4   C3=Constants(3);
          5
          6   Lambda_Sum=0.001:0.002:1;
Constants 7
          8   %Friction Coefficient in Braking at Slip Angle = 0
          9   SlipAngle_0=0*pi/180;
         10   LambdaResultant_B_0=((2*(1-Lambda_Sum)*(1-cos(SlipAngle_0)))+Lambda_Sum.^2).^(0.5);
         11   LongitudinalSlip_B_0=cos(SlipAngle_0)-(1-Lambda_Sum);
         12   LongitudinalFrictionCoefficient_B_0=(C1*(1-exp(-C2*LambdaResultant_B_0))./LambdaResultant_B_0-C3).*LongitudinalSlip_B_0;
         13   LateralSlip_B_0=sin(SlipAngle_0);
         14   LateralFrictionCoefficient_B_0=(LongitudinalFrictionCoefficient_B_0.*LateralSlip_B_0)./LongitudinalSlip_B_0;
         15
         16   %Friction Coefficient in Braking at Slip Angle = 2
         17   SlipAngle_2=2*pi/180;
         18   LambdaResultant_B_2=((2*(1-Lambda_Sum)*(1-cos(SlipAngle_2)))+Lambda_Sum.^2).^(0.5);
         19   LongitudinalSlip_B_2=cos(SlipAngle_2)-(1-Lambda_Sum);
         20   LongitudinalFrictionCoefficient_B_2=(C1*(1-exp(-C2*LambdaResultant_B_2))./LambdaResultant_B_2-C3).*LongitudinalSlip_B_2;
         21   LateralSlip_B_2=sin(SlipAngle_2);
         22   LateralFrictionCoefficient_B_2=(LongitudinalFrictionCoefficient_B_2.*LateralSlip_B_2)./LongitudinalSlip_B_2;
         23
         24   %Friction Coefficient in Braking at Slip Angle = 7
         25   SlipAngle_7=7*pi/180;
         26   LambdaResultant_B_7=((2*(1-Lambda_Sum)*(1-cos(SlipAngle_7)))+Lambda_Sum.^2).^(0.5);
         27   LongitudinalSlip_B_7=cos(SlipAngle_7)-(1-Lambda_Sum);
         28   LongitudinalFrictionCoefficient_B_7=(C1*(1-exp(-C2*LambdaResultant_B_7))./LambdaResultant_B_7-C3).*LongitudinalSlip_B_7;
         29   LateralSlip_B_7=sin(SlipAngle_7);
         30   LateralFrictionCoefficient_B_7=(LongitudinalFrictionCoefficient_B_7.*LateralSlip_B_7)./LongitudinalSlip_B_7;
         31
         32   %Friction Coefficient in Braking at Slip Angle = 15
         33   SlipAngle_15=15*pi/180;
         34   LambdaResultant_B_15=((2*(1-Lambda_Sum)*(1-cos(SlipAngle_15)))+Lambda_Sum.^2).^(0.5);
         35   LongitudinalSlip_B_15=cos(SlipAngle_15)-(1-Lambda_Sum);
         36   LongitudinalFrictionCoefficient_B_15=(C1*(1-exp(-C2*LambdaResultant_B_15))./LambdaResultant_B_15-C3).*LongitudinalSlip_B_15;
         37   LateralSlip_B_15=sin(SlipAngle_15);
         38   LateralFrictionCoefficient_B_15=(LongitudinalFrictionCoefficient_B_15.*LateralSlip_B_15)./LongitudinalSlip_B_15;
```

FIGURE A1.3 MathScript of the equations for $\mu_{x,B}$-curve simulation.

A.2 LabVIEW PROGRAM OF EXAMPLE 4.2

The implementation for $\mu_{x,T}$-curve simulation in LabVIEW is set up as follows. Figure A2.1 display the front panel of the program while Figure A2.2 shows the block diagram. In the front panel, the road surface can be selected, and each surface number corresponds to a different sets of parameters C1, C2 and C3. Those parameters are the inputs to the block diagram so that the associated $\mu_{x,T}$-curve is sumulated. Figure A2.3 lists the MathScript codes of the equations, which the calculation of $\mu_{x,T}$-curve is based on.

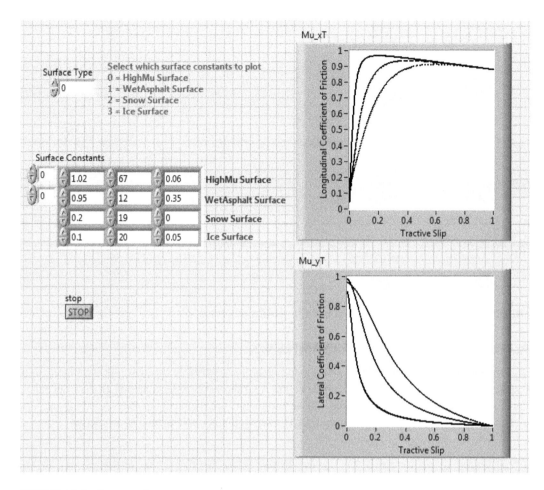

FIGURE A2.1   Program front-panel for $\mu_{x,T}$-curve simulation.

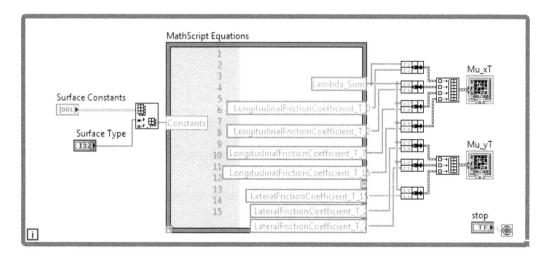

FIGURE A2.2   Program block-diagram for $\mu_{x,T}$-curve simulation.

Equations in MathScript

```
C1=Constants(1);
C2=Constants(2);
C3=Constants(3);

Lambda_Sum=0.001:0.002:1;

%Friction Coefficient in Traction at Slip Angle = 0
SlipAngle_0=0*pi/180;
LambdaResultant_T_0=((2*(1-Lambda_Sum)*(1-cos(SlipAngle_0)))+Lambda_Sum.^2).^(0.5);
LongitudinalSlip_T_0=1-(1-Lambda_Sum)*cos(SlipAngle_0);
LongitudinalFrictionCoefficient_T_0=((((1-exp(-C2.*LambdaResultant_T_0))-(C3.*LambdaResultant_T_0))./(LambdaResultant_T_0))-C3).*LongitudinalSlip_T_0;
LateralSlip_T_0=(1-Lambda_Sum)*sin(SlipAngle_0);
LateralFrictionCoefficient_T_0=(LongitudinalFrictionCoefficient_T_0.*LateralSlip_T_0)./LongitudinalSlip_T_0;

%Friction Coefficient in Traction at Slip Angle = 2
SlipAngle_2=2*pi/180;
LambdaResultant_T_2=((2*(1-Lambda_Sum)*(1-cos(SlipAngle_2)))+Lambda_Sum.^2).^(0.5);
LongitudinalSlip_T_2=1-(1-Lambda_Sum)*cos(SlipAngle_2);
LongitudinalFrictionCoefficient_T_2=((((1-exp(-C2.*LambdaResultant_T_2))-(C3.*LambdaResultant_T_2))./(LambdaResultant_T_2))-C3).*LongitudinalSlip_T_2;
LateralSlip_T_2=(1-Lambda_Sum)*sin(SlipAngle_2);
LateralFrictionCoefficient_T_2=(LongitudinalFrictionCoefficient_T_2.*LateralSlip_T_2)./LongitudinalSlip_T_2;

%Friction Coefficient in Traction at Slip Angle = 7
SlipAngle_7=7*pi/180;
LambdaResultant_T_7=((2*(1-Lambda_Sum)*(1-cos(SlipAngle_7)))+Lambda_Sum.^2).^(0.5);
LongitudinalSlip_T_7=1-(1-Lambda_Sum)*cos(SlipAngle_7);
LongitudinalFrictionCoefficient_T_7=((((1-exp(-C2.*LambdaResultant_T_7))-(C3.*LambdaResultant_T_7))./(LambdaResultant_T_7))-C3).*LongitudinalSlip_T_7;
LateralSlip_T_7=(1-Lambda_Sum)*sin(SlipAngle_7);
LateralFrictionCoefficient_T_7=(LongitudinalFrictionCoefficient_T_7.*LateralSlip_T_7)./LongitudinalSlip_T_7;

%Friction Coefficient in Traction at Slip Angle = 15
SlipAngle_15=15*pi/180;
LambdaResultant_T_15=((2*(1-Lambda_Sum)*(1-cos(SlipAngle_15)))+Lambda_Sum.^2).^(0.5);
LongitudinalSlip_T_15=1-(1-Lambda_Sum)*cos(SlipAngle_15);
LongitudinalFrictionCoefficient_T_15=((((1-exp(-C2.*LambdaResultant_T_15))-(C3.*LambdaResultant_T_15))./(LambdaResultant_T_15))-C3).*LongitudinalSlip_T_15;
LateralSlip_T_15=(1-Lambda_Sum)*sin(SlipAngle_15);
LateralFrictionCoefficient_T_15=(LongitudinalFrictionCoefficient_T_15.*LateralSlip_T_15)./LongitudinalSlip_T_15;
```

FIGURE A2.3   MathScript of the equations for $\mu_{x,T}$-curve simulation.

A.3   LabVIEW PROGRAM OF EXAMPLE 9.2

The implementation for transient-response simulation in LabVIEW is set up as follows. Figure A3.1 display the front panel of the program while Figure A3.2 shows the block diagram. In the front panel, the model parameters are to be entered. Those parameters are the inputs to the block diagram. The bicycle model is programmed in MathScript. The step input is implemented as a function.

❏

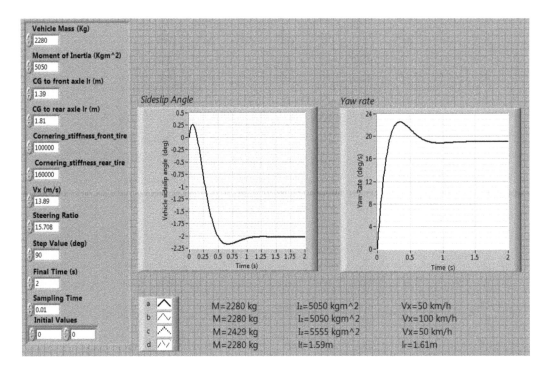

FIGURE A3.1   Program front-panel for the transient response.

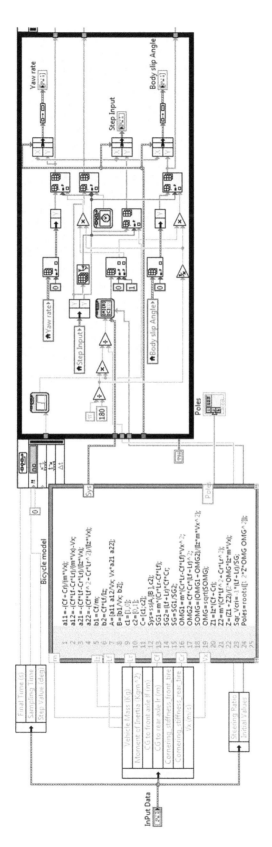

FIGURE A3.2 Program block-diagram for transient response.

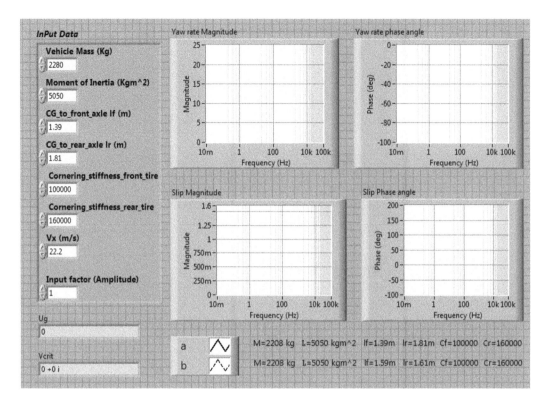

FIGURE A4.1 Program front-panel for the frequency response.

A.4 LabVIEW PROGRAM OF EXAMPLE 9.3

The implementation for frequency-response simulation in LabVIEW is set up as follows. Figure A4.1 display the front panel of the program while Figure A4.2 shows the block diagram. In the front panel, the model parameters are to be entered. Those parameters are the inputs to the block diagram. The bicycle model is programmed in MathScript. The step input is implemented as a function. ❏

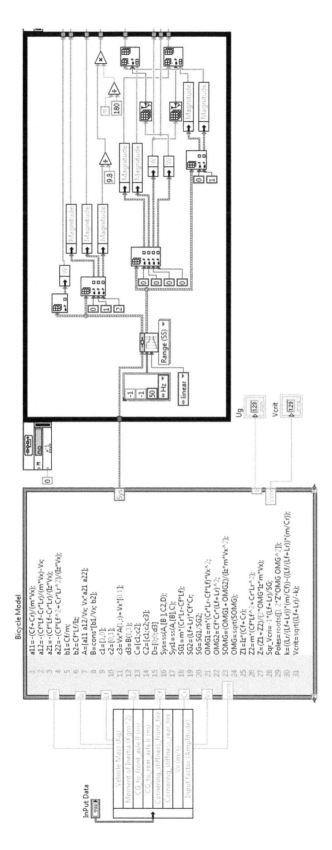

FIGURE A4.2 Program block-diagram for the frequency response.

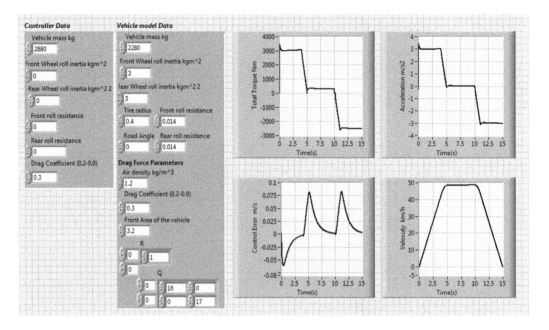

FIGURE A5.1   Program front-panel of the speed control.

A.5   LabVIEW PROGRAM OF EXAMPLE 13.1

The front panel is shown in Figure A5.1. The vehicle model data is the input field for the vehicle model parameters while the controller data includes the vehicle parameters assumed for the controller.

From Figure 13.2, more details can be shown in Figure A5.2. The vehicle model is implemented in the MathScript format on the left side of the block diagram. The controller is programmed in MathScript as well and placed on the right side of the block diagram. As seen in Figure 13.2, CD LQR provides controller gains using Riccati equation. Figure A5.3 shows the created LQR VI receiving the controller gains and calculates equation (13.7).

❏

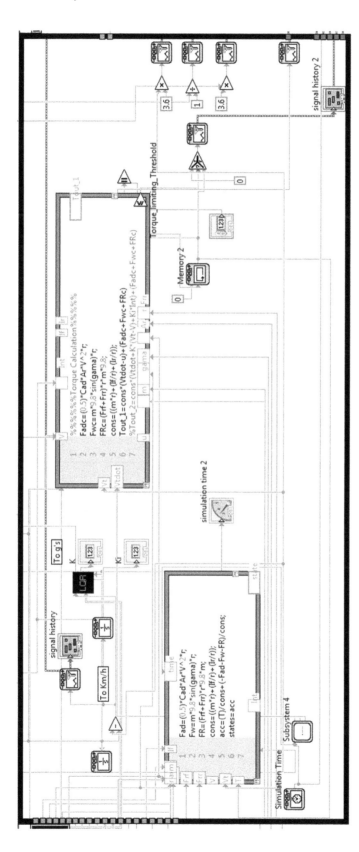

FIGURE A5.2 MathScript of the controller and the vehicle model.

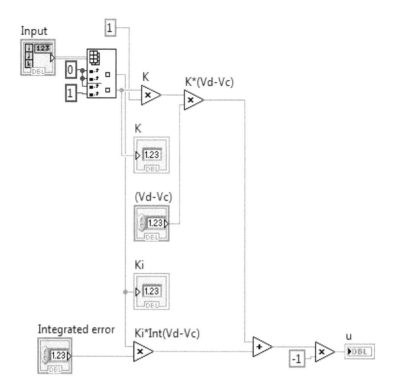

FIGURE A5.3   LQR VI for the speed control.

A.6   LabVIEW PROGRAM TO FIGURE 14.26

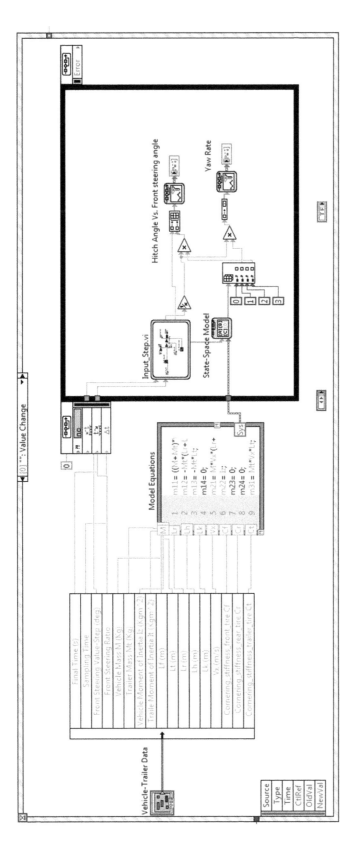

FIGURE A6.1 Block diagram to Figure 14.26.

Model Equations in MathScript

```
1   Xm11= ((M+Mt)*Vx); m12= -Mt*(Lr+Lh+Lt); m13= -Mt*Lt; m14= 0;
2   Xm21= M*Vx*(Lr+Lh); m22= Iz; m23= 0; m24= 0;
3   Xm31= Mt*Vx*Lt; m32= -Mt*Lt*(Lr+Lh+Lt)-It; m33= -((Mt*Lt^2)+It); m34= 0;
4     m41= 0; m42= 0; m43= 0; m44= 1;
5   Xa11= -(Cf+Cr+Ct); a12= -((M+Mt)*Vx)-(((Cf*Lf)-(Cr*Lr)-(Lr+Lh+Lt+Lk)*Ct)/Vx); a13= ((Lt+Lk)*Ct)/Vx; a14= Ct;
6   Xa21= -Cf*(Lf+Lr+Lh)-Cr*Lh; a22= -((Cf*Lf*(Lf+Lr+Lh)-(Cr*Lr*Lh))/(Iz*Vx))-(M*Vx*(Lr+Lh)); a23= 0; a24= 0;
7   Xa31= -Ct*(Lt+Lk); a32= ((Ct*(Lr+Lh+Lt+Lk)*(Lt+Lk))/(Vx))-(Mt*Vx*Lt); a33= (((Lt+Lk)^2)*Ct)/(Vx); a34= (Lt+Lk)*Ct;
8     a41= 0; a42= 0; a43= 1; a44= 0;
9   Xb11= Cf; b12= Cr;
10  Xb21= Cf*(Lf+Lr+Lh); b22= Cr*Lh;
11    b31= 0; b32= 0;
12    b41= 0; b42= 0;
13    M=[m11 m12 m13 m14; m21 m22 m23 m24; m31 m32 m33 m34; m41 m42 m43 m44]
14    A=[a11 a12 a13 a14; a21 a22 a23 a24; a31 a32 a33 a34; a41 a42 a43 a44];
15    B=[b11 b12; b21 b22; b31 b32; b41 b42];
16    N=inv(M);
17    AA=N*A;
18    BB=N*B;
19    C=[0 1 0 0];
20    D=[0 0];
21    Sys=ss(AA,BB,C);
```

FIGURE A6.2   MathScript codes of the vehicle-trailer model equations.

BIBLIOGRAPHY

A. Bedford and W. Fowler, Engineering Mechanics: Dynamics, New Jersey: Prentice Hall, 2002.

J. Reimpell, Fahrwerktechnik: Grundlagen, Würzburg: Vogel Verlag, 1995.

J. C. Dixon, Tires, Suspension and Handling, Warrendale: SAE, 1996.

M. Mitschke, Dynamik der Kraftfahrzeuge, Band A: Antrieb und Bremsung, Berlin: Springer-Verlag, 1982.

M. Burckhardt, Radschlupf-Regelsysteme, Würzburg: Vogel Verlag, 1993.

A. Zomotor, Fahrwerktechnik: Fahrverhalten, Würzburg: Vogel Verlag, 1991.

M. Mitschke, Dynamik der Kraftfahrzeuge, Band C: Fahrverhalten, Berlin: Springer-Verlag, 1990.

T. D. Gillespie, Fundamentals of Vehicle Dynamics, Warrendale: SAE, 1992.

J. E. Duffy and C. Johanson, Auto Drive Trains Technology, Illinois: The Goodheart-Willcox Company, 1995.

U. Kiencke and L. Nielsen, Automotive Control Systems, Warrendale: SAE, , 2000.

Vehicle Dynamics Terminology, SAE Standard J670, revised in 2008.

H. Fennel, R. Gutwein, A. Kohl, M. Latarnik and G. Roll, "Das modulare Regler- und Regelkonzept beim ESP von ITT Automotive," in *7th Achener Kolloquium Fahrzeug- und Motortechnik*, Achen, Okt. 1998.

A. van Zanten, R. Erhardt and G. Pfaff, "FDR-Die Fahrdynamikregelung von Bosch," *ATZ Automobiltechnische Zeitschrift*, vol. 96, no. 11, pp. 674–689, 1994.

Bosch, Automotive handbook, 8th ed., Plochingen: Robert Bosch GmbH, 2011.

E. C. Glasner von Ostenwall, Beitrag zur Auslegung von Kraffahrzeugbremsanlagen, Dissertation, Universität Stuttgart, Germany, 1973.

M. Mitschke and E. Sagan, Fahrdynamik von Pkw-Wohnnanhängerzügen, Köln: Verlag TÜV Rheinland, 1998.

Y. Gao, L. Chen and M. Ehsani, Investigation of the Effectiveness of Regenerative Braking for EV and HEV, SAE International Conference, Costa Mesa, California, 1999.

Road vehicles - Lateral transient response test methods, ISO Standard 7401, 2011.

J. Y. Wong, Theory of Ground Vehicles, New Jersey: John Wiley & Sons, 2008.

A. Alleyne and J. K. Hedrick, "Nonlinear Adaptive Control of Active Suspensions," *IEEE Transactions on Control Systems Technology*, vol. 3, no. 1, 1995.

R. Rajamani and J. K. Hedrick, "Adaptive Observers for Automotive Suspensions: Theory and Experiment," *IEEE Transactions. on Control Systems Technology*, vol. 3, no. 1, 1995.

P. C. Müller and W. O. Schiehlen, Linear Vibrations, Dordrecht: Martinus Nijhoff Publishers, 1985.

D. J. M. Sampson, Active Roll Control of Articulated Heavy Vehicles, Ph.D. thesis, University of Cambridge, UK, 2000.

K. T. Feng, H. S. Tan and M. Tomizuka, "Automatic Steering Control of Vehicle Lateral Motion with the Effect of Roll Dynamics," in *Proceedings of the American Control Conference*, Philadelphia, Pennsylvania, 1998.

K. Ogata, Modern Control Engineering, 3rd ed., New Jersey: Prentice Hall, 1997.

B. Friedland, Control System Design, New York: Dover Publications, 2005.

R. G. Brown, Introduction to Random Signal Analysis and Kalman Filtering, New Jersey: Wiley, 1983.

F. L. Lewis, C. T. Abdallah and D. M. Dawson, Control of Robot Manipulators, New York: Macmillan Publishing Company, 1993.

F. P. Beer and E. R. Johnson Jr., Vector Mechanics for Engineers: Dynamics, Boston: McGraw-Hill, 1997.

J. J. Craig, Introduction to Robotics, 4th ed., New Jersey: Pearson Education, 2017.

M. Gopal, Modern Control System Theory, New Delhi: New Age International Ltd., 1993.

J. Lu, D. Messih and A. Salib, "Roll rate based stability control-the roll stability control system," in *Proceedings of the 20th Enhanced Safety of Vehicles Conference*, 2007.

P. C. Müller, Class Manuscript, Topic: Regelungstheorie, University of Wuppertal, Germany, 1990.

J. Yu, Lateral Control in Path-tracking of Autonomous Vehicles, US Patent Application 17330536 , 2021.

H. P. Willumeit, Modelle und Modellierungsverfahren in der Fahrzeugdynamik, Stuttgart: B.G. Teubner, 1998.

A. Hughes, Electric Motors and Drives, 3rd ed., Amsterdam: Elsevier Ltd., , 2006.

I. Boldea and S. A. Nasar, Electric Drives, Boca Raton, London, New York: CRC Press, 2017.

W. L. Brogan, Modern Control Theory, Upper Saddle River, NJ: Prentice Hall, 1991.

N. M. Enache, S. Mammar, M. Netto and B. Lusetti, "Driver Steering Assistance for Lane-Departure Avoidance Based on Hybrid Automata and Composite Lyapunov Function," *IEEE Transactions on Intelligent Transportation Systems*, vol. 11, p. 28, 2010.

J. Yu, Verfahren und Vorrichtung zum Angaben der auf ein Rad wirkenden Stoermomente. Patent DE 10104599A1, Feburary 2001.

P. C. Müller, "Indirect Measurement of Nonlinear Effects by State Observers," in *Nonlinear Dynamics in Engineering Systems, IUTAM Symposium, Stuttgart, Germany,* 1989.

D. Söffker, J Bajkowski and P.C. Müller, "Detection of Cracks in Turborotors - A New Observer Based Method," *ASMS Journal of Dynamics Systems, Measurement and Control*, pp. 518–524, Sept. 1993.

Alexandr F. Andreev, Viachaslau Kabanau, Vladimir Vantsevich, Driveline Systems of Ground Vehicles: Theory and Design, Boca Raton, London, New York: CRC Press, 2010.

Dean C. Karnopp, Donald L. Margolis, Engineering Applications of Dynamics, Hoboken, New Jersey: John Wiley & Sons, Inc. 2008.

Masato Abe, Vehicle Handling Dynamics: Theory and Application, Boston, New York, London: Elsevier, 2009.

Georg Rill, Abel Arrieta Castro, Road Vehicle Dynamics: Fundamentals and Modeling, Second Edition, Boca Raton, London, New York: CRS Press, Taylor & Francis Group, 2020.

A. Galip Ulsoy, Huei Peng, Melih Cakmakci, Automotive Control Systems, Cambridge: Cambridge University Press, 2012.

Rolf Johansson, Anders Rantzer (eds.), Nonlinear and Hybrid Systems in Automotive Control, London: SAE International, Springer Verlag, 2003.

Michael Blundell, Damian Harty, The Multibody Systems Approach to Vehicle Dynamics, Boston, New York, London: Elsevier, 2014.

Elzbieta Jarzebowska, Model-based Tracking Control of Nonlinear Systems, Boca Raton, London, New York: CRS Press, Taylor & Francis Group, 2012

Iqbal Husain, Eletric and HYbrid Vehicles: Design Fundamentals, 2nd ed., Boca, London, New York: CRS Press, Taylor & Francis Group, 2011

J. Ackermann, J. Guldner, W. Sienel, R. Steinhauser and V.I. Utkin, "Linear and Nonlinear Controller Design for Robust Automatic Steering," IEEE Transactions on Control Systems Technology, vol. 3, No. 4, March 1995

W. Deng and X. Kang, "Parametric Study on Vehicle-Trailer Dynamics for Stability Control," SAE World Congress, Detroit, Michigan, USA, 2003

Chr. Chatzikomis, A. Sorniotti, P. Gruber, M. Zanchetta, D. Willans and B. Balcombe, "Comparison of Path Tracking and Torque-Vectoring Controllers for Autonomous Electric Vehicles," IEEE Transactions on Intelligent Vehicles, vol. 3, No. 4, December 2018

R. Attia, R. Orjuela, and M. Basset, "Combined Longitudinal and Lateral Control for Automated Vehicle Guidance," Veh. Syst. Dyn., vol. 52, No. 2, 2014

C. M. Filho, D. F. Wolf, V. Grassi, and F. S. Osorio, "Longitudinal and Lateral Control for Autonomous Ground Vehicles," IEEE Intelligent Vehicles Symposium, Dearborn, MI, USA, 2014

Index

For Product Safety Concerns and Information please contact our EU
representative GPSR@taylorandfrancis.com
Taylor & Francis Verlag GmbH, Kaufingerstraße 24, 80331 München, Germany

www.ingramcontent.com/pod-product-compliance
Ingram Content Group UK Ltd.
Pitfield, Milton Keynes, MK11 3LW, UK
UKHW050926180425
457613UK00003B/35